Studies in Big Data

Volume 16

Series editor

Janusz Kacprzyk, Polish Academy of Sciences, Warsaw, Poland
e-mail: kacprzyk@ibspan.waw.pl

About this Series

The series "Studies in Big Data" (SBD) publishes new developments and advances in the various areas of Big Data-quickly and with a high quality. The intent is to cover the theory, research, development, and applications of Big Data, as embedded in the fields of engineering, computer science, physics, economics and life sciences. The books of the series refer to the analysis and understanding of large, complex, and/or distributed data sets generated from recent digital sources coming from sensors or other physical instruments as well as simulations, crowd sourcing, social networks or other internet transactions, such as emails or video click streams and other. The series contains monographs, lecture notes and edited volumes in Big Data spanning the areas of computational intelligence incl. neural networks, evolutionary computation, soft computing, fuzzy systems, as well as artificial intelligence, data mining, modern statistics and operations research, as well as self-organizing systems. Of particular value to both the contributors and the readership are the short publication timeframe and the world-wide distribution, which enable both wide and rapid dissemination of research output.

More information about this series at http://www.springer.com/series/11970

Nathalie Japkowicz · Jerzy Stefanowski
Editors

Big Data Analysis: New Algorithms for a New Society

 Springer

Editors
Nathalie Japkowicz
University of Ottawa
Ottawa, ON
Canada

Jerzy Stefanowski
Institute of Computing Sciences
Poznań University of Technology
Poznań
Poland

ISSN 2197-6503
Studies in Big Data
ISBN 978-3-319-80053-0
DOI 10.1007/978-3-319-26989-4

ISSN 2197-6511   (electronic)

ISBN 978-3-319-26989-4   (eBook)

Springer Cham Heidelberg New York Dordrecht London
© Springer International Publishing Switzerland 2016
Softcover re-print of the Hardcover 1st edition 2016

Printed on acid-free paper

Springer International Publishing AG Switzerland is part of Springer Science+Business Media (www.springer.com)

Preface

This book is dedicated to Stan Matwin in recognition of the numerous contributions he has made to the fields of machine learning, data mining, and big data analysis to date. With the opening of the Institute for Big Data Analytics at Dalhousie University, of which he is the founder and the current Director, we expect many more important contributions in the future.

Stan Matwin was born in Poland. He received his Master's degree in 1972 and his Ph.D. in 1977, both from the Faculty of Mathematics, Informatics and Mechanics at Warsaw University, Poland. From 1975 to 1979, he worked in the Institute of Computer Science at that Faculty as an Assistant Professor. Upon immigrating to Canada in 1979, he held a number of lecturing positions at Canadian universities, including the University of Guelph, York University, and Acadia University. In 1981, he joined the Department of Computer Science (now part of the School of Electrical Engineering and Computer Science) at the University of Ottawa, where he carved out a name for the department in the field of machine learning over his 30+ year career there (he became a Full Professor in 1992, and a Distinguished University Professor in 2011). He simultaneously received the State Professorship from the Republic of Poland in 2012.

He founded the Text Analysis and Machine Learning (TAMALE) lab at the University of Ottawa, which he led until 2013. In 2004, he also started cooperating as a "foreign" professor with the Institute of Computer Science, Polish Academy of Sciences (IPI PAN) in Warsaw. Furthermore, he was invited as a visiting researcher or professor in many other universities in Canada, USA, Europe, and Latin America, where in 1997 he received the UNESCO Distinguished Chair in Science and Sustainable Development (Universidad de Sao Paulo, ICMSC, Brazil).

In addition to his position as professor and researcher, he served in a number of organizational capacities: former president of the Canadian Society for the Computational Studies of Intelligence (CSCSI), now the Canadian Artificial Intelligence Society (CAIAC), and of the IFIP Working Group 12.2 (Machine Learning), Founding Director of the Information Technology Cluster of the Ontario Research Centre for Electronic Commerce, Chair of the NSERC Grant Selection

Committee for Computer Science, and member of the Board of Directors of Communications and Information Technology Ontario (CITO).

Stan Matwin is the 2010 recipient of the Distinguished Service Award of the Canadian Artificial Intelligence Society (CAIAC). He is Fellow of the European Coordinating Committee for Artificial Intelligence and Fellow of the Canadian Artificial Intelligence Society.

His research spans the fields of machine learning, data mining, big data analysis and their applications, natural language processing and text mining, as well as technological aspects of e-commerce. He is the author and co-author of over 250 research papers.

In 2013, he received the Canada Research Chair (Tier 1) in Visual Text Analytics. This prestigious distinction and a special program funded by the federal government allowed him to establish a new research initiative. He moved to Dalhousie University in Halifax, Canada, where he founded, and now directs, the Institute for Big Data Analytics.

The principal aim of this Institute is to become an international hub of excellence in Big Data research. Its second goal is to be relevant to local industries in Nova Scotia, and in Canada (with respect to applications relating to marine biology, fisheries and shipping). Its third goal is to develop a focused and advanced training program that covers all aspects of big data, preparing the next generation of researchers and practitioners for research in this field of study.

On the web page of his Institute, he presents his vision on Big Data Analytics. He stresses, "Big data is not a single breakthrough invention, but rather a coming together and maturing of several technologies: huge, inexpensive data harvesting tools and databases, efficient, fast data analytics and data mining algorithms, the proliferation of user-friendly data visualization methods and the availability of affordable, massive and non-proprietary computing. Using these technologies in a knowledgeable way allows us to turn masses of data that get created daily by businesses and the government into a big asset that will result in better, more informed decisions."

He also recognizes the potential transformative role of big data analysis, in that it could support new solutions for many social and economic issues in health, cities, the environment, oceans, education access, personalized medicine, etc. These opinions are reflected in the speech he gave at the launch of his institute, where his recurring theme was "Make life better." His idea is to use big data (i.e., large and constantly growing data collections) to learn how to do things better. For example, he proposes to turn data into an asset by, for instance, improving motorized traffic in a big city or ship traffic in a big port, creating personalized medical treatments based on a patient's genome and medical history, and so on.

Notwithstanding the advantages of big data, he also recognizes its risks for society, especially in the area of privacy. As a result, since 2002, he has been engaged in research on privacy preserving data mining.

Other promising research directions, in his opinion, include data stream mining, the development of new data access methods that incorporate sharing ownership mechanisms, and data fusion (e.g., geospatial applications).

We believe that this book reflects Stan Matwin's call for careful research on both the opportunities and the risks of Big Data Analytics, as well as its impact on society.

Nathalie Japkowicz
Jerzy Stefanowski

Acknowledgments

We take this opportunity to thank all contributors for submitting their papers to this edited book. Their joint efforts and good co-operation with us have enabled to successfully finalize the project of this volume.

Moreover, we wish to express our gratitude to the following colleagues who helped us in the reviewing process: Anna Kobusińska, Ewa Łukasik, Krzysztof Dembczyński, Miłosz Kadziński, Wojciech Kotłowski, Robert Susmaga, Andrzej Szwabe on the Polish side and Vincent Barnabe-Lortie, Colin Bellinger, Norrin Ripsman and Shiven Sharma on the Canadian side.

Continuous guidance and support of the Springer Executive Editor Dr. Thomas Ditzinger and Springer team are also appreciated. Finally, we owe a vote of thanks to Professor Janusz Kacprzyk who has invited us to start the project of this book and has supported for our efforts.

Contents

A Machine Learning Perspective on Big Data Analysis

Nathalie Japkowicz and Jerzy Stefanowski

Abstract This chapter surveys the field of Big Data analysis from a machine learning perspective. In particular, it contrasts Big Data analysis with data mining, which is based on machine learning, reviews its achievements and discusses its impact on science and society. The chapter concludes with a summary of the book's contributing chapters divided into problem-centric and domain-centric essays.

1 Preliminaries

In 2013, Stan Matwin opened the Institute for Big Data Analytics at Dalhousie University. The Institute's mission statement, posted on the website is, "To create knowledge and expertise in the field of Big Data Analytics by facilitating fundamental, interdisciplinary and collaborative research, advanced applications, advanced training and partnerships with industry." In another position paper [46] he posited that Big Data sets the new problems they come with and represent the challenges that machine learning research needs to adapt to. In his opinion, Big Data Analytics will significantly influence the field with respect to developing new algorithms as well as in the creation of applications with greater societal importance.

The purpose of this edited volume, dedicated to Stan Matwin, is to explore, through a number of specific examples, how the study of Big Data analysis, of which his institute is at the forefront, is evolving and how it has started and will most likely continue to affect society. In particular, this book focuses on newly developed algorithms affecting such areas as business, financial forecasting, human mobility, the Internet of Things, information networks, bioinformatics, medical systems and life science.

N. Japkowicz (✉)
School of Electrical Engineering & Computer Science, University of Ottawa,
Ottawa, ON, Canada
e-mail: nat@site.uottawa.ca

J. Stefanowski
Institute of Computing Sciences, Poznań University of Technology,
Poznań, Poland
e-mail: Jerzy.Stefanowski@cs.put.poznan.pl

© Springer International Publishing Switzerland 2016
N. Japkowicz and J. Stefanowski (eds.), *Big Data Analysis: New Algorithms for a New Society*, Studies in Big Data 16, DOI 10.1007/978-3-319-26989-4_1

Moreover, this book will provide methodological discussions about the principles of mining Big Data and the difference between traditional statistical data analysis and newer computing frameworks for processing Big Data.

This chapter is divided into three sections. In Sect. 2, we define Big Data Analysis and contrast it with traditional data analysis. In Sect. 3, we discuss visions about the changes in science and society that Big Data brings about, along with all of their benefits. This is countered by warnings about the negative effects of Big Data Analysis along with its pitfalls and challenges. Section 4 introduces the work that will be presented in the subsequent chapters and fits it into the framework laid out in Sects. 2 and 3. Conclusions about the research presented in this book will be presented in the final chapter along with a review of Stan Matwin's contributions to the field.

2 What Do We Call Big Data Analysis?

For a traditional Machine Learning expert, "Big Data Analysis" can be both exciting and threatening. It is threatening in that it makes a lot of the research done in the past obsolete since previously designed algorithms may not scale up to the amount of new data now typically processed, or they may not address the new problems generated by Big Data Analysis. In addition, Big Data analysis requires a different set of computing skills from those used in traditional research. On the other hand, Big Data analysis is exhilarating because it brings about a multitude of new issues, some already known, and some still to be discovered. Since these new issues will need to be solved, Big Data analysis is bringing a new dynamism to the fields of Data Mining and Machine Learning.

Yet, what is Big Data Analysis, really? In this section and this introductory chapter, in general, we try to figure out what Big Data Analysis really is, at least as far as Machine Learning scientists are concerned, whether it is truly different from what Machine Learning scientists have been doing in the past, and whether it has the potential, heralded by many, to change society in a dramatic way or whether the changes will be incremental and relatively small. Basically, we are trying to figure out what all the excitement is about! We begin by surveying some general definitions of mining Big Data, discussing characteristic features of these data and then move on to more specific Machine Learning issues. After discussing a few well-known successful applications of Big Data Analysis, we conclude this section with a survey of the specific innovations in Machine Learning and Data Mining research that have been driven by Big Data Analysis.

2.1 General Definitions of Big Data

The original and much cited definition from the Gartner project [10] mentions the "three *Vs*": Volume, Velocity and Variety. These "*V* characteristics" are usually explained as follows:

Volume—the huge and continuously increasing size of the collected and analyzed data is the first aspect that comes to mind. It is stressed that the magnitude of Big Data is much larger than that of data managed in traditional storage systems. People talk about terabytes and petabytes rather than gigabytes. However, as noticed by [35], the size of Big Data is "a constantly moving target- what is considered to be Big today will not be so years ahead".

Velocity—this term refers to the speed at which the data is generated and input into the analyzing system. It also forces algorithms to process data and produce results in limited time as well as with limited computer resources.

Variety—this aspect indicates heterogeneous, complex data representations. Quite often analysts have to deal with structured as well as semi-structured and unstructured data repositories.

IBM added a fourth "*V*" which stands for "**Veracity**". It refers to the quality of the data and its trustworthiness. Recall that some sources produce low quality or uncertain data, see e.g. tweets, blogs, social media. The accuracy of the data analysis strongly depends on the quality of the data and its pre-processing.

In 2011, IDC added, yet, another dimension to Big Data analysis: "**Value**". Value means that Big Data Analysis seeks to economically extract value from very large volumes of a wide variety of data [16]. In other words, mining Big Data should provide novel insights into data, application problems, and create new economical value that would support better decision making; see some examples in [22].

Another "*V*", still, is for "**Variability**". Authors of [23] stress that there are changes in the structure of the data, e.g. inconsistencies which can be shown in the data as time goes on, as well as changes in how users want to interpret that data.

These are only a few of the definitions that have previously been proposed for Big Data Analysis. For an excellent survey on the topic, please see [20].

Note, however, that most of these definitions are general and geared at business. They are not that useful for Machine Learning Scientists. In this volume, we are more interested in a view of Big Data as it relates to Machine Learning and Data Mining research, which is why we explore the meaning of Big Data Analysis in that context next.

2.2 Machine Learning and Data Mining Versus Big Data Analysis

One should remember that data mining or more generally speaking the field of Knowledge Discovery from Databases started in the late 1980s [50]—before the appearance of Big Data applications and research. Machine Learning is an older research discipline that provided many algorithms for carrying out data mining steps or inspired more specialized and complex solutions. From a methodological point of view, it strongly intersects with the field of data mining. Some researchers even identify traditional data mining with Machine Learning while others indicate differences, see e.g., discussions in [33, 41, 42].

It is not clear that there exists a simple, clear and concise Big Data definition that applies to Machine Learning. Instead, what Machine Learning researchers have done is list the kinds of problems that may arise with the emergence of Big Data. We now present some of these problems in the table below (due to its length it is actually split into two Tables 1 and 2). The table is based on discussions in [23, 24], but we organized them by categories of problems. We also extended these categories according to our own understanding of the field. Please note that some of the novel aspects of Big Data characteristics have already been discussed in the previous subsection, so here, we only mention those that relate to machine learning approaches in data mining.

Please note, as well, that this table is only an approximation: as with the boundary between machine learning and data mining, the boundary between the traditional data mining discipline and the Big Data analysis discipline is not clear cut. Some issues listed in the Big Data Analysis column occurred early on in the discipline, which could still be called traditional data mining. Similarly, some of the issues listed in the Data Mining column may, in fact, belong more wholly to the Big Data Analysis column even if early, isolated work on these problems had already started before the advent of the Big Data Analysis field. This is true at the data set level too: the distinction between a Data Mining problem and a Big Data Analysis problem is not straightforward. A problem may encounter some of the issues described in the "Big Data Analysis" column and still qualify as a data mining problem. Similarly, a problem that qualifies as a "Big Data" problem may not encounter all the issues listed in the "Big Data Analysis" column.

Once again, this table serves as an indication of what the term "Big Data Analysis" refers to in machine learning/data mining. The difference between traditional data mining and Big Data Analysis and most particularly, the novel elements introduced by Big Data Analysis, will be further explored in Sect. 2.4 where we look at specific problems that can and must now be considered. Prior to that, however, we take a brief look at a few successful applications of Big Data Analysis in the next section.

2.3 Some Well-Known and Successful Applications of Big Data Analysis

Big Data analysis has been successfully applied in many domains. We limit ourselves to listing a few well known applications, though these applications and many others are quite interesting and would have been given greater coverage if space restrictions had not been a concern:

Table 1 Part A—Traditional data mining versus big data analysis with respect to different aspects of the learning process

	Traditional data mining	Big data analysis
Memory access	The data is stored in centralized RAM and can be efficiently scanned several times	The data may be stored on highly distributed data sources In case of huge, continuous data streams, data is accessed only in a single scan and limited subsets of data items are stored in memory
Computational processing and architectures	Serial, centralized processing is sufficient A single-computer platform that scales with better hardware is sufficient	Parallel and distributed architectures may be necessary Cluster platforms that scale with several nodes may be necessary
Data types	The data source is relatively homogeneous The data is static and, usually, of reasonable size	The data may come from multiple data sources which may be heterogeneous and complex The data may be dynamic and evolving. Adapting to data changes may be necessary.
Data management	The data format is simple and fits in a relational database or data warehouses. Data management is usually well-structured and organized in a manner that makes search efficient. The data access time is not critical	Data formats are usually diverse and may not fit in a relational database The data may be greatly interconnected and needs to be integrated from several nodes Often special data systems are required that manage varied data formats (NoSQL databases, Hadoop or Spark platforms, etc.) The data access time is critical for scalability and speed
Data quality	The provenance and pre-processing steps are relatively well documented Strong correction techniques were applied for correcting data imperfection Sampling biases can, somehow, be traced back The data is relatively well tagged and labeled	The provenance and pre-processing steps may be unclear and undocumented There is a large amount of uncertainty and imprecision in the data Sampling biases are unclear Only a small number of data are tagged and labeled

Table 2 Part B—Traditional data mining versus big data analysis with respect to different aspects of the learning process

	Traditional data mining	Big data analysis
Data handling	Security and Privacy are not of great concern Policies about data sharing are not necessary	Security and Privacy may matter Data may need to be shared and the sharing must be done appropriately
Data processing	Only batch learning is necessary Learning can be slow and off-line The data fits into memory All the data has some sort of utility The curse of dimensionality is manageable No compression and minimal sampling is necessary Lack of sufficient data is a problem	Data may arrive in a stream and need to be processed continuously Learning may need to be fast and online The scalability of algorithms is important The data may not fit in memory The useful data may be buried in a mass of useless data The curse of dimensionality is disproportionate Compression or sampling techniques must be applied Lack of sufficient data of interest remains a problem
Result analysis and integration	Statistical significance results are meaningful Many visualization tools have been developed Interaction with users is well developed The results do not usually need to be integrated with other components	With massive data sets, non-statistically significant results may appear statistically significant Traditional visualization software may not work well with massive data The results of the Big Data analysis may need to be integrated with other components

- Google Flu Trends—Researchers at Google and the Center for Disease Control (CDC) teamed together to build and analyse a surveillance system for early detection of flu epidemics, which is based on tracking a different kind of information from flu-related web search queries [29].[1]
- Predicting the Next Deadly Manhole Explosion in New York Electric Network— In 2004 the Con Edison Company began a proactive inspection program, with the goal of finding the places in New York's network of electrical cables where trouble was most likely to strike. The company co-operated with a research team at

---

[1] While this application was originally considered a success, it subsequently obtained disappointing results and is now in the process of getting improved [4].

Columbia University to develop an algorithm that predicts future manhole failure and could support the company's inspection and repair programs [55, 56].

- Wal-Mart's use of Big Data Analytics—Wal-Mart has been using Big Data Analysis extensively to achieve a more efficient and flexible pricing strategy, better-managed advertisement campaigns and a better management of their inventory [36].
- IBM Watson—Watson is the famous Q&A Computer System that was able to defeat two former winners of the TV game show *Jeopardy!* in 2011 and win the first prize of one million dollars. This experiment shows how the use of large amounts of computing power can help clear up bottlenecks when constructing a sophisticated language understanding module coupled with an efficient question answering system [63].
- Sloan Sky Digital Survey—The Sloan Sky Digital Survey has gathered an extensive collection of images covering more than a quarter of the sky. It also created three-dimensional maps of over 930,000 galaxies and 120,000 quasars. This data is continuously analyzed using Big Data Analytics to investigate the origins of the universe [60].
- FAST—Homeland Security FAST (Future Attribute Screening Technology) intends to detect whether a person is about to commit a crime by monitoring the contractions of their facial muscles, which are, believed to reflect seven primary emotions and emotional cues linked to hostile intentions. Such a system would be deployed at airports, border crossings and at the gates of special events, but it is also the subject of controversy due to its potential violation of privacy and the fact that it would probably yield a large number of false positives [25].

The reviews of additional applications where Big Data Analysis has already proven itself worthy can be found in other articles such as [15, 16, 67].

2.4 Machine Learning Innovations Driven by Big Data Analysis

In this subsection, we look at the specific new problems that have emanated from the handling of Big Data sets, and the type of issues they carry with them. This is an expansion of the information summarized in Tables 1 and 2.

We divide the problems into two categories:

1. The completely new problems which were never considered, or considered only in a limited range, prior to the advent of Big Data analysis, and originate from the format in which data is bound to present itself in Big Data problems.
2. The problems that already existed but have been disproportionately exaggerated since the advent of Big Data analysis.

2.4.1 New Problems Caused by the Format in Which Big Data Presents Itself

We now briefly discuss problems that were, perhaps, on the mind of some researchers prior to the advent of Big Data, but that either did not need their immediate attention since the kind of data they consider did not seem likely to occur immediately, or were already tackled but need new considerations given the added properties of Big Data sets. The advent and fast development of the Internet and the capability to store huge volumes of data and to process them quickly changed all of that. New forms of data have emerged and are here to stay, requiring new techniques to deal with them. These new kinds of data and the problems associated with them are now presented.

Graph Mining Graphs are ubiquitous and can be used to represent several kinds of networks such as the World Wide Web, citation graphs, computer networks, mobile and telecommunication networks, road traffic networks, pipeline networks, electrical power-grids, biological networks, social networks and so on. The purpose of graph mining is to discover patterns and anomalies in graphs and use these discoveries to useful means such as fraud detection, cyber-security, or social network mining (which will be discussed in more detail below). There are two different types of graph analyses that one may want to perform [37]:

- Structure analysis (which allows the discovery of patterns and anomalies in connected components, the monitoring of the radius and diameter of graphs, their structure, and their evolution).
- Spectral Analysis (which allows the analysis of more specific information such as tightly connected communities and anomalous nodes).[2]

Structure analysis can be useful for discovering anomalously connected components that may signal anomalous activity; it can also be used to give us an idea about the structure of graphs and their evolution. For example, [37] discovered that large real-world graphs are often composed of a "core" with small radius, "whiskers" that still belong to the core but are more loosely connected to it and display a large radius, and finally "outsiders" which correspond to disconnected components each with a small radius.

Spectral Analysis, on the other hand, allows for much more pinpointed discoveries. The authors of [37] were able to find that adult content providers create many twitter accounts and make them follow each other so as to look more popular. Spectral Analysis can thus be used to identify specific anomalous (and potentially harmful) behaviour that can, subsequently, be eliminated.

Mining Social Networks The notion of social networks and social network analysis is an old concept that emanated from Sociology and Anthropology in the 1930s [7]. The earlier research was focused on analysing sociological aspects of personal relationships using rigorous data collection and statistical analysis procedures. However,

---

[2]Please note that graphs were sometimes considered in traditional data mining (e.g., as structures of chemical compounds), but the graphs in question were of much smaller size than those considered today.

with the advent of social media, this kind of study recently took a much more concrete turn since all traces of social interactions through these networks are tangible.

The idea of Social Network Analysis is that by studying people's interactions, one can discover group dynamics that can be interesting from a sociological point of view and can be turned into practical uses. That is the purpose of mining social networks which can, for example, be used to understand people's opinions, detect groups of people with similar interests or who are likely to act in similar ways, determine influential people within a group and detect changes in group dynamics over time [7].

Social Network Mining tasks include:

- Group detection (who belongs to the same group?),
- Group profiling (what is the group about?),
- Group evolution (understanding how group values change),
- Link prediction (predict when a new relationship will form).

Social Network Analysis Applications are of particular interest in the field of business since they can help products get advertised to selected groups of people likely to be interested, can encourage friends to recommend products to each other, and so on.

Dealing with Different and Heterogeneous Data Sources Traditional machine learning algorithms are typically applied to homogeneous data sets, which have been carefully prepared or pre-processed in the first steps of the knowledge discovery process [50]. However, Big Data involves many highly heterogeneous sources with data in different formats. Furthermore, these data may be affected by imprecision, uncertainty or errors and should be properly handled. While dealing with different and heterogeneous data sources, two issues are at stake:

1. How are similar data integrated to be presented in the same format prior to analysis?
2. How are data from heterogeneous sources considered simultaneously in the analysis?

The first question belongs primarily to the area of designing data warehouses. It is also known as the problem of data integration. It consists of creating a unified database model containing the data from all the different sources involved.

The second question is more central to data mining as it may lead researchers to abandon the construction of a single model from integrated and transformed data in favour of an approach that builds several models from homogeneous subsets of the overall data set and integrates the results together (see e.g., [66]).

Combining the questions on graph mining discussed in the previous subsections and on heterogeneous sources discussed here leads to a commonly encountered problem: that of analysing a network of heterogeneous data (e.g., the nodes of the network represent people, documents, photos, etc.). This started a new sub-field called heterogeneous information network analysis [61], which consists of using a network scheme listing meta-information about the nodes and the links.

Data Stream Mining In the past, most of machine learning research and applications were focused on batch learning from static data sets. These, usually not massive, data sets were efficiently stored in databases or file systems and, if needed, could be accessed by algorithms multiple times. Moreover, the target concepts to be learned were well defined and stable. In some recent applications, learning algorithms have had to act in dynamic environments, where data are continuously produced at high speed. Examples of such applications include sensor networks, process monitoring, traffic management, GPS localizations, mobile and telecommunication call networks, financial or stock systems, user behaviour records, or web log analysis [27]. In these applications, incoming data form a data stream characterized by a huge volume of instances and a rapid arrival-rate which often requires a quick, real-time response.

Data stream mining, therefore, assumes that training examples arrive incrementally one at a time (or in blocks) and in an order over which the learning algorithm has no control. The learning system resulting from the processing of that data must be ready to be applied at any time between the arrivals of two examples, or consecutive portions (blocks) of examples [11]. Some earlier learning algorithms, like Artificial Neural Networks or Naive Bayes, were naturally incremental. However, the processing of data streams imposes new computational constraints for algorithms with respect to memory usage, limited learning and testing time, and single scanning of incoming instances [21]. In practice, incoming examples can be inspected briefly, cannot all be stored in memory, and must be processed and discarded immediately in order to make room for new incoming examples. This kind of processing is quite different from previous data mining paradigms and has new implications on constructing systems for analysing data streams.

Furthermore, with stream mining comes an important and not insignificant challenge: these algorithms often need to be deployed in dynamic, non-stationary environments where the data and target concepts change over time. These changes are known as concept drifts and are serious obstacles to the construction of a useful stream-mining system [39].

Finally, from a practical point of view, mining data streams is an exciting area of research as it will lead to the deployment of ubiquitous computing and smart devices [26].

Unstructured or Semi-Structured Data Mining Most Big Data sets are not highly structured in a way that can be stored and managed in relational databases. According to many reports, the majority of collected data sets are semi-structured, like in the case of data in HTML, XML, JSON or bibtex format, or unstructured, like in the case of text documents, social media forums or sound, images or video format [1]. The lack of a well-defined organization for these data types may lead to ambiguities and other interpretation problems for standard data mining tools.

The typical way to deal with unstructured data sets is to find ways to impose some structure on them and/or transform them into another representation, in order to be able to process them with existing data mining tools. In text mining, for example, it is customary to find a representation of the text using Natural Language Processing and Text Analytic tools. These include tools for removing redundancies and inconsistencies, tokenization, eliminating stop words, stemming, identification of terms based on unigrams, bigrams, phrases or other features of the text which could lead to vector space models [43]. Some of these methods may also require collecting reference corpora of documents. Similar approaches are used for images or sound where high-level features are defined and used to describe the data. These features can then be processed by traditional learning systems.

Spatio-Temporal Data Mining Spatio-temporal data corresponds to data that has both temporal and spatial characteristics. The temporal characteristics refer to the fact that over time certain changes apply to the object under consideration and these changes are recorded at certain time intervals. The spatial aspect of the data refers to the location and shape of the object. Typical spatio-temporal applications include environment and climate (global change, land-use classification monitoring), the evolution of an earthquake or a storm over time, Public Health (monitoring and predicting the spread of disease), public security (finding hotspots of crime), geographical maps and census analysis, geo-sensor measurement networks, transportation (traffic monitoring, control, traffic planning, vehicle navigation), tracking GPS/mobile and localization-based services [54, 57, 58].

Handling spatio-temporal data is particularly challenging for different reasons. First, these data sets are embedded in continuous spaces, whereas typical data are often static and discrete. Second, classical data mining tends to focus on discovering the global patterns of models while in spatio-temporal data mining there is more interest on local patterns. Finally, spatio-temporal processing also includes aspects that are not present with other kinds of data processing. For example, geometric and temporal computations need to be included in the processing of the data, normally implicit spatial and temporal relationships need to be explicitly extracted, scale and granularity effects in space and time need to be considered, the interaction between neighbouring events has to be considered, and so on [65]. Moreover, the standard assumption regarding sample independence is generally false because spatio-temporal data tends to be highly correlated.

Issues of Trust/Provenance Early on, data mining systems and algorithms were typically applied to carefully pre-processed data, which came from relatively accurate and well-defined sources, thus trust was not a critical issue. With emerging Big Data, the data sources have many different origins, which may be less known and not all verifiable [15]. Therefore, it is important to be aware of the provenance of the data and establish whether or not it can be trusted [17]. Provenance refers to the path that the data has followed before arriving at its destination and Trust refers to whether both the source and the intermediate nodes through which the database passed are trustworthy.

Typically, data provenance explains the creation process and origins of the data records as well as the data transformations. Note that provenance may also refer to the type of transformation [58] that the data has gone through, which is important for people analysing it afterwards (in terms of biases in the data). Additional metadata, such as conditions of the execution environment (the details of software or computational system parameters), are also considered as provenance.

Data provenance has previously been studied in the database, workflow and geographical information systems communities [18]. However, the world of Big Data is much more challenging and still not sufficiently explored. The main challenges in Big Data Provenance come from working with:

- massive scales of sources and their inter-connections as well as highly unstructured and heterogeneous data (in particular, if users also apply ad-hoc analytics, then it is extremely difficult to model provenance [30]);
- complex computational platforms (if jobs are distributed onto many machines, then debugging the Big Data processing pipeline becomes extremely difficult because of the nature of such systems);
- data items that may be transformed several times with different analytical pieces of software;
- extremely long runtimes (even with more advanced computational systems, analysing provenance and tracking errors back to their sources may require unacceptably long runtimes);
- difficulties in providing sufficiently simple and transparent programming models as well as high dynamism and evolution of the studied data items.

It is therefore an issue to which consideration must be given, especially if we expect the systems resulting from the analysis to be involved in critical decision making.

Privacy Issues Privacy Preserving Data Mining deals with the issue of performing data mining, i.e., drawing conclusions about the entire population, while protecting the privacy of the individuals on whose information the processing is done. This imposes constraints on the regular task of data mining. In particular, ways have to be found to mask the actual data while preserving its aggregate characteristics. The result of the data mining process on this constrained data set needs to be as accurate as if the constraint were not present [45].

Although privacy issues had been noticed earlier, they have become extremely important with the emergence of mining Big Data, as the process often requires more personal information in order to produce relevant results. Instances of systems requiring private information include localization-based and personalized recommendations or services, targeted and individualized advertisements, and so on. Systems that require a user to share his geo-location with the service provider are of particular concern since even if the user tries to hide his personal identity, without hiding his location, his precautions may be insufficient—the analysts could infer a missing identity by querying other location information sources. Barabasi et al. have, indeed, shown that there is a close correlation between people's identities and their movement patterns [31].

In social data sets, the privacy issue is particularly problematic since such sets usually contain many highly interconnected pieces of personal information. Even if the basic records could, somehow, be blocked from public view, a lot of personal information can be found and mined out when links to other data are found. At this point, all the pieces of information about a given person will be integrated and privacy compromised. Cukier and Mayer-Schoenberger describe several such case studies in their book [47]; see, for example, the surprising results obtained by an experimental analysis of old queries provided by AOL. Although the personal names and IP were anonymized, researchers were able to correctly identify a single person by looking at associations between particular search phrases and additional data [6]. A similar situation occurred in the Netflix Prize Datasets, where researchers discovered correlations of ranks similar to those found in data sets from other services that used the users' full names [49]. This allowed them to clearly identify the anonymized users of the Netflix data.

This concludes our review of new problems that stemmed from the emergence of Big Data sets. We now move to existing problems that were amplified by the advent of Big Data.

2.4.2   Existing Problems Disproportionately Exaggerated by Big Data

Although the learning algorithms derived in the past were originally developed for relatively small data sets, it is worth noting that machine learning researchers have always been aware of the computational efficiency of their algorithms and of the need to avoid data size restrictions. Nonetheless, these efforts are not sufficient to deal with the flood of data that Big Data Analysis brought about. The two main problems with Big Data analysis, other than the emergence of new data format as discussed in previous subsections, consequently, are that:

1. The data is too big to fit into memory and is not sufficiently managed by typical analytical systems using databases.
2. The data is not, currently, processed efficiently enough.

The first problem is addressed by the design of distributed platforms to store the data and the second, by the parallelization of existing algorithms [15]. Some efforts have already been made in both directions and these are, now, briefly presented.

Distributed Platforms and Parallel Processing There have been several ventures aimed at creating distributed processing architectures. The best known one, currently, is the pioneering one introduced by Google. In particular, Google created a programming model called MapReduce which works hand in hand with a distributed file system called Google File System (GFS). Briefly speaking, MapReduce is a framework for processing parallelizable problems over massive data sets using a large number of computer nodes that construct a computational cluster. The programming consists of two steps: map and reduce. At the general level, map procedures read data from the distributed file system, process them locally and generate intermediate results, which are aggregated by reduce procedures into a final output. The framework also provides the distributed shuffle operations (which manage communication and data transfers), the orchestration of running parallel tasks, and deals with redundancy and fault tolerance.

Yahoo and other companies emulated the MapReduce architecture in an opensource framework. That Apache version of MapReduce is called Hadoop MapReduce and uses the Hadoop Distributed File System (HDFS), which is the open-source Apache equivalent of GFS [32]. The term Hadoop also refers to the collection of additional software wrappers that can be installed on top of Hadoop and MapReduce, and can provide programmers with a better environment, see, for example, Apache Pig (SQL-like environment), Apache Hive (Hive is a warehouse system that conquers and analyses files stored in HDFS) and Apache HBase (a massive scale database management system) [59].

Hadoop and MapReduce are not the only platforms around. In fact, they have several limitations: most importantly, MapReduce is inefficient for running iterative algorithms, which are often applied in data mining. A few new fresh platforms have recently been developed to deal with this issue. The Berkeley Data Analytics Stack (BDAS) [9] is the next generation open-source data analysis tool for computing and analysing complex data. In particular, the BDAS component, called Spark, represents a new paradigm for processing Big Data, which is an alternative to Hadoop and should overcome some of its I/O limitations and eliminate some disk overhead in running iterative algorithms. It is reported that for some tasks it is much faster than Hadoop. Several researchers claim that Spark is better designed for processing machine learning algorithms and has much better programming interfaces. There are also several Spark wrappers such as Spark Streaming (large scale real time stream processing), GraphX (distributed graph system), and MLBase/Mlib (distributed machine learning library based on Spark) [38]. Other competitive platforms are ASTERIX or SciDB. Furthermore, specialized platforms for processing data streams include Apache S4 and Storm.

The survey paper [59] discusses criteria for evaluating different platforms and compares their application dependent characteristics.

Parallelization of Existing Algorithms In addition to the Big Data platforms that have been developed by various companies and, in some cases, made available to the public through open source platforms, a number of machine learning algorithms have been parallelized and placed in software packages made available to the public through open source channels.

Here is a list of some of the most popular open source packages:

- Apache's Mahout [40] which includes many implementations of distributed or otherwise scalable machine learning algorithms focused primarily on the areas of collaborative filtering, clustering and classification. Many of the implementations originally used the Apache Hadoop and MapReduce framework. However, some researchers judged that the implementations are too slow and the package not user-friendly [15]. In April 2014 the Mahout community decided to move its codebase onto newer data processing systems, such as Apache Spark, that offer a richer programming model and more efficient execution than Hadoop and MapReduce.
- BC-PDM (Big Cloud-Parallel Data Mining) is a cloud based series of implementations also based on Hadoop. It also supports parallel ETL (Extraction Transformation Load) processes and is more applicable to industrial Business Intelligence.
- MOA is an open source software package for stream data mining and contains implementations of classifiers, regression, clustering and frequent set mining [11]. Another newer, related project for distributed stream mining is the SAMOA project [48].

- NIBLE is yet another portable toolkit for implementing parallel ML-DM algorithms and runs on top of Hadoop [28].
- VowpalWabbit was developed by Yahoo and Microsoft Research. Its main aim is to provide efficient scalable implementations of online machine learning and support for a number of machine learning reductions, importance weighting, and a selection of different loss functions and optimization algorithms.
- h2o is the most recent open source mathematical and machine learning software for Big Data, released by Oxdata in 2014 [62]. It offers distribution and parallelism to powerful algorithms and allows programmers to use the R and JSON languages as APIs. It can be run on the top of either Hadoop or Spark.
- Graph mining tools are often used in mining Big Data. PEGASUS (Peta-scale Graph Mining System) is an open source package specifically designed for graph mining and also based on Hadoop. Giraph and GraphLab are two other such systems for Graph Mining.

A comprehensive survey of the various efforts made to scale up algorithms for parallel and distributed platforms can be found in the book entitled "Scaling Up Machine Learning. Parallel and Distributed Approaches" [8].

This concludes our general overview of Big Data Analysis from a Machine Learning point of view. The next section will discuss the scientific and societal changes that Big Data Analysis has led to.

3 Is Big Data Analysis a Game Changer?

In this section, we discuss the visions of people who believe that the foundations of science and society are fundamentally changing due to the emergence of Big Data. Some people see it as a natural and positive change, while others are more critical as they worry about the risks Big Data Analysis pose to Science and Society. We begin by surveying the debate concerning potential changes in the way scientific research is or will be conducted, and move on to the societal effects of Big Data Analysis.

3.1 Big Data Analysis and the Scientific Method

A few prominent researchers have recently suggested that there is a revolution underway in the way scientific research is conducted. This argument has three main points:

- Traditional statistics will not remain as relevant as it used to be,
- Correlations should replace models, and
- Precision of the results is not as essential as it was previously believed to be.

These arguments, however, are countered by a number of other scientists who believe that the way scientific research is conducted did not and should not change as radically as advocated by the first group of researchers. In this section, we look at the arguments for and against these statements.

Arguments in support of the Big Data revolution

The four main proponents of this vision are Cukier, Mayer-Shoenberger, Anderson and Pentland [3, 47, 52]. Here are the rationales they give for each issue:

Traditional Statistics Will Not Remain as Relevant as It Used to Be: With regard to this issue, Cukier and Mayer-Schoenberger [47] point out that humans have always tried to process data in order to understand the natural phenomena surrounding them and they argue that Big Data Analysis will now allow them to do so better. They believe that the reason why scientists developed Statistics in the 19th century was to deal with small data samples, since, at that time, they did not have the means to handle large collections of data. Today, they argue, the development of technology that increases computer power and memory size, together with the so-called "datafication" of society makes it unnecessary to restrict ourselves to small samples.

This view is shared, in some respect, by Alex Pentland who believes that more precise results will be obtainable, once the means to do so are derived. He bases his argument on the observation that Big Data gives us the opportunity not to aggregate (average) the behaviour of millions, but instead to take it into consideration at the micro-level [52]. This argument will be expanded further in a slightly different context in the next subsections.

Correlations Should Replace Models: This issue was advocated by Anderson in his article provocatively titled "The End of Theory" [3] in which he makes the statement that theory-based approaches are not necessary since "with enough data the numbers speak for themselves". Cukier and Mayer-Schoenberg agree as all three authors find that Big Data Analysis is changing something fundamental in the way we produce knowledge. Rather than building models that explain the observed data and show what causes the phenomena to occur, Big Data forces us to stop at understanding how data correlates with each other. In these authors' views, abandoning explanations as to why certain phenomena are related or even occur can be justified in many practical systems as long as these systems produce accurate predictions. In other words, they believe that "the end justifies the means" or, in this case, that "the end can ignore the means". Anderson even believes that finding correlations rather than inducing models in the traditional scientific way is more appropriate. This, he believes, leads to the recognition that we do not know how to induce correct models, and that we simply have to accept that correlations are the best we can do. He further suggests that we need to learn how to derive correlations as well as we can since, despite them not being models, they are very useful in practice.

Precision of the Results Is Not as Essential as It Was Previously Believed to Be: This issue is put forth by Cukier and Mayer-Schoenberger who assert that "looking at vastly more data (...) permits us to loosen up our desire for exactitude" [47]. It is, once again, quite different from traditional statistical data analysis, where samples had to be clean and as errorless as possible in order to produce sufficiently accurate results. Although they recognize that techniques for handling massive amounts of unclean data remain to be designed, they also argue that less rigorous precision is acceptable as Big Data tasks often consists of predicting trends at the macro level. In the Billion Price Project, for example, the retail price index based on daily sales data in a large number of shops is computed from data collected from the internet [12]. Although these predictions are less precise than results of systematic surveys carried out by the US Bureau of Labour Statistics, they are available much faster, at a much lesser cost and they offer a sufficient accuracy for the majority of users. The next part of this subsection considers the flip-side of these arguments.

Arguments in denial of the Big Data revolution

There have been a great number of arguments denying that a Big Data revolution is underway, or at least, warning that the three main points just discussed are filled with misconceptions and errors. The main proponents of these views are: Danah Boyd and Kate Crawford, Zeynep Tufekci, Tim Harford, Wolfgang Pietsch, Gary Marcus and Ernest Davis, Michael Jordan, David Ritter, and Alex Pentland (who participates in both sides of the argument). Once again, we examine each issue separately.

Traditional Statistics Will Not Remain as Relevant as It Used to Be: The point suggesting a decline in the future importance of traditional Statistics in the world of Big Data Analysis raises three sets of criticisms. The first one comes with a myriad of arguments that will now be addressed:

- Having access to massive data sets does not mean that there necessarily is a sufficient amount of appropriate data to draw relevant conclusions from without having recourse to traditional statistics tools
  In particular,

  - **Sample and selection biases will not be eliminated:** The well-known traps of traditional statistical analysis will not be eliminated by the advent of Big Data Analysis. This important argument is made by Danah Boyd and Kate Crawford as well as Tim Harford and Zeynep Tufekci. Tufekci, in particular, looks at this issue in the context of Social Media Analysis [64]. She notes, for example, that most Social Media research is done with data from Twitter. The reasons are that Twitter data is accessible to all (Facebook data, on the other hand, is proprietary) and has an easy structure. The problem with this observation is that not only is Twitter data not representative of the entire population, but by the features it presents it forces the users to behave in certain ways that would not necessarily happen on different platforms.
  - **Careful Variable Selection is still warranted:** The researchers that argue that more data is better and that better knowledge can be extracted from large data sets are not necessarily correct. For example, the insights that can be extracted from a qualitative study using only a handful of cases and focusing on a few carefully selected variables may not be inferable from a quantitative study using thousands of cases and throwing in hundreds of variables simultaneously, see, e.g. Tim Harford's essay [34].

- **Unknowns in the data and errors are problematic:** These are other problems recognized by both Boyd and Crawford and Tufekci [13, 14, 64]. An example of unknowns in the data is illustrated as follows: a researcher may know who clicked on a link and when the click happened, based on the trace left in the data, but he or she does not know who saw the link and either *chose* not to click it or *was not able* to click it. In addition, Big Data sets, particularly those coming from the Internet, are messy, often unreliable, and prone to losses. Boyd, Crawford and Tufekci believe that these errors may be magnified when many data sets are amalgamated together. Boyd and Crawford thus postulate that the lessons learned from the long history of scientific investigation, which include asking critical questions about the collection of data and trying to identify its biases, cannot be forgotten. In their view, Big Data Analysis still requires an understanding of the properties and limits of the data sets. They also believe that it remains necessary to be aware of the origins of the data and the researcher's interpretation of it. A similar opinion is, in fact, presented in [44, 51].
- **Sparse data remains problematic:** Another very important statistical limitation, pointed out by Marcus and Davis in [44], is that while Big Data analysis can be successful on very common occurrences it will break down if the data representing the event of interest is sparse. Indeed, it is not necessarily true that massive data sets improve the coverage of very rare events. On the contrary, the class imbalance may become even more pronounced if the representation of common events increases exponentially, while that of rare events remains the same or increases very slowly with the addition of new data.

- **The results of Big Data Analysis are often erroneous:** Michael Jordan pulled the alarm on Big Data Analysis by suggesting that a lot of results that have been and will continue to be obtained using Big Data Analysis techniques are probably invalid. He bases his argument on the well-known statistical phenomenon of spurious correlations. The more data is available, the more correlations can be found. With current evaluation techniques, these correlations may look insightful, when, in fact, many of them could be discarded as white noise [2]. This observation is related to older statistical lessons on dealing with other dangers, such as the multiple comparison problems and false discovery.
- **Computing power has limitations:** [24] points out that even if computational resources improve, as the size of the data sets increases, the processing tools may not scale up quickly enough and the computations necessary for data analysis may quickly become infeasible. This means that the size of the data sets cannot be unbounded since even if powerful systems are available they can quickly reach their limit. As a result, sampling and other traditional statistical tools are not close to disappearing.

Correlations Should Replace Models: This issue is, once again, countered by three arguments:

- **Causality cannot be forgone:** In their article, Boyd and Crawford completely disagree with the provocative statement by Chris Anderson that Big Data Analysis will supersede any other type of research and will lead to a new theory-free perspective. They argue, instead, that Big Data analysis is offering a new tool in the scientific arsenal and that it is important to reflect on what this new tool adds to the existing ones and in what way it is limited. In no way, do they believe, however, that Big Data analysis should replace other means of knowledge acquisition since they believe that causality should not be replaced by correlations. Each has their place in scientific investigation. A similar discussion concerning the need to appreciate causality is expressed by Wolfgang Pietsch in his philosophical essay on the new scientific methodology [51]

- **Correlations are not always sufficient to take action:** In his note entitled "When to act on a correlation, and when not to", Ritter considers the dilemma of whether one can intervene on the basis of discovered correlations [53]. He recommends caution while taking actions. However, he also claims that the choice of acting or not depends on balancing two factors: (1) confidence that the correlation will re-occur in the future and (2) trade-off between risk and reward of acting. Following this, if the risk of acting and being wrong is too high, acting on strong correlations may not be justified. In his opinion, confidence in a correlation is a function of not only the statistical frequency but also the understanding of what is causing that correlation. He calls it the "clarity of causality" and shows that the fewer possible explanations there are for a correlation, the higher the likelihood that the two events are really linked. He also says that causality can matter tremendously as it can drive up the confidence level of taking action. On the other hand, he also distinguishes situations where, if the value of acting is high, and the cost of wrong decisions is low, it makes sense to act based on weaker correlations. So, in his opinion a better understanding of the dynamics of the data and working with causality is still critical in certain conditions, and researchers should better identify situations where correlation is sufficient to act on and what to do when it is not.

- **Big Data Analysis will allow us to understand causality much better:** Unlike Anderson and Cukier and Mayer-Schoenberger, Alex Pentland does not believe in a future without causality. On the contrary, in line with his view that Big Data Analysis will lead to more accurate results, he believes that Big Data will allow us to understand causalities much more precisely than in the past, once new methods for doing so are created. His argument, as seen earlier, is that up to now, causalities were based on averages. Big Data, on the other hand, gives us the opportunity not to aggregate the behaviour of millions, but instead to take it into consideration at the micro-level [52].

Precision of the Results Is Not as Essential as It Was Previously Believed to Be: This argument in favour of decreasing the rigour of the results is countered by two arguments as follows:

- **Big Data Analysis yields brittle systems:** When considering the tools that can be constructed from Big Data analysis engines, Marcus and Davis [44] point out that these tools are sometimes based on very shallow relationships that can easily be guessed and defeated. That is obviously undesirable and needs to be addressed in the future. They illustrate their point by taking as an example a tool for grading student essays, which relies on sentence length and word sophistication that were found to correlate well with human scores. A student knowing that such a tool will be used could easily write long non-sense sentences peppered with very sophisticated words to obtain a good grade.
- **Big Data Analysis yields tools that lack in robustness:** Because Big Data Analysis based tools are often built from shallow associations rather than provable deep theories, they are very likely to lack in robustness. This is exactly what happened with Google Flu Trends, which appeared to work well based on tests conducted on one flu season, but over-estimated the incidence of the flu the following year [4].

3.2 Big Data Analysis and Society

Kenneth Cukier and Viktor Mayer-Schoenberger as well as Alex Pentland believe that Big Data Analysis is on its way to changing society and that it is doing so for the better. Others wonder whether that is indeed the case and warn against the dangers of this changed society. After summarizing Pentland and Cukier and Mayer-Schoenberger's positive vision, we survey the issues that have come up against the changes that Big Data Analysis is bringing to Society.

The Benefits of Big Data Analysis for Society

Alex Pentland is a great believer in the societal changes that Big Data Analysis can bring about. He believes that the management of organizations such as cities or governments can be improved using Big Data analysis and develops a vision for the future in his article entitled the "Data Driven Society" [52]. In particular, he believes, from his research on social interactions, that free exchanges between entities (people, organizations, etc.) improve productivity and creativity. He would, therefore, like to create societies that permit the flow of ideas between citizens, and believes that such activity could help prevent major disasters such as financial crashes, epidemics of dangerous diseases and so on. Cukier and Mayer-Schoenberger agree that Big Data Analysis applications can improve the management of organizations or the effectiveness of certain processes. However, they do not go as far as Pentland who implemented the idea of an open-data city in an actual city (Trento, Italy), which is used as a living lab for this experiment.

The Downside of Big Data Analysis for Society

In this section, we discuss the perception of negative societal repercussions that have been discussed since the advent of Big Data Analysis. First, however, we would like to mention that not everyone is convinced that Big Data Analysis is as significant as it is made up to be. Marcus and Davis, for example, wonder whether the hype given to Big Data analysis is justified [44]. Big Data analysis is held as a revolutionary advance, and as Marcus and Davis suggest, it is an important innovation, but they wonder how tools built from Big Data such as Google Flu Trends compare to advances such as the discovery of antibiotics, cars or airplanes. This consideration aside, it is clear that Big Data Analysis causes a number of changes that can affect society, and some, in a negative way, as listed below:

- **Big Data Analysis yields a carefree/dangerous attitude toward the validity of the results:** Traditional Statistical tools rely on assumptions about the data characteristics and the way it was sampled. However, as previously discussed, such assumptions are more likely to be violated when dealing with huge data sets whose provenance is not always known, and which have been assembled from disparate sources [24]. Because of these data limitations, Boyd and Crawford, as well as Tufekci, caution scientists against wrong interpretations and inferences from the observed results. Indeed, massive data makes the researchers less careful about what the data set represents: instances and variables are just thrown in with the expectation that the learning system will spit out the relevant results. This danger was less present in carefully assembled smaller data collections.

- **Big Data Analysis causes a mistaken semblance of authority:** Marcus and Davis [44] note that answers given by tools based on Big Data analysis may give a semblance of authority when, in fact, the results are not valid. They cite the case of tools that search large databases for an answer. In particular, they cite the example of two tools that searched for a ranking of the most important people in history from Wikipedia documents. Unfortunately, the notion of "importance" was not well defined and because the question was imprecise, the tools were allowed to go in unintended directions. For example, although the tools correctly retrieved people like Jesus, Lincoln and Shakespeare, one of them also asserted that Francis Scott Key, whose claim to fame is the writing of the US National Anthem, "The Star-Spangled Banner", was the 19th most important poet in history. The tools seem authoritative because they are exhaustive (or at least, they search a much larger space than any human could possibly search), however, they suffer from the same "idiot savant" predicament as the 1970s expert systems.

- **Data Privacy and Transparency are compromised by Big Data Analysis:** Many Big Data studies concern personal data. Some personal data are submitted by individuals on their own initiative (as in social networks or as a result of gaining access to free services), others may be collected automatically (by using some devices or specific services) or may be shared with external sources to enrich data sets. Finally, some data may be inferred from other data, and the apparent anonymity may get lost, as was previously discussed. Therefore, privacy or data protection are more serious challenges than they have ever been before. While a number of computational and legal solutions have been proposed, this problem is far from resolved and will continue to cause great concern in society. In his open-data city experiment, Pentland proposes a solution to this issue in which people would keep ownership of their data the way they do of money in a bank, and, likewise, would control how this data is used by choosing to share it or not, on a one-to-one basis. Another solution is proposed by Kord Davis, the author of [19], who believes in the need for serious conversations among the Big Data Analysis community regarding companies' policies and codes of ethics related to data privacy, identifiable customer information, data ownership and allowed actions with data results. In his opinion, transparency is a key issue and the data owners need to have a transparent view of how personal data is being used. Transparent rules should also refer to the case of how data is sold or transferred to other, third parties [5]. In addition, transparency may also be needed in the context of algorithms. For instance, Cukier and Mayer-Schoenberger, in Chap. 9 of their book [47], call for the special monitoring of algorithms and data, especially if they are used to judge people. This is another critical issue since algorithms may make decisions concerning bank credits, insurance or job offers depending on various individual data and indicators of individual behaviour.

- **Big Data Analysis causes a new digital divide:** As previously mentioned, and noted by Boyd and Crawford and Tufekci, everyone has access to most of Twitter's data, but not everyone can access Google or Facebook data. Furthermore, as discussed by Boyd and Crawford, Big Data processing requires access to large computers, which are available in some facilities but not others. As well, Big Data research is accessible to people with the required computational skills but not to others. All these requirements for working in the field of Big Data Analysis create a divide that will perpetuate itself since students trained in top-class universities where large computing facilities are available, and access to Big Data may have been paid for, will be the ones publishing the best quality Big Data research and be invited to work in large corporations, and so on. As a result, the other less fortunate individuals will be left out of these interesting and lucrative opportunities.

This concludes our discussion of the effect of Big Data Analysis on the world, as we know it. The next section takes a look at the various scientific contributions made in the remainder of this volume and organizes them by themes and applications.

4   Edited Volume's Contributions

The contributed chapters of this book span the whole framework established in this introduction and enhance it by providing deeper investigations and thoughts into a number of its categories. The papers can be roughly divided into two groups: the problem-centric contributions and the domain-centric ones. Though most papers span both groups, they were found to put more emphasis toward one or the other one and are, therefore, classified accordingly.

In the problem-centric category, we present four chapters on the following topics:

1. The challenges of Big Data Analysis from a Statistician's viewpoint
2. A framework for Problem-Solving Support tools for Big Data Analysis

3. Proposed solutions to the Concept Drift problem
4. Proposed solutions to the mining of complex Information Networks

In the domain-centric category, we present seven chapters that fit in the areas of Business, Science and Technology, and Life Science. More specifically, the papers focus on the following topics:

1. Issues to consider when using Big Data Analysis in the Business field
2. Dealing with data uncertainties in the Financial Domain
3. Dealing with Capacity issues in the Insurance Domain
4. New issues in Big Data Analysis emanating from the Internet of Things
5. The mining of complex Information Networks in the Telecommunication Sector
6. Issues to consider when using Big Data Analysis for DNA sequencing
7. High-dimensionality in Life Science problems

We now give a brief summary of each of these chapters in turn, and explain how they fit in the framework we have created. A deeper discussion of each of these contributions along with their analysis will be provided in the conclusion of this edited volume. The next four paragraphs pertain to the problem-centric type of papers.

4.1 Problem Centric Contributions

In the problem centric contribution, four types of problems were considered: the relationship between Big Data Analysis and Statistics, the creation of supporting tools for Big Data Analysis, the problem of concept drift, and the issues surfacing when handling information networks.

Big Data Analytics and Statistics Chapter "An Insight on Big Data Analytics" by Ross Sparks, Adrien Ickowicz, and Hans J. Lenz, gives an excellent discussion of the statistical problems raised by a careless application of algorithmic tools to Big Data sets without prior statistical considerations. The chapter begins by discussing the statistical issues that come up when several data sets originating from different sources are joined together. It also questions whether all the data available is, in fact, really needed. This leads to a discussion of how to manage the size of the data. The solution to this problem can take a couple of forms: a series of tools and techniques for decreasing the size of the data sets is presented and a discussion on how to decompose problems into manageable chunks is also proposed. After commenting on the fact that Big Data Analysts and Statisticians take different views on the question of Big Data and need to collaborate rather than ignore each other, the chapter tackles the question of whether correlations without an underlying model are sufficient. This chapter fits perfectly and expands on our Big Data Analysis framework. It addresses a lot of the questions raised in Sect. 3.1 which discusses "Big Data Analysis and the Scientific Method", and brings additional issues for the reader to consider. It also briefly touches upon the question of data ownership that was raised in Sect. 3.2.

Supporting Tools for Big Data Analysis Chapter "Toward Problem Solving Support based on Big Data and Domain Knowledge: Interactive Granular Computing based Computing and Adaptive Judgment" by Andrzej Skowron, Andrzej Jankowski and Soma Dutta, presents an innovative framework for developing support tools for Big Data Analysis using ideas from the field of Interactive Agents. More specifically, the chapter introduces a framework for modelling interactive agents that can be deployed in decision support tools for helping users deal with their problem solving tasks in the context of Big Data Analysis. The idea of Interactive Granular Computing is put forth which extends basic Granular Computing with the notion of complex granules. A new kind of reasoning called Adaptive Judgment is also introduced to help control, predict and bound the behaviour of the system when large scales of data are involved. This chapter expands our Big Data Analysis framework by considering the construction of support tools to help a user solve the complex tasks he or she encounters when dealing with Big Data. It is most related to the "Result Analysis and Integration" entry of Table 2 which it illustrates in an interesting way. It is not a chapter that discusses learning from the data per se. Instead, it looks at how the interactive agents learn and adapt as learning from the data progresses.

Concept Drift Chapter "An overview of concept drift applications" by Indrė Žliobaitė, Mykola Pechenizkiy, and João Gama, proposes a very nice framework for classifying problems in which concept drifts occur in terms of the task they solve, the kind of drift they encounter and the regimen by which the data and its characteristics become available. They simultaneously classify the problems with respect to their type: monitoring and control, information management, analytics and diagnostics; and within different industrial sectors. For each of these types, they identify solutions that have previously been proposed and illustrate the type with a concrete example. This chapter provides an excellent point of departure for researchers interested in delving into the concept-drift problem. The chapter fits well within our Big Data Analysis framework as it covers an important aspect of Big Data analysis mentioned in the "Data Types" entry of Table 1. It also fits closely within the discussion on "Data Stream mining" in Sect. 2.4 where concept drifts are very likely to occur.

Information Networks Chapter "Analysis of text-enriched heterogeneous information networks" by Jan Kralj, Anita Valmarska, Miha Grcar, Marko Robnik-Sikonja and Nada Lavrac provides an up-to-date introduction to information network analysis distinguishing between three types of information networks: homogeneous, heterogeneous and text-enriched heterogeneous information networks. They then survey the various tasks commonly performed in each type of information network. These include various kinds of classifications, rankings, link predictions, graph extraction, etc. Next, they present a specific method for mining text-enriched information networks which combines text mining as well as previously proposed mining from text-enriched heterogeneous information networks techniques. This chapter provides, once again, an excellent point of departure for researchers interested in working with information networks together with a concrete example of one such advanced study. The chapter also fits well within our Big Data Analysis framework as it covers an important aspect of Big Data analysis mentioned in the "Data man-

agement" entry of Table 1. It also fits closely within Sect. 2.4 which discusses "graph mining" and "social network mining", respectively.
We now move on to the description of the domain-centric papers.

4.2 Domain Centric Contributions

Three broad domains were considered in the domain centric contributions of this book: the business sector, the science and technology sector, and the life sciences sector.

4.2.1 Business

A Framework for Big Data Analysis in Business Chapter "Implementing Big Data Analytics Projects in Business" by Françoise Fogelman-Soulié and Wenhuan Lucan can be viewed as a concise, yet exhaustive, guide for introducing Big Data Analysis to a company. The chapter begins by taking the reader through the series of steps that form the task of Big Data Analysis. They include data collection, data cleaning, feature engineering, modelling, evaluation and deployment. The chapter emphasizes the type of skills required by employees of a firm involved in Big Data Analysis. These are Statistical skills to build, evaluate and analyze models; Information Technology skills to collect data, engineer features and deploy models; and Business skills to ask the right questions, identify critical issues, and evaluate the models' ultimate value for the business. The infrastructure needed to perform Big Data Analysis is also discussed. In this context, the chapter first introduces the notion of Data Lakes which keep a large collection of data available for different Big Data Analysis projects the company may, at different times, want to engage in given the strategic value of such moves. Data Lakes are the successors of Data Warehouses which have become too small given the scale of Big Data sets and cannot adapt easily to dynamic data. The chapter also touches upon Big Data platforms and Big Data Analysis software available for Business projects. It overviews virtually all aspects discussed in Tables 1 and 2, but does so with a business application in mind. It is meant to introduce company executives to the realities of dealing with Big Data in their business. The discussion on infrastructure is related to the "Data management" entry of Table 1 and it addresses some of the questions raised in Sect. 2.4 on "Distributed Platforms and Parallel Processing". As a matter of fact, it brings new elements to that discussion, while also discussing other management issues of which company executives should be aware before embarking on the Big Data bandwagon.

Big Data Models in Finance Chapter "Data Mining in Business: Current Advances and Future Challenges" by Eric Paquet, Herna Viktor, and Hongyu Guo addresses some of the important issues that come up in the Finance sector. In that domain, the data presents characteristics that are different from those found in other domains, making the traditional non-parametric approaches non-applicable. These characteristics include highly fluctuating data, data arriving at a fast rate, late-arriving data, and data including a lot of randomness with parameters that are difficult to estimate (i.e., modelling the unknown), handling conflicting information, and integrating boundary conditions such as the price of stocks when bought or sold. The chapter takes the reader through the various techniques that have been proposed to handle these kinds of data and points out the strengths and weaknesses of each approach. This chapter focuses on the topics covered in the "Data types" entry of Table 1. More specifically, it is in line with Sect. 2.4 on "Data Streams Mining" which it expands so as to discuss the specific types of issues encountered in the Finance domain.

Risk Analysis for Reinsurance Companies Chapter "Industrial-Scale Ad Hoc Risk Analytics Using MapReduce" by Andrew Rau-Chaplin, Zhimin Yao, and Norbert Zeh discusses solutions for the risk analysis problem that insurance companies need to assess. The particular problem considered in this chapter is the problem of risk analysis for reinsurance companies which are (secondary) insurance companies that insure other (primary) insurance companies. The idea for this model is that the reinsurance company would share in an agreed-upon percentage of the cost born by the primary insurance company in case where a claim is made to the primary insurance company, and the terms of this claim is covered by the agreement between the primary and secondary (re)insurance companies. The chapter focuses specifically on the amount of computing power necessary to respond to ad-hoc risk-analysis queries. The authors make it clear that such an application could not be carried out without a parallel architecture to support the computation. They argue that closed-form solutions to these queries cannot succeed given the amount of data involved in these estimations and that, instead, risk analysts have recourse to Monte-Carlo simulations. These are both data-intensive and time-consuming. The authors present a system implemented using MapReduce as well as other features of Apache Hadoop that can be used by risk analysts, with good knowledge of the field but little computer background, to answer these queries. A very nice feature of their paper is the time analysis of the system that they provide. The chapter's application is closer to the database side than the machine learning side of Big Data Analysis, but it is quite relevant. It pertains to the "computational processing and architectures" entry of Table 1 as well as to Sect. 2.4 on "Distributed Platforms" and "Parallel Processing and Parallelization of existing algorithms", respectively.

4.2.2 Science and Technology

The Internet of Things Chapter "Big Data and the Internet of Things" by Mohak Shah presents an excellent introduction to the concept of the Internet of Things (IoT) and the challenges that accompany it, and it discusses what aspects of Big

Data Analysis are particularly important to solve these challenges. The specific challenges anticipated in the context of Big Data Analysis are in the tasks of data integration and management, in the provision of an appropriate computing infrastructure, and in the development of new analytical tools. The variety of domains in which the IoT is expected to have a very big impact include the manufacturing sector, asset and fleet management, operations management, resource exploration and others. Although Big Data Analysis is viewed as the enabler of the IoT, there are a number of concerns that arise from its development: privacy and security issues, data quality issues and interpretability of the models. In addition, the author foresees other problems including validation of the models, human-analytics interaction and reconciliation of the models with the human understanding of the domain, potential for errors and failures, and over-personalization. Once again, this chapter spans and expands upon many of the topics introduced in our framework including the "Data management", "Data quality", "Data Handling", "Data Processing" and "Result Analysis and Integration" entries of Table 2. It expands upon many subsections while also considering some of the issues discussed in the last section.

Telecommunication Chapter "Social Network Analysis in Streaming Call Graphs" by Rui Sarmento, Marcia Oliveira, Mario Cordeiro, and João Gama describes some of the problems that are encountered in the particular sector of telecommunications services. The problem consists of analyzing the data generated by telecommunication providers. Three specific issues faced by these companies are that their data is typically represented by graphs where, for example, each node represents a phone number and the directed edges represent a phone call initiated from one node and directed at another; the graphs change quickly over time; and the amount of information generated by the company is enormous. They cast this problem as one of mining data streams where the data stream consists of a succession of information networks. Within this context they describe techniques that have previously been proposed to sample from such networks, a problem that is common to all cases of large network analysis but which is compounded here by the dynamic nature of the network. They also describe visualization techniques as well as network analysis such as centrality detection and community detection, which again are different in dynamic networks. This chapter discusses issues that belong to the "Data type" and "Data management" categories of Table 1. It is particularly interesting because it merges two issues that are already very difficult to handle—"Graph and Social Networks Mining" discussed in several sub-sections of 2.4 on the one hand, and "Data Streams mining" discussed in the later parts of Sect. 2.4, on the other hand to derive an even more challenging type of data.

4.2.3 Life Sciences

DNA Sequencing Chapter "Scalable cloud-based data analysis software systems for Big Data from next generation sequencing" by Monika Szczerba, Marek S.

Wiewiórka, Michał J. Okoniewski and Henryk Rybiński explains the computational challenge caused by the advent of the next generation sequencing technology, a new generation of machines that permits fast as well as cheap sequencing of DNA. While this is a great development for medicine since it will allow the development of improved diagnostic and personalized treatment, it causes great challenges in terms of both data storage and data analysis efficiency. This chapter presents the tools that are currently used by or developed for genomic data analysis. These tools are cloud-based and generally come from the Hadoop environment. The chapter begins by presenting two kinds of problems that come up in genomic data analysis: searching for genome variants and RNA expression profiling. It then describes two tools particularly useful for implementing solutions to these problems: Apache Hadoop, MapReduce and Apache Spark. The second part of the chapter presents a particular software tool available for the analysis of next generation sequencing data called SparkSeq. The performance of the system is assessed in terms of various criteria such as data access efficiency and data protection. This chapter covers a number of entries from Tables 1 and 2, namely, "Memory access", "Computational processing and architectures", "Data management" and "Data handling". It expands upon the discussion of these parts of Sect. 2.4 on "Data Privacy", "Distributed Platforms" and "Parallel Processing and Parallelization of existing algorithms", respectively.

Feature Selection for Life Science Problems Chapter "Discovering networks of interdependent features in high-dimensional problems" by Michał Draminski, Michał J. Dąbrowski, Klev Diamanti, Jacek Koronacki, and Jan Komorowski presents a new methodology for selecting features and discovering their interactions in extremely high dimensional problems such as those encountered in the field of Life Sciences. Using their previously designed Monte-Carlo Feature Selection algorithm to rank the features, they then proceed to construct a directed graph that models the interactions between these features and the strengths of their interdependencies. Rather than focusing on features that provide similar information, they attempt to discover features that cooperate together in making a decision about a data sample. They test their Inter Dependent Graph (or ID Graph) construction approach by feeding its results into software tools for building classifiers and extracting rules (ROSETTA and Ciruvis, respectively). They assess the effectiveness of their ID Graph approach on the task of understanding the ancestry influence on certain aspects of the immune system development. This chapter fits into the "Data Processing" entry of Table 2. It discusses issues that are considered in Sect. 2.4 about existing problems disproportionately exaggerated by Big Data: feature selection has been applied to data since the earliest days of machine learning; yet, the dimensionality and the type of interactions between features that occurs in Life Science problems are on a very different scale from what has previously been observed in data.

This concludes our brief overview of the chapters of this volume. Each of the studies just mentioned will now be presented in great detail by their authors, and we will draw conclusions from these discussions and present them in our concluding chapter. Stan Matwin's contributions to the field of Big Data Analysis throughout the years will also be overviewed in the concluding chapter.

References

1. Abiteboul, S.: Querying semi-structured data. In: ICDT '97 Proceedings of the 6th International Conference on Database Theory, pp. 1–18 (1997)
2. An interview with Michal Jordan—Why Big Data Could Be a Big Fail. IEEE Spectrum. (Posted by Lee Gomes, 20 Oct 2014)
3. Anderson, C.: The end of Theory. The data deluge makes the scientific method obsolete, Wired Magazine, 16/07 (2008, June 23)
4. Auerbach, D.: The Mystery of the Exploding Tongue. How reliable is Google Flu Trends? Slate Web page. http://www.slate.com/articles/technology/bitwise/2014/03/google_flu_trends_reliability_a_new_study_questions_its_methods.html (2014)
5. Azzara, M.: Big Data Ethics: Transparency, Privacy, and Identity. Blog cmo.com. (Retrieved 2015)
6. Barbaro, M., Zeller, Jr, T.: A Face Is Exposed for AOL Searcher No. 4417749. The New York Times Magazine. (August 9, 2006)
7. Barbier, G., Liu, H.: Data Mining in Social Media. In: Aggarwal, C. (eds.) Social Network Data Analytics, pp. 327–352. Kluwer Academic Publishers, Springer (2011)
8. Bekkerman, R., Bilenko, M., Langford, J.: Scaling Up Machine Learning. Parallel and Distributed Approaches. Cambridge University Press, Cambridge (2011)
9. Berkeley Data Analysis Stack. https://amplab.cs.berkeley.edu/software/
10. Beyer, M.A., Laney, D.: The importance of "Big Data": a definition. Gartner Publications, pp. 1–9 (2012). See also: http://www.gartner-com/it-glosary/big-data
11. Bifet, A., Holmes, G., Kirkby, R., Pfahringer, B.: MOA: massive online analysis. J. Mach. Learn. Res. **11**, 1601–1604 (2010)
12. Billion Price Project. http://bpp.mit.edu/
13. Boyd, D., Crawford, K.: Six provocations for Big Data. Presented at "A Decade in Internet Time: Symposium on the Dynamics of the Internet and Society" Oxford Internet Institute, Sept 21 (2011)
14. Boyd, D., Crawford, K.: Critical questions for big data. Inf. Commun. Soc. **15**(5), 662–679 (2012)
15. Che, D., Safran, M., Peng, Z.: From big data to big data mining: challenges, issues and opportunities. In: Hong, B, et al. (eds.) DASFAA Workshops, Springer LNCS 7827, pp. 1–15 (2013)
16. Chen, M., Mao, S., Liu, Y.: Big data: a survey. Mobile New Appl. **19**, 171–209 (2014)
17. Dai, C., Lin, D., Bertino, E., Kantarcioglu, M.: An approach to evaluate data trustworthiness based on data provenance. In: Proceedings of the 5th VLDB Workshop on Secure Data Management, pp. 82– 98 (2008)
18. Davidson, S., Freire, J.: Provenance and scientific workflows: challenges and opportunities. In: Proceedings of the SIGMOD'08 (2008)
19. Davis, K.: Ethics of Big Data. Balancing Risk and Innovation. O'Reily (2012)
20. De Mauro, A., Greco, M., Grimaldi, M.: What is big data? a consensual definition and a review of key research topics. In: Proceedings of 4th Conference on Integrated Information (2014)
21. Domingos, P., Hulten, G.: Mining high-speed data streams. In: Proceedings of the 6th ACM SIGKDD International Conference on Knowledge Discovery and Data Mining, pp. 71–80 (2000)
22. Einav, L., Levin, J.D.: The data revolution and economic analysis. National Bureau of Economic Research Working Paper, no. 19035 (2013)
23. Fan, W., Bifet, A.: Mining big data: current status, and forecast to the future. SIGKDD Explor. Newsl. **12**(2), 1–5 (2013)
24. Frontiers in Massive Data Analysis. The National Research Council, the National Academy of Sciences, USA (2013)
25. Future Attribute Screening Technology. Wikipedia article. https://en.wikipedia.org/wiki/Future_Attribute_Screening_Technology
26. Gaber, M., Zaslavsky, A., Krishnaswamy, S.: Mining data streams: a review. ACM Sigmod Record **34**(2), 18–26 (2005)

27. Gama, J.: Knowledge Discovery from Data Streams, 1st ed. Hall/CRC, (2010)
28. Ghoting, A., Kambadur, P., Pednault, E., Kannan, R.: NIMBLE: A toolkit for the implementation of parallel data mining and machine learning algorithms on mapreduce. In: Proceedings of the 17th ACM SIGKDD International Conference on Knowledge Discovery and Data Mining KDD 2011, pp. 334–342 (2011)
29. Ginsberg, J., Mohebbi, M. H., Patel, Rajan S., Brammer, L., Smolinski, M.S., Brilliant, L.: Detecting influenza epidemics using search engine query data. Nature **457**(7232), 1012–1014 (19 Feb 2009)
30. Glavic, B.: Big Data provenance: challenges and implications for benchmarking. In: Specifying Big Data Benchmarks, pp. 72–80. Springer (2014)
31. Gonzalez, M.C., Hidalgo, C.A., Barabasi, A.L.: Understanding individual human mobility patterns. Nature **453**, 779–782 (2008)
32. Hadoop. http://hadoop.apache.org
33. Han, J., Kamber, M.: Data Mining: Concepts and Techniques, 2nd edn. San Francisco, Morgan Kaufmann (2005)
34. Harford, T.: Big Data: are we making a big mistakes? Financial Times, March 28 (2014)
35. Hashem, I., Yaqoob, I., Anuor, N., Mokhter, S., Gani, A., Khan, S.: The rise of bog data on cloud computing. Review and open research issues. Inf. Syst. **47**, 98–115 (2015)
36. How big data analysis helped increase Walmart's sales turnover. DeZyre Web page (23 May 2015)
37. Kang, U., Faloutsos, C.: Big graph mining: algorithms and discoveries. ACM SIGKDD Explor. Newsl. **14**(2), 29–36 (2012)
38. Kraska, T., Talwalkar, A., Duchi, J.C., Griffith, R., Franklin, M.J., Jordan, M.I. MLbase: A distributed machine-learning system. In: Proceedings of Sixth Biennial Conference on Innovative Data Systems Research (2013)
39. Krempl, G., Zliobaite, I., Brzezinski, D., Hullermeier, E., Last, M., Lemaire, V., Noack, T., Shaker, A., Sievi, S., Spiliopoulou, M., Stefanowski, J.: Open challenges for data stream mining research. ACM SIGKDD Explor. **16**(1), 1–10 (2014). June
40. Mahout software. http://mahout.apache.org/
41. Maimon, O., Rokach, L. (eds.): The Data Mining and Knowledge Discovery Handbook. Springer (2005)
42. Mannila, H.: Data mining: machine learning, statistics, and databases, In: Proceedings of the Eight International Conference on Scientific and Statistical Database Management. Stockholm June 18–20, pp. 1–8 (1996)
43. Manning C., Schutze H. Foundations of Statistical Natural Language Processing. MIT Press (1999)
44. Marcus, G., Davis, E.: Eight (No, Nine!) Problems With Big Data. New York Times (Apr 6, 2014)
45. Matwin, S.: Privacy-preserving data mining techniques: survey and challenges. In: Custers, B., Calders, T., Schermer, B., Zarsky T. (eds.) Discrimination and Privacy in the Information Society. Springer Series on Studies in Applied Philosophy, Epistemology and Rational Ethics, vol. 3, pp. 209–221 (2013)
46. Matwin, S.: Machine learning: four lessons and what is next? Bull. Pol. AI Soc. **2**, 2–7 (2013)
47. Mayer-Schonberger, V., Cukier, K.: Big Data: A Revolution That Will Transform How We Live, Work and Think. Eamon, Dolan/Houghton Mifflin Harcourt (2013)
48. Morales, G., Bifet, A.: SAMOA: scalable advanced massive online analysis. J. Mach. Learn. Res. **16**, 149–153 (2015)
49. Narayanan, A., Shmatikov, V.: Robust De-anonymization of Large Datasets (How to Break Anonymity of the Netflix Prize Dataset). In: Proceedings of the 2008 IEEE Symposium on Security and Privacy SP'08, pp. 111–125 (2008)
50. Piatetsky-Shapiro, G., Matheus, C. (eds): Knowledge discovery in databases. AAAI/MIT Press (1991)
51. Pietsch, W.: Big Data? The New Science of Complexity. In: 6th Munich-Sydney-Tilburg Conference on Models and Decisions (Munich; 10–12 April 2013)

52. Reinventing Society in the Wake of Big Data—Edge's interview with Alex "Sandy" Pentland (Posted August 30, 2012)
53. Ritter, D.: When to act on a correlation and when no to. Harward Business Review, March 19 (2014)
54. Roddick, J., Hornsby, K., Spiliopoulou, M.: An updated bibliography of temporal, spatial, and spatio-temporal data mining research. Lect. Notes Comput. Sci. **2007**, 147–163 (2001)
55. Rudin, C., Passonneau, R., Radeva, A., Jerome, S., Issac, D.: 21st century data miners meet 19-th century electrical cables. IEEE Comput. 103–105 (June 2011)
56. Rudin, C., et al.: Machine learning for the New York city power grid. IEEE Trans. Pattern Anal. Mach. Intell. **34**(2), 328–345 (2012)
57. Shekhar, S.: What is special about mining spatial and spatio-temporal datasets? Tutorial (2014)b. http://www-users.cs.umn.edu/~shekhar/talk/sdm2.html
58. Simmhan, Y., Plale, B., Gannon, D.: A survey on data provenance techniques. Technical Report Indiana University, IUB-CS-TR618 (2005)
59. Singh, D., Reddy, C.: A survey on platforms for Big Data analytics. J. Big Data **1**(8), 2–20 (2014)
60. Sloan Digital Sky Survey. Wikipedia article. https://en.wikipedia.org/wiki/loan_Digital_Sky_Survey
61. Sun, Y., Han, J.: Mining Heterogeneous Information Networks: Principles and Methodologies. Morgan & Claypool Publishers (2012)
62. The h2o software. http://0xdata.com/h2o
63. Thomson, C.: What Is IBMs Watson? The New York Times Magazine, June 16 (2010)
64. Tufekci, Z.: Big Data: Pitfalls, methods and concepts for an emergent field. SSRN (March 2013). http://dx.doi.org/10.2139/ssrn.2229952
65. Venkateswara Rao, K., Govardhan, A., Chalapati, Rao K.V.: Spatiotemporal data mining: issues, tasks and applications. Int. J. Comput. Sci. & Eng. Surv. (IJCSES) **3**(1) (Feb 2012)
66. Vucetic S., Obradovis, Z.: Discovering homogeneous regions in spatial data through competition. In: Proceedings of the 17th International Conference of Machine Learning ICML, pp. 1095–1102 (2000)
67. Zhou, Z.H., Chavla, N., Jin, Y., Williams, G.: Big Data opportunities and challenges: discussions from data analytics perspectives. IEEE Comput. Intell. Mag. **9**(4), 62–74 (2014)

An Insight on Big Data Analytics

Ross Sparks, Adrien Ickowicz and Hans J. Lenz

Abstract This paper discusses the opportunities big data offers decision makers from a statistical perspective. It calls for a multidisciplinary approach by computer scientists, statisticians and domain experts to providing useful big data solutions. Big data calls for us to think in new ways and communicate effectively within such teams. We make a plea for linking data-driven and model-driven analytics, and stress the role of cause-effect models for knowledge enhancement in big data analytics. We remember Kant's statement that theory without data is blind, but facts without theories are meaningless. A case is made for each discipline to define the contribution they offer to big data solutions so that effective teams can be formed to improve inductions. Although new approaches are needed much of the past learning related to small data are valuable in providing big data solutions. Here we have in mind the long-term academic training and field experience of statisticians concerning reduction of dataset volumes, sampling in a more general setting, data depreciation and quality, model design and validation, visualisation, etc. We expect that combining the present approaches will give incentives for increasing the chances for "real big solutions".

1 Introduction

Generally Big Data involves routinely collected data that is integrated from different sources and joined together. The theory is that the combined data hold more information in it than analysing the separate datasets independently. Combined datasets do not always have all the variables of interest but generally hold more variables of interest than the separate datasets. This suggests that Big Data has the potential to solve many problems we could not by analysing these datasets separately. Generally observation studies need to be carefully planned for them make causal inferences

R. Sparks · A. Ickowicz
CSIRO Computational Informatics, North Ryde, NSW 1670, Australia

H.J. Lenz (✉)
Institut Für Statistik und Ökonometrie, Inst. Für Wirtschaftsinformatik,
Freie Universität Berlin, Boltzmannstr.20, K30 D-14195 Berlin, Germany
e-mail: hans-j.lenz@fu-berlin.de

© Springer International Publishing Switzerland 2016
N. Japkowicz and J. Stefanowski (eds.), *Big Data Analysis: New Algorithms for a New Society*, Studies in Big Data 16, DOI 10.1007/978-3-319-26989-4_2

and unless this happens with Big Data it is not going to solve many of the problems we are interested in as statisticians. More effort is needed in designing Big Data collection processes for specific analytical purposes before its value can be broadened in making reliable inferential judgments.

The quality of the information in Big Data collection processes is an issue. There is the issue of gross recording errors that need to be addressed and the quality of the information used to join datasets from different sources need to be considered in the analytical approach. If joined datasets are used and there are selection bias issues with each dataset used in the join, then the combined Big Data will have compounded selection bias issues if the join is carried out using the intersection principle. If the join includes all the data as well as all information that is missing using the union principle in joining the datasets, then we could be dealing with a massive missing data problem, but the selection bias issue will generally be reduced. There are significant challenges when dealing with such situations. For example if a probabilistic join is used then the join has some uncertainty, and this requires a change to the analytical methods to deal with this uncertainty [7]. This adds to the challenge. All of this fits into the section of the paper that looks at the issue of whether the Big Data are fit for purpose.

Section 3 will discuss some general tools that may be useful in reducing the size of the analytical effort in analysing big datasets. These are often useful in taking the original dataset that may be in peta scale or tera scale say down to the more manageable giga scale. This section is by no means complete but it documents what we have found to be useful.

Section 4 of the paper looks at the issue of analysing massive datasets when it is impossible to include all the data in the routines for their analysis. This will use the divide and rule principle, that is, the big dataset will be divided up into manageable pockets in a way that helps improve the analytical purpose, e.g., inference or predictions (or forecasts). How to divide the datasets up is an open research question which will not be answered in this paper, but some general principles will be discussed in Sect. 3.

Section 5 has another look at ways of reducing the volume of data in certain instances. Section 6 finishes with comments about the tension between data mining and statisticians and a call for a collaborative approach to building knowledge that will help us better manage the future. Section 7 examines the question of whether theory is essential. Section 8 briefly examines intellectual property issues. Section 9 finishes with a discussion of the issues and summarises

2   Is Big Data Fit for Purpose?

2.1   Do We Need Big Data?

It is fashionable to talk about the opportunities that Big Data offer decision makers. Big Data is attracting the interests of industry and resulting in their preparedness to invest money and resources in achieving the related business gains. Business gains can be achieved using both Big Data and small data. Therefore industry should think carefully about what they would like to achieve, and then establish whether the appropriate data are available for achieving their objectives or making their decisions. That is, they should think about investing better not investing more. Often we require the appropriate data to achieve unbiased solutions. Before the Big Data focus showed up, statisticians addressed their problems by carefully thinking whether the available data are adequate for the purpose or whether new data needed to be collected by an efficient experimental design.

Statistics has long been the avenue for answering important research questions. However the computer has increased our ability to deal with larger and larger datasets, and in some sense answering more complex questions. However ensuring that the data is fit for purpose is even more important in the Big Data context. Before computers were available, inverting a 4×4 matrix of reals took a considerable amount of time, while now this is trivial. Therefore statistics has evolved with the advent of computers. The growth in computational statistics research has allowed statisticians to fit more complex models than previously was possible by using MCMC methods, and improving inference using bootstrap and cross-validation methods.

Our view is that Big Data increases opportunities, but much that has been learned in the past is also relevant in the Big Data space, and in fact we argue that it is even more important. Our view is that answering the right question is more important than the appropriate data, but a close second is having the appropriate data to answer these questions. Big Data is sold as the means of solving all questions but we feel this perception is misguided. Savage's book [10], links the development of statistics in the late twentieth century to the British-American school and its view of probability as objectivistic theory of knowledge. According to this view, the mathematical concept (model) by which we understand our problems must be obtained by observing repetition of events, and *from no other source whatsoever*. This is quite enlightening in the Big Data settings. The first point made, is that the modern statistics (as defined by [10]) referred to as statistical inference, is the daughter of the probability theory. Accurate inference lies in the construction of a model to understand the data. We will explore that point further in Sect. 7. The second point is that any information other than the repetition of the event remains clueless in regard to the application of statistical techniques. Big Data implies more data, but it may not imply more information. Big Data may not build on our current knowledge or answer our important questions. An excess of non-relevant information is likely to be misleading or may create confusion with what is important or add to our spurious/false 'discoveries'. However Big Data that is built on a theoretical framework of knowledge discovery (see [3], p. 106) is likely to improve our understanding and build on our current knowledge. The view that Big Data offers all the answer to our quests for knowledge, and all we need to do is discover where it is embedded in the Big Data is dangerous.

2.2   What About Big Data Do We Need?

Big Data is unlikely to solve all problems of interest to the data custodians unless it has been designed to achieve this aim. Most routine datasets collect the measures that are easy to accumulate mostly because they are necessary administrative data such as revenues and expenditures, because they are easy to measure and collect or simply "open data" ready for downloading free of any fees. A typical example is data from social networks. But the question arises whether what we got is what we need or is "N = ALL" perhaps a seductive illusion, Harford [6]?

The first step before using any dataset is to decide whether the dataset is fit for purpose. We break the fit for purpose evaluation down into the answering following questions:

1. Are all the appropriate variables available?
2. Are these variables measured accurately enough to answer these question? Are there potential recording errors?
3. Does the data represent the population we wish to make inferences about or wish to predict? What selection biases are there?
4. Does the data cover the appropriate time frames for the purpose? Is the time between measures and the duration of collection appropriate?
5. Are there any redundancies in the dataset that are worth removing?
6. Are all measures well defined and consistently measured over time?
7. Has measurement accuracy improved over time and therefore what historical data are useful for the purpose?
8. Is there any missing data and if there is, then what is the nature of the missingness?
9. Do any of the measurement suffer detection limits? For example, is the measurement process incapable of measuring values either below or above a certain limit?
10. Is the spatial information adequate for the purpose?

Some of these fit well with the five V's raised by Megahed and Jones-Farmer [8] as volume, variety, velocity, veracity and value. Veracity refers to the trustworthiness of the data in terms of creating knowledge relating to the purpose. This calls for data management processes for maintaining the veracity of the data. For example in large scale sensor networks, where many measures are collected every 5 min over long periods of time, requires real-time checks on the spatio-temporal consistency of measures as well as checking whether the measures are consistent with related measures collected at the same site (e.g., see [11]). Therefore Big Data increases the need for the appropriate level of data management. Improved accuracy can sometimes be forced by a certain level of aggregation either over space/geography or by temporal aggregation. For example considering the average measurement per 5 min when the data are recorded every minute or averaging measurements made within a spatial grid. This certainly has advantages when 1 min measures are highly autocorrelated and neighbouring measures are almost measuring the same entity. On the other hand, this can result in a loss of either spatial or temporal resolution when aggregating over too large space or too large time periods, respectively. It is therefore better to build in the appropriate level of accuracy into measures by using the appropriate data management techniques and controls on the measurement process.

The challenges with sensor networks is whether consistency of measures checking be done at the location of each sensor before sending the information back to the root node in the network (thus not checking for spatial consistency) or send the information to the root node first and then do the multivariate-spatio-temporal consistency checking. Such decision may not depend on which approach delivers greater accuracy but in wireless solar operation sensors this may be based on power considerations. Nevertheless accuracy of measurement will impact on what analytical approach will be used to analyse the data.

3   Basic Toolbox for Analysing Big Data

Datasets are increasing in size and purchasing memory space in this digital age is becoming cheaper. Therefore the size and complexity of datasets is growing nearly exponentially. Having the appropriate tools for dealing with such complexity is important with both n (sample size) and p (number of variables) being large in the n by p data matrix. The following methods are useful in managing the computational complexity:

1. **Aggregation and Grouping**: There are many common examples of aggregations that are common place to-day:

   - The billions of market transactions per second in the world involving over 1000 TB per annum (PB/a) is aggregated into GDP per year (USD/a) published in the UNO Yearbook by the National Accounts Group of UNO, New York (8 Bytes/a).
   - Instead of singletons like screws, nails etc. these are combined into one category/class called hardware as a larger.
   - It is fairly common to bin peoples ages into groups, e.g. age intervals [0, 18], [18, 65], [65, 120], and to study behaviour within cohorts.

2. **Blocking**: Semantic keys are built so that users can find certain information very fast. As an example the Administrative Record Census 2011, Germany, used attribute 'address' for household generation as a main blocking variable. Privacy concerns often result in the lowest level of geography that is released on individuals is postal code, and in many analyses this is used as a blocking variable. This is at times used to define people who are similar in some way, e.g., with similar social disadvantage index.

3. **Compression and Sparsity exploitation**: An example is the sparse matrix storage of images such as that used by 'jpeg'. Dimension reduction techniques of data compression are fairly common. Examples are multi-dimensional scaling (MDS), Projection Pursuit, PCA, non linear PCA, radial basis functions or

wavelets. Examples of application are image reconstruction using wavelets or PCA.

4. **Sufficient statistics**: Another very common data compression approach is to only store the sufficient statistics for later analysis, such as is commonly used in Meta Analysis. This reduces the full data by only storing and using statistical functions of the data, e.g., the sample mean and sample standard deviation for Gaussian data. In modern control theory this principle is applied by signal filtering techniques like the Kalman Filter.

5. **Fragmentation and Divisibility (divide et impera)**: We fragment a feature in such a way that it preserves its essential features for analysis. For example, a company made up of different stores at different location around a country. Keeping the total sales at each store allows us to calculate the total sales for the company. The maximum or minimum sales at each store still allows us to calculate in minimum or maximum sale for the company. The top ten sales at each store allows us to calculate the top ten sales for the company. Where this fails is with the median sales at each store; this does not allow us to estimate the median sale for the company.

   A good example of divisibility is that a joint multivariate density can be preserved by factorization of densities say using Markov fields or Markov chains/processes, e.g., example if $X \rightarrow Y \rightarrow Z$ is a Markov chain, then $f(x, y, z) = f_x(x) f_{y|x}(y|x) f_{z|y}(z|y)$ where $f(x, y, z)$ is the joint density of x, y and z, $f_x(x)$ is the marginal density of x, $f_{y|x}(y|x)$ is the conditional density of y given the value of x, and $f_{z|y}(z|y)$ is the conditional density of z given the value of y.

6. **Recursive versus global Estimation (parameter learning) procedures/ algorithms**: This could involve Generalised Least Squares (GLS) or Ordinary Least Squares (OLS) estimation versus Kalman Filtering or recursive GLS/OLS. For example: the recursive arithmetic mean estimator is given by

$$\bar{x}_n = (1 - \lambda_n)\bar{x}_{n-1} + \lambda_n x_n$$

   where $\lambda_n = 1/n$, while the Kalman filter includes a signal to noise (variance) ratio, υ, leading to $\lambda_n = 1/(1/\upsilon + n)$.

7. **Algorithms**: One Pass Algorithm (like Greedy Algorithm) versus Multi Pass Algorithms (cf. backtracking, Iteration)

8. **Type of Optimum**: Local optimum/Pareto optimum/global optimum. Heuristic optimisation often delivers a "practical useful" local optimum with strongly bounded computational efforts, the proof of its optimality may be very CPU-time consuming.

9. **Solution types of combinatorial problems**: Limited enumeration, branch and bound methods or full enumeration. Example: Traversing or exploring game trees or social/technical networks.

10. **Sequencing of operations (for additive or coupled algebraic operations) or parallelisation** Examples: Linking of stand-alone programs for solving one (separable) problem in 1- memory-1 CPU machine. Dividing the task up into parallel streams that can be run in parallel to each other.

11. **Invariant Embedding**: Instead of sampling with a given frequency (time window) we record the time stamp, event and value. An example is the measuring of electricity consumption of private households either using a fixed sampling frequency or recording the triple (time stamp, load (kw), type of electric appliance).

Many of the methods mentioned in this section are used to divide the analytical task into more manageable chunks.

4 Dividing the Analytical Task Up into Manageable Chunks

This section will focus on two applications both involving forecasting. The first application deals with forecasting or inferential generalised linear models with a unique defined response variable. The second deals with forecasting counts in complex tabular settings. As the sample size increases generally the proportion of the error due to the systematic error reduces but the proportion of model error starts to increase. Therefore much more attention needs to be devoted to establishing the appropriate model for Big Data applications.

4.1 Generalised Linear Models Example

When datasets get too large to include all observations in the analysis phase, then subdividing the data is important in managing the analytical task. This has computational advantages and information advantages as well (see the results later). Even in smaller datasets it makes statistical sense to divide the data into a test (learning) sample and a validation sample, cf. cross-validation. This is particularly true of model building which is the main focus of this section. The test sample is used to formulate a useful model for prediction/forecasting or inference where selecting: the model, the explanatory variables, and transformations of the response or explanatory variables (e.g., see projection pursuit by [4]). Furthermore, the splitting helps avoiding overfitting and biased estimates of goodness of fit criteria. In addition the test data are used to validate whether any assumptions may hold approximately. After we have settled on a useful operating model with the test data, then we validate the selected model using the new validation dataset. Here we recheck assumptions and assess the goodness-of-fit for the selected model. In other words, the validation dataset is used to assess the usefulness of the selected model. If there are two comparable models selected at the test phase, then the validation dataset can be used to differentiate them and select the better one, or decide to use both and apply ensemble forecasts or inferences.

Different tasks would involve different ways of dividing up the work and so this section is not going to do justice in providing advice for all different tasks. We will consider the forecasting task as one option and we start by looking at forecasting a continuous variable. The model formulation stage would use test dataset (generally about two thirds to half the data). However since this test data may still be excessively large we split this test data into say 100 datasets that are selected randomly without replacement of roughly equal size (n). This process divides the test data into 100 subsets that are non-overlapping but exhaustive of the test dataset. Assume that the ith test sample subset has response variable observations given by vector y_i with related predictor variables that include the same number of observations as in y_i (some of these explanatory variables could be lag response variables). This matrix of predictor variables is denoted by X_i. Consider the generalised linear model structure as an example where

$$g(E(y_i)) = X_i \beta_i, \quad i = 1, 2, \ldots, 100$$

where g is the link function and β_i is the coefficient for the ith test sample and E is the expectation operator. We expect that if the model was appropriate then $\beta_i = \beta$ for all i. We may want to compare either several g link function options or several distribution options for the response y_i. Assume that the fitted models for the ith test dataset resulting in an estimated model formed by substituting β_i by its estimate $\widehat{\beta_i}$ in the equation above. Then since $\beta_i = \beta$ the ensemble estimate for the component β_j is $\bar{\beta}_j = \sum_{i=1}^{100} \widehat{\beta}_{ij}/100$ where $\bar{\beta}_j$ is the generalised linear model estimate of the regression coefficient derived from the partitioned test dataset. The model fitting algorithms produce estimates of model standard errors for each $\widehat{\beta_i}$ which are denoted s_w and interpreted as the within sample uncertainty in the estimate of the coefficient. However the sample estimated standard errors (s_j) for the between test data subsets estimates of the jth regression parameter in the model is given by

$$s_j^2 = \sum_{i=1}^{100} (\widehat{\beta}_{ij} - \bar{\beta}_j)^2/100$$

which assesses how much the individual estimate differs on average from the ensemble estimate. In addition, the distribution of estimated parameters $\widehat{\beta}_{ij}$ for all $i = 1, 2, \ldots, 100$ would be useful in determining the consistency in the jth regression parameters across the various test subset samples. The s_j^2 value is a reflection of the stability of the model across different random samples and measures the robustness of the model parameter estimates. With highly collinear explanatory variables the regression parameter estimates can be unstable, but prediction is usually stable in such cases. We therefore can compare the variation in model prediction errors by calculating

$$S_i^2 = \sum_{k=1}^{n} (y_{ik} - g^{-1}(X_i \widehat{\beta}_{ik}))^2/n$$

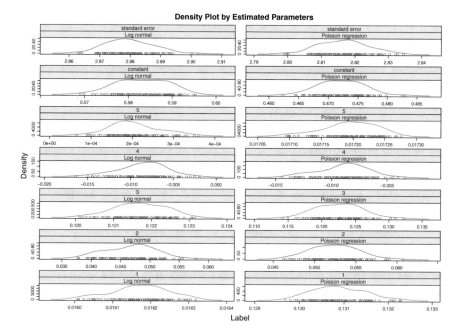

Fig. 1 Comparison of competing models: distribution of estimated parameters and validation standard errors

across all validation samples and test samples. The variation in S_i^2 values provide evidence for the robust performance of the model predictions. These between sample variations could be useful in comparing the robustness of competing models, and therefore help make a decision on the appropriate approximating model (denoted the operating model).

A simulated example is presented in Fig. 1. The data contains 20 million observations generated using the following Poisson regression model $\mu = \exp$ $(0.15 \times x_1 + 0.02 \times x_3 - 0.01 \times x_4) \times as.factor(x_2) \times (\exp(0.06), \exp(0.12),$ $\exp(0.2))$ where $x_1 \sim N(4, 4)$, $as.factor(x_2)$ is a ordinal factor having three levels, $x_3 \sim U(0, 25)$ and $x_4 \sim U(0, 100)$ i.e., uniformly distributed. The response variables were simulated as Poisson with mean μ. The data is split into two 100 validation samples of n = 100,000 observations and the same as training data. The model is fitted using each 100,000 observations in test samples and then the prediction are validated using 100,000 validation dataset. This cycled through each of the 100 training and validation sets. The distributions of the estimated parameters of the model and the prediction standard errors are reported in Fig. 1. Two models were fitted based on no knowledge of the true Poisson regression model used to simulate the data. The Poisson regression model for the counts with expected value:

$$\mu = \exp(\beta_0 + x_1\beta_1 + as.factor(x_2)_{\text{level } 2}\beta_2 + as.factor(x_2)_{\text{level } 3}\beta_3 + x_3\beta_4 + x_4\beta_5)$$

The linear model that is fitted is $log(y + 1) = \alpha_0 + x_1\alpha_1 + x_2\alpha_2 + x_3\alpha_3 + x_4\alpha_4 + x_2 \times x_4\alpha_5 + error$. These models are compared in Fig. 1. The regression coefficients in Fig. 1 are in number order (β for the Poisson regression model and α otherwise). Looking at the distribution of the estimated regression parameters and the validation standard errors for the fitted models; the Poisson regression is the better model, and therefore this model is preferred. The estimated regression coefficients generally vary less in the Poisson regression model and the standard errors are on average smaller. The evidence is more clear if the density plot of the differences between the two model matched validation standard errors are plotted, which indicated that the Poisson regression always had a smaller validation standard error. In this way competing models can be compared when faced with large data sets.

A similar approach to the above can be used for fitting Bayesian Hierarchical models (for example). Here we have established credible intervals for model parameters (and forecasts if that is the purpose) for each sample i. These credible intervals could be plotted for all $i = 1, 2, \ldots, 100$ as a way of assessing the validity of the model and the consistency of these intervals. Combining of the Bayesian parameter estimates as mentioned before could provide ensemble estimates for parameters, and the variation of these from the ensemble estimate could be a way of validating the robustness of the model. In addition such empirical evidence can be used to compare different Bayesian hierarchical models and select the model which show the better properties. We believe that in the case of Big Data a validation sample is still necessary because model decisions are still made based on it. This same approach could also be used to compare different burn-in and iteration estimation strategies.

With forecasts, using very large datasets, we wish to avoid refitting the model using all the data each time a new data value is observed. In linear models this can largely be avoided by using some recursive estimation procedure such as the Kalman filter and some state space models [13]. Bolt and Sparks adopted a simpler approach of using a moving window of the same size and exponential weights to give the most recent observation a greater weight, but their approach is only reasonable for one-step-ahead forecasts.

4.2   Forecasting Counts in Complex Tabular Settings

If we are trying to forecast the daily social service needs within a country, then the challenge is a little different. We could still follow the approach designed above, but it is our view that this would not be as efficient as defining cohorts of the population with similar needs and temporal trends. For example, all university students apply for similar support for their university education at the same time of the year. Dividing the population into m different cohorts which have very similar temporal trends and seasonal variation for their demands on the country's social services or geographical regions whose population has homogeneous services needs across time and with the same longitudinal influences seems sensible. The divide of the population into non-overlapping and exhaustive population cohorts is likely to improve the forecasts of needs within cohorts and thus improve the forecasts of the national needs by aggregating up from these cohorts. This approach is not only likely to help make the task more manageable but it will also help improve forecasts.

On the other hand if our interests were in forecasting particular cohort needs and we notice that many cohorts have similar temporal trends, then it may be helpful to decide which cohort counts would be better predicted by forecasting the total counts from the cohorts with similar trends and then proportionally allocate these forecast counts to the respective cohorts. This simplifies the task by aggregating counts to a more manageable level and at times delivers more robust predictions if the cohorts aggregated over all have the same trends.

5   Reducing the Size of the Data that Needs to Be Modeled

The very basic way of reducing the size of the data in space-time applications is by either temporal aggregation thus reducing the number of measures within a unit of time, or spatial aggregation reducing the spatial resolution of the data. An example is the sea surface temperature measured at a fine grid all around Australia with these measures having high spatio-temporal correlations. Assume we were trying to predict the insured costs of floods at 20 locations around Australia given the sea's surface temperatures as explanatory variables. There are several ways of tackling this problem. One is to use technology which exploits Lasso type technology [5, 12] exploiting sparsity, boosting and use ensemble methods. The other approach which we prefer is to create latent variables from sea surface temperature that have physical meaning to the climatologists and are good predictors of flood insured costs at each of the locations of interest. This latent variable (or latent variables) takes the place of these many temperature measurements and therefore reduces the size of the data needed for forecasts.

When we are trying to forecast multi-way tabular counts, e.g., a large array of counts, then at times a drastic reduction in number of cell counts that require forecasts is needed. In such cases it may be worth modelling aggregated cell counts over several dimensions and then proportionally allocate counts to cells that were aggregated over in a way that preserves all interactions. This could be achieved by establishing the cells with the same temporal trends and model the aggregations over these cells counts and then proportionally allocate these forecast totals to the individual cells used to get these total to derive cell forecasts. An example of this is presented in Bolt and Sparks [1]. The only issue with this is if any covariate interacts with time then this model is unlikely to be adequate. Such local errors can quite easily be fixed using temporal smoothing adjustments. Bolt and Sparks [1] approach to forecasting large volumes of counts suited their monitoring applications where early detections of interactions with time were important. Hence this modelling approach will not generally be useful for forecasting applications involving a large number of cells. Another way of reducing the size of the problem is by conditioning, for example, if we condition on age group j and modeled only those in age group j, and repeat this for all other age groups. This could be made more complex by conditioning on age and ethnicity, or by conditioning on three variables. Once the aggregated counts for the conditioned space is found this can be modeled and forecasts established. Forecasts for the whole space is achieved by aggregating over the entire conditional space that makes up the 'whole'. All of these examples lend themselves very well to parallel processing.

6 The Tension Between Data Mining and Statistics

Deming ([3], p. 106) said that "Knowledge comes from theory. Without theory, there is no way to use the information that comes to us on the instant". The Deming quote relating to knowledge may not sit that well with many data mining approaches that search for something interesting in the data. Theory we think is formulated by past observations generating beliefs that are tested by well planned studies, and only then integrated into knowledge when the belief has been "proven" to be true. Data is certainly not information—it has to be turning into information. Many data mining methods are rather short on theory but they still aim to turn data into information. We believe that data mining plays an important role in generating beliefs that needed to be integrated into a theoretical frame which we will call knowledge. When modelling data statisticians sometimes find these theoretical frameworks are too restrictive. At times statisticians make assumptions that have theoretical foundations which are practically unrealistic. This is generally used to make progress towards solving a problem and it is a step in the right direction, but not the appropriate solution. Eventually over time someone builds on this idea and the problem can then be solved without unrealistic assumptions. This is how the theoretical framework is extended to solving the more difficult problems. Non-statistically trained data-miners we believe too often drop the theoretical considerations. Some data-miners attempt to transform data into information using common sense and make judgments about knowledge called learning from the data—sometimes they may get it wrong but often they may be right. Have we statisticians got too hung-up about theory? We do not think so. We may assume too much at first in trying to solve a problem but our foundations are the theory. The current Big Data initiatives are mostly based on the assumption that Big Data is going to drive knowledge (without a theoretical framework). We disagree with this assertion and believe the solution is for data-miners and statistician to collaborate in the process of generating knowledge within a sound theoretical framework. We believe that statisticians should stop making assumptions that remain unchecked and data-miners should work with statisticians in helping discover knowledge that will help manage the future. It is knowledge that helps us improve the management of the future and this should be our focus.

In risk assessment statisticians are generally good at estimating the likelihood, they are trained to evaluate beliefs or hunches and they are trained to building efficient empirical models, but generally they are not adequate trained in the efficient manipulations of massive datasets. Data-miners and computer scientists have the advantage in mining very large volumes of data and extracting features of interest. However there are many issues that data-miners may ignore, e.g., defining the population under study with respect to time, region and subject, defining problem *adequate* variables, utilising background information ("meta data"), paying attention to selection biases when collecting data, the efficient design of observational studies caring about randomness and test/control groups etc.

7 Does the New Big Data Initiative Need No Theory?

The view of Savage on modern statistics raises the question of whether Big Data offers us more information. An interesting question is: does the Big Data current thrust lie outside the modern statistics theory and practice. Alternatively should we define post-modern statistics with Big Data as the main driver. The introduction of this paper questions the current Big Data focus. The ensemble approach of aggregating over the predictions of different models to achieve better predictions may deliver more accurate predictions, but it may not lead to a better understanding than one model. This highlights the importance in selecting the appropriate analytical approach relating to the aim or purpose. However an important question is whether a well thought out model or theory are needed at all?

Statisticians use empirical models to approximate the "real data model" and integrate this with mathematical theory to understand processes and build knowledge. The focus is to understand the sources of variation, and then make conclusions that are supported by the data. Statistical modelling align with Popper [9] view, "*the belief that we can start with pure observations alone, without anything in the nature of a theory, is absurd; as may be illustrated by the story of the man who dedicated his life to natural science, wrote down everything he could observe, and bequeathed his priceless collection of observations to the Royal Society to be used as inductive evidence. This story should show us that though beetles may profitably be collected, observations may not*". This was true for most of the data we have access to. The model shapes the data in trying to best fit it, and the data shapes the model in that it helps us use models with the appropriate assumptions. The less data we have, the more the appropriate model will help in drawing unbiased-low variance estimators/predictions for our problem. However is the assertion that Big Data reduces the need for developing an operating model? Alternatively can every problem be solved by constructing an appropriate empirical model? Like Breiman [2] we believe statisticians need to be more pragmatic. Breiman [2] notes the existence of two parallel cultures in statistical modelling. The first one assumes the data are generated by a given stochastic data model. The second culture uses algorithmic models and treats the data mechanism as unknown. Breiman [2] accuses the statistical community of having focused too much on appropriate empirical models, leading to the development of "*irrelevant theory and questionable scientific conclusions*". Luckily, since 2001, the discipline evolved through this, and made better use of the available computational resources available. Techniques like Gaussian Processes, Bayesian Non-parametric statistic and machine learning deliver successful outcomes (see [14]). It is probably safe to say that modern statisticians have nowadays a toolbox full of machine learning tricks and data-miners similarly have modern statistical tools in their toolbox. However, as a mathematical discipline, it is unlikely that statisticians will move too far away from their theory-driven techniques to full black-box algorithms.

8   Who Owns Big Data?

Another question of interest is the current shift in the intellectual property from the scientific methodology to the data itself. Until recently the major intellectual property was in building the model/technique/algorithm to extract/infer valuable information from the data. Protection was controlled through patents and publications and ownership was recognised by law. Now there is a view that the intellectual property resides in the data. Companies may trust scientists to use their data to answer research questions, but not without protecting the ownership of their data with confidentiality agreements. Big Data is by essence collected from everywhere. The danger is in every corporate entity protecting their data and this lack of data sharing limits the amount value that integrating data from different sources can offer us in understanding our world. For example understanding the consequence of changes in climate requires insurance companies and companies to share their data on insured costs and losses respectively.

9   Discussion

Big data offers us scientists with numerous challenges, and therefore it demands contributions from computer scientists, data-miners, mathematicians, and statisticians. The greatest difficulty is deciding on what value our various skills offer in solving problems and answering questions using Big Data. We feel that collaboration and co-teaching across each of these disciplines is the best way of deciding on the value we each offer.

The big advantage is that all these disciplines have added to the tools that are needed to manipulate and analyse Big Data. As datasets increase in size we statisticians are going to need to lean on the tools developed by computer scientists and data-miners more and more. In addition, new theoretical frameworks may be needed to ensure that judgment mistakes are not made. The Big Data challenge is extracting information in real-time decision making situations where both n and p are large and there is a real-time dimension to the problem. Often people use simple statistical methods to analyse such data and limit their inference to answering fairly simple questions. However, the challenge for both data-miners and statistician working in this area is to move the questions and analytical methods up to the more complex questions with a particular emphasis on avoiding giving biased solutions.

For large data sets, it is well known now that testing for statistical significance is of limited value and the challenges are more aligned with accurate estimates and confidence intervals. Clearly Big Data research demands diverse skills recognizing that the problems are too difficult and large to be "owned" by one discipline area. Statisticians are lacking in the skills necessary for manipulating these large data sets efficiently, but statisticians have the skills that avoid biases and help divide the analytical tasks into manageable chunks without the loss of information.

The general view is that Big Data is the data miners domain and statistics does not play a key role. However, this view is narrow for the following reasons:

- The data quality challenges for ensuring the data are fit-for-purpose are enormous. It requires statistical skills involving outlier detection that avoids masking and swamping. These would involve:

1. The need for prospective robust statistical quality control methods involving the multivariate spatio-temporal consistency checking of data. The aim being that the measurement process is accurate and that the data are free from influential errors.
2. Planning of the dimension reduction process in a way that preserves all the sufficient statistics for future decisions. In other words, design the aggregation process and data compression process to maximize the information needed for its purpose.
3. Plan for future studies using the data—stratify the population into homogeneous groups to help with sample designs for future analyses. Think about how the data can be used for future longitudinal studies.
4. Propensity score matching should be used to avoid biases in observational studies and planning for potential future designed trials.
5. The whole aspect of assuring that the data are fit for purpose needs careful statistical thought and planning.

- Compressing data is not just about selecting a window over which to aggregate values—it is about compressing the data in a way that retains as much of the necessary information as possible. It is about preserving the sufficient statistics.
- Inference becomes more about mathematical significance (the size of the influence of a variable) and less about statistical significance. Estimation and prediction is all about avoiding biases—there may be selection bias issues.

The challenges listed above are statistical in nature and by no means are complete, but it is important to decide what part each discipline plays in the future development of analytical techniques for large data sets, and what parts are best done in partnership with others.

A quick summary of needs are:

1. Fast and efficient exploratory data analysis
2. Intelligent ways of reducing dimensions (both in the task and the data).
3. Intelligent ways of exploiting sparsity.
4. Intelligent ways of breaking up the analytical task (e.g., stratification and the parallel processing of different strata).

5. Intelligent and efficient visualisation, anomaly detection, feature extraction, pattern recognition.
6. Commitment to unbiased estimation and prediction/forecasting analytics.
7. Effective design—supported by starting with a thinking about what data to collect, how to collect it, and then how to analyse it.
8. Efficient designs for breaking the data into training and validation samples.
9. Real-time challenges—fast processing—estimation, forecasting, feature extraction, anomaly detection, clustering, etc.

References

1. Bolt, S., Sparks, R.: Detecting and diagnosing hotspots for the enhanced management of hospital emergency departments in Queensland, Australia. Med. Inform. Dec. Making **13**, 134 (2013)
2. Breiman, L.: Statistical modeling: the two cultures (with comments and a rejoinder by the author). Statis. Sci. **16**(3), 199–231 (2001)
3. Deming, W.E.: The new economics: for industry, government, education, 2nd edn. The MIT Press, Cambridge (2000)
4. Friedman, J.H., Stuetzle, W.: Projection pursuit regression. J. Am. Statis. Assoc. **76**, 817–823 (1981)
5. Friedman, J.H.: Fast sparse regression and classification. Int. J. Forecast. **28**, 722–738 (2012)
6. Harford, T.: Big data: are we making a big mistake? Significance **11**(5), 14–19 (2014)
7. Lahiri, P., Larsen, M.: Regression analysis with linked data. J. Am. Statis. Assoc. **100**, 222–230 (2005)
8. Megahed, F.M., Jones-Farmer, L.A.: A statistical process monitoring perspective on big data. In: XIth International Workshop on Intelligent Statistical Quality Control, CSIRO, Sydney (2013)
9. Popper, K.: Science as falsification. Conject. Refutat. Readings in the Philosophy of Science, 33–39 (1963)
10. Savage, L.J.: The Foundations of Statistics, Dover edn, 352pp (1972)
11. Sparks, R.S., Okugami, C.: Data quality: algorithms for automatic detection of unusual measurements. Front. Statis. Proc. Control **10**, 385–400 (2012)
12. Tibshirani, R.: Regression shrinkage and selection via the lasso. J. Royal Statis. Soc. Series B (Methodological), **24**, 267–288 (1996)
13. West, M., Harrison, P.J.: Bayesian Forecasting and Dynamic Models. Springer, New York (1997)
14. Williams, C., Rasmussen, C.: Gaussian processes for regression (1996). http://eprints.aston.ac.uk/651/1/getPDF.pdf

Toward Problem Solving Support Based on Big Data and Domain Knowledge: Interactive Granular Computing and Adaptive Judgement

Andrzej Skowron, Andrzej Jankowski and Soma Dutta

Big Data is defined by the three V's:
1. Volume — large amounts of data
2. Variety — the data comes in different forms, including traditional databases, images, documents, and complex records
3. Velocity — the content of the data is constantly changing, through the absorption of complementary data collections, through the introduction of previously archived data or legacy collections, and from streamed data arriving from multiple sources

—Jules J. Berman [1]

Abstract Nowadays efficient methods for dealing with Big Data are urgently needed for many real-life applications. Big Data is often distributed over networks of agents involved in complex interactions. Decision support for users, to solve problems using

This work by Andrzej Skowron and Andrzej Jankowski was partially supported by the Polish National Science Centre (NCN) grants DEC-2011/01/D/ST6/06981, DEC-2012/05/B/ST6/03215, DEC-2013/09/B/ST6/01568 as well as by the Polish National Centre for Research and Development (NCBiR) under the grant O ROB/0010/03/001. Soma Dutta was supported by the ERCIM postdoc fellowship.

A. Skowron (✉) · S. Dutta
Institute of Mathematics, Warsaw University, Banacha 2,
02-097 Warsaw, Poland
e-mail: skowron@mimuw.edu.pl

S. Dutta
e-mail: somadutta9@gmail.com

A. Skowron
Systems Research Institute, Polish Academy of Sciences, Newelska 6,
01-447 Warsaw, Poland

A. Jankowski
Knowledge Technology Foundation, Nowogrodzka 31, 00-511 Warsaw, Poland
e-mail: andrzej.adgam@gmail.com

© Springer International Publishing Switzerland 2016
N. Japkowicz and J. Stefanowski (eds.), *Big Data Analysis: New Algorithms for a New Society*, Studies in Big Data 16, DOI 10.1007/978-3-319-26989-4_3

49

Big Data, requires to develop relevant computation models for the agents as well as methods for incorporating changes in the reasoning of the computation models themselves; these requirements would enable agents to control computations for achieving the target goals. It is to be noted that users are also agents. Agents are performing computations on complex objects of very different natures (e.g., (behavioral) patterns, classifiers, clusters, structural objects, sets of rules, aggregation operations, reasoning schemes etc.). One of the challenges for systems based on Big Data is to provide the systems with high-level primitives of users for composing and building complex analytical pipelines over Big Data. Such primitives are very often expressed in natural language, and they should be approximated using low-level primitives, accessible from raw data. In Granular Computing (GrC), all such constructed and/or induced objects are called granules. To model interactive computations, performed by the agent in complex systems based on Big Data, we extend the existing approach to GrC by introducing *complex granules* (c-*granules* or *granules*, for short). Many advanced tasks, concerning complex systems based on Big Data may be classified as control tasks performed by agents aiming at achieving the high quality trajectories (defined by computations) relative to the considered target tasks and quality measures. Here, new challenges are to develop strategies to control, predict, and bound the behavior of the system based on Big Data at scale. We propose to investigate these challenges using the GrC framework. The reasoning, which aims at controlling the computational schemes from time-to-time, in order to achieve the required target, is called an *adaptive judgement*. This reasoning deals with granules and computations over them. Adaptive judgement is more than a mixture of reasoning based on deduction, induction and abduction. Due to the uncertainty the agents generally cannot predict exactly the results of actions (or plans). Moreover, the approximations of the complex vague concepts initiating actions (or plans) are drifting with time. Hence, adaptive strategies for evolving approximation of concepts with respect to time are needed. In particular, the adaptive judgement is very much needed in the efficiency management of granular computations, carried out by agents, for risk assessment, risk treatment, cost/benefit analysis. The approach, discussed in this paper, is a step towards realization of the Wisdom Technology (WisTech) program [2, 3], and is developed over years of experiences, based on the work on different real-life projects.

Keywords Rough set · (Interactive) granular computing · Interactive computation · Adaptive judgement · Efficiency management · Risk management · Cost/benefit analysis · Big data technology · Cyber-physical system · Wisdom web of things · Ultra-large system

1 Introduction

Nowadays we observe a flood of data [1]:

> Data pours into millions of computers every moment of every day. It is estimated that the total accumulated data stored on computers worldwide is about 300 exabytes (that's 300 billion gigabytes). Data storage increases at about 28 % per year. The data stored is peanuts compared to data that is transmitted without storage. The annual transmission of data is estimated at about 1.9 zettabytes (1900 billion gigabytes) [...] From this growing tangle of digital information, the next generation of data resources will emerge.

Presenting a characterization of Big Data is the motto of this article. Many topics related to Big Data Technology (BDT) for systems based on Big Data [1, 4–11], are getting great attention nowadays. Efficient methods for dealing with data of this type are important for many real-life applications. The importance of such applications is characterized in [6] (see Foreword, p. ix) in the following way:

> Big data and analytics promise to change virtually every industry and business function over the next decade. Any organization that gets started early with big data can gain a significant competitive edge. Just as early analytical competitors in the "small data" era (including Capital One bank, Progressive Insurance, and Marriott hotels) moved out ahead of their competitors and built a sizable competitive edge, the time is now for firms to seize the big data opportunity.

One can also consider the following sentence [1]:

> Perhaps the greatest potential benefit of Big Data is the ability to link seemingly disparate disciplines, for the purpose of developing and testing hypotheses that cannot be approached within a single knowledge domain.

Big Data may be embedded in Cyber-Physical Systems (CPSs) [12] and/or systems based on Wisdom Web of Things (W2T) [13]. CPSs [9, 12, 14] are characterized by a high degree of coupling between computations and physical processes, with the collaboration of computational elements and their respective physical entities. In [13], W2T is characterized by a data cycle, *viz.*, "from things to data, information, knowledge, wisdom, services, humans, and then back to things" making it possible to realize the harmonious symbiosis of humans, computers, and things in the emerging hyper world. Such systems based on Big Data may be treated as special cases of *Ultra-Large-Scale* (ULS) systems [9, 14]. ULS systems are interdependent webs consisting of software-intensive systems, people, policies, cultures, and economics. ULS are characterized by properties such as: (i) decentralization, (ii) inherently conflicting, unpredictable, and diverse requirements, (iii) continuous evolution and deployment, (iv) heterogeneous, inconsistent, and changing elements, (v) erosion of the people/system boundary, and (vi) routine failures [9, 14].

It is predicted that applications based on the above mentioned systems will have enormous societal impact and economic benefit. However, there are many challenges related to such systems. In this article, we claim that further development of such systems, in particular systems based on Big Data for supporting users in problem solving, should be based on the relevant computation models.

There are several important issues which should be taken into account in developing such a model. Among them some are as follows.

- Computations are performed on complex objects with very different structures, where the structures themselves are constructed and/or induced from data and domain knowledge.
- Computations are performed in an open world where interactions of physical objects are unavoidable.
- Due to uncertainty, the properties and results of interactions can be perceived by agents only partially.
- Computations are realized in the societies of interacting agents including humans.
- Agents are aiming at achieving their tasks by controlling computations.
- Agents can control computations by using *adaptive judgement*, in which all of deduction, induction and abduction are used.

For BDT, we propose to base on the model of the Granular Computing (GrC) framework.

Granulation of information is inherent in human thinking and reasoning processes. It is often realized that precision is sometimes expensive and not very meaningful in modeling and controlling complex systems. When a problem involves incomplete, uncertain, and vague information, it may be difficult to discern distinct objects, and one may find it convenient to consider granules for tackling the problem of concern. Granules are composed of objects that are drawn together by indiscernibility, similarity, and/or functionality among the objects [15]. Each of the granules according to its structure and size, with a certain level of granularity, may reflect a specific aspect of the problem, or form a portion of the system's domain. GrC is considered to be an effective framework in the design and implementation of intelligent systems for various real life applications.

The systems based on GrC, e.g., for pattern recognition, exploit the tolerance for imprecision, uncertainty, approximate reasoning and partial truth of soft computing framework and are capable of achieving tractability, robustness, and close resemblance with human-like (natural) decision-making [16–19].

In GrC, computations are performed on granules of different structures, where granularity of information plays an important role. Granules should be constructed and/or induced from data and domain knowledge during computations realized by systems based on Big Data. In particular, when one would like to build a system for supporting users in solving problems using Big Data, granules may represent computational building blocks for approximating (or inducing models of) high-level primitives used by users in order to compose complex analytical pipelines over Big Data [20]. In this way, granules are enabling to make such primitives "comprehensible" to the system based on Big Data. Let us note that these primitives, used by the users, are often expressed by complex vague concepts represented in a natural language.

Let us note that agents or teams of agents may be treated in GrC as complex c-granules, too. It is worthwhile mentioning that the idea of coupling among brain (control), body, and environment presented in [21] (p. 1088) is related to some high level c-granules:

> [...] First, [...] system's behavior is also affected by the ecological niche in which the system is physically embedded, by its morphology (the shape of its body and limbs, as well as the type and placement of sensors and effectors), and by the material properties of the elements composing the morphology. [...] Second, physical constraints shape the dynamics of the interaction of the embodied system with its environment... Third, [...] Coupled sensory-motor activity and body morphology induce statistical regularities in sensory input and within the control architecture and therefore enhance internal information processing.

The size of ULS and their hybrid nature (consisting of (i) physical elements as well as quasicontinuous and discrete controls, (ii) communication channels, and (iii) local and system-wide optimization algorithms as well as management systems), imply that hierarchical and multi-domain approaches for their simulation, analysis and design are needed [14]. We propose to develop such approaches based on granulation and/or de-granulation of constructed and/or induced granules.

The usefulness of GrC in the context of hierarchical and multi-domain approaches for ontology approximation and domain knowledge transfer is evident from (see, e.g., [22–24]). While working with complex real-life projects such as control of UAV, risk assessment in medical diagnosis, algorithmic trading, fraud detection, and real-time decision support of the commander of firefighters, we realized the need for a new methodology for approximating of complex vague concepts (e.g., safe situation on the road, risk of losing life by a child admitted to emergency room or risk of fire spreading or explosion etc.). In all these cases, the existing methods are proved to be infeasible in inducing the relevant granules (e.g., patterns and/or models) directly from data. Fortunately, it was possible to acquire ontologies from domain experts. These acquired ontologies along with the concepts approximated by cooperation with experts were used for enriching the construction of the decision system. Concepts and relations present in these ontologies are often vague. However, it was possible gradually to approximate them by induced granules (e.g., patterns and/or models) — starting from the lowest level of ontology to that at the highest level. Approximations were constructed using methods based on a combination of rough sets and fuzzy sets. Using the decision systems, enriched by the induced approximations of concepts and relations from ontologies, a very good performance on real-life data has been observed (see, e.g., [22–24]).

It is important to note that the agents in the discussed systems are linked with the physical objects and very often the aim is to control the performance of computations in the physical world for achieving the desired goals. Hence, in developing our computation model we have no choice but to recognize the dependence of our mathematical knowledge on laws of physics and consider computations as physical processes (see, e.g., [25]).

Information granules (infogranules, for short) in GrC are widely discussed in the literature [17]. In particular, let us mention here the rough granular computing approach based on the rough set approach and its combination with other approaches to soft computing, such as fuzzy sets. However, the issues related to the interactions of infogranules with the physical world, and perception of interactions in the physical world by means of infogranules are not well elaborated yet. The understanding of interactions is the critical issue of complex systems [26]. For example, the ULS are autonomous or semiautonomous systems, and cannot be designed as closed systems that operate in isolation; rather, the interaction and potential interference among smart components, among CPSs, and among CPSs and humans, require to be modeled by coordinated, controlled, and cooperative behavior of agents representing components of the system [14]. We extend the existing approach to GrC by introducing *complex granules* (c-*granules*, for short) [2] and making it possible to model interactive computations carried out by agents and their teams in complex systems working in an open world.

Any agent operates on a local world of c-granules. The agent aims at controlling computations performed on c-granules from this local world for achieving the target goals. In our approach, computations in systems based on Big Data proceed through complex interactions among physical objects. Some results of such interactions are perceived by agents.

Some important aspects of BDT, related to providing decision support to users, in the context of problem solving with the usage of Big Data, concern reasoning methods on computations performed by such systems. These reasoning methods may be based on formal languages, i.e., the expressions from such languages are used as labels (syntax) of granules. However, there are also paradigms such as Computing With Words (CWW), due to Professor Lotfi Zadeh [27–32], where labels of granules are *words* (i.e., words or expressions from a relevant fragment of natural language), and computations are performed on *words* (http://www.cs.berkeley.edu/~zadeh/presentations.html):

> Manipulation of perceptions plays a key role in human recognition, decision and execution processes. As a methodology, computing with words provides a foundation for a computational theory of perceptions - a theory which may have an important bearing on how humans make - and machines might make - perception-based rational decisions in an environment of imprecision, uncertainty and partial truth. [...] computing with words, or CW for short, is a methodology in which the objects of computation are words and propositions drawn from a natural language.

In GrC it is necessary to develop new methods extending the approaches for approximating vagues concepts (e.g., the high-level primitives of users corresponding to these concepts in the context of BDT) expressed in natural language for *approximating* reasoning on such concepts. It is also important to note that information granulation plays a key role in implementation of the strategy of divide-and-conquer in human problem-solving [32, 33]. Hence, it is important to develop methods which could (by using primitives of the system based on Big Data) perform approximate reasoning along such decomposition schemes delivered by the strategy of divide-and-conquer in human problem-solving, and induce the relevant granules as computational building blocks for constructing the solutions for the problem.

In case of systems based on Big Data, the users are often specifying problems in fragments of a natural language with the requirement, that their solutions will satisfy specifications to some satisfactory degrees. Hence, methods for approximation of domain ontology (i.e., ontology on which a fragment is based) as well as approximations of constructions representing solutions based on concepts from the domain ontology should be developed. These approximations may help the system to follow, in an approximate sense, the judgement schemes expressed in the relevant (for considered problems) fragment of a natural language. We also emphasize the importance of dialogues between users and system in the process of obtaining the relevant approximations.

Very often the problems related to BDT are related to control tasks. Examples of control tasks may be found in different areas, such as the medical therapy support, management of large software projects, algorithmic trading or control of unmanned vehicles, to name a few. Such projects are typical for Cyber-Physical Systems (CPSs) or Wisdom Web of Things (W2T). Any of such exemplary projects is supported by Big Data and domain knowledge distributed over computer networks and/or Internet. Moreover, interactions of agents with the physical world, which are often unpredictable, are unavoidable. Computations performed by agents are aiming at constructing, learning, or discovering granules, which in turn makes it possible to understand the concerned situation (state) to a satisfactory degree. The relevant controlling of computations based on this understanding is realized using approximations of complex vague concepts playing the role of guards, responsible for initiation of actions (or plans) by agents. In particular, for constructing these approximations different kinds of granules, discovered from Big Data, are used. The main processes, namely granulation and degranulation, characterize respectively the synthesis and decomposition of granules in the process of obtaining relevant resultant granules.

The efficiency management in controlling the computations [2] in BDT are of great importance for the successful behavior of individuals, groups or societies of agents. In particular, such efficiency management is important for constructing systems based on Big Data for supporting users in problem solving. The efficiency management covers risk assessment, risk treatment, and cost/benefit analysis. The tasks related to this management are related to control tasks aiming at achieving the high quality performance of (societies of) agents. The novelty of the proposed approach is in interpreting the complex vague concepts as the guards of control actions (or plans) performed by agents. These vague concepts are represented using domain ontologies. The rough set approach in combination with other soft computing approaches is used for approximation of the vague concepts used in the process of judgement involved in the efficiency management.

One of the challenges, here, is to develop methods and strategies for adaptive reasoning, called adaptive judgement, e.g., for adaptive control of computations. In particular, adaptive judgement is very much needed in the efficiency management. The efficiency management in decision systems requires tools to discover, represent, and access approximate reasoning schemes (ARSs) (over domain ontologies) representing the judgement schemes [23, 34, 93]. ARSs are approximating, in a sense, judgement expressed in relevant fragments of simplified natural language. Methods for inducing of ARSs are still under development. The systems for problem solving are enriched not only by approximations of concepts and relations from ontologies but also by ARSs.

The discussed approach is a step towards one way of realization of the Wisdom Technology (WisTech) program [2, 3]. The approach was developed over years of work on different real-life projects.

This chapter is organized as follows. Agent language for basic tasks in BDT is discussed in Sect. 2. In Sect. 3, we present some basic postulates concerning agent's behaviors in the physical world. In Sect. 4, an introduction to Interactive Granular Computing (IGrC) is presented. Interactive computations on complex granules realized by agents are discussed in Sect. 5. Section 6 is devoted to the issues related to decision support of users in problem solving with the use of BDT. This section includes comments on problem specification in BDT (Sect. 6.1). Next, some strategies for construction and discovery of new granules are presented (Sect. 6.2). In particular, such strategies can be based on (i) aggregation by joining information systems with constraints, (ii) inducing a hierarchy of satisfiability relations, (iii) self-organization of agents, or (iv) communications and dialogue among agents. The approach to efficiency analysis, in particular to risk management, in controlling computations over granules in BDT, is presented in Sect. 6.3. The role of reasoning based on adaptive judgement is discussed in Sect. 6.4. Section 7 concludes the chapter.

The chapter summarizes as well as extends the work developed in [35–38].

2   Agent Language for Basic Tasks in BDT

Agents realize their goals by performing actions. Hence, it is very important to discover some measures for evaluating the correctness of a selection of a given action in a given situation. For any action a, one can consider a complex vague concept Q_a representing such a measure. For a particular situation s, the value of $Q_a(s)$, is a c-granule representing the degree to which $Q_a(s)$ is satisfied at s, i.e., the correctness degree of selection of the action a at s. The c-granule $Q_a(s)$ consists of two main c-subgranules representing arguments *for* and *against* the satisfiability of $Q_a(s)$. These arguments are derived from the judgement based on the estimation that, if a potentially can be initiated in situation s with respect to the efficiency management [2]. For example, in the risk assessment [39] the goal of the judgement is to identify the main risks. On the basis of the risk degrees another judgement, called the risk treatment, is performed. Some modifications of performed actions, called controls (or new controls), are considered against existing (or possible) vulnerabilities. These new controls could suggest in favor of avoiding the risk, reducing the risk, removing the source of the risk, modifying consequences, changing probabilities, sharing the risks with other agents, retaining the risk or even increasing the risk to pursue the opportunity (see www.praxiom.com/iso-31000-terms.htm and Fig. 14).

In a relevant fragment of natural language, one should judge the degrees of satisfiability of $Q_a(s)$ for all relevant actions. One should also judge conflict among

Fig. 1 Judgement on satisfiability degrees of guards, i.e., complex vague concepts used for initiation of actions. Agents are using actions to control interactions and, in consequence, the computations over granules progressing due to interactions [37]

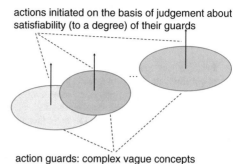

actions initiated on the basis of judgement about satisfiability (to a degree) of their guards

action guards: complex vague concepts

the degrees corresponding to different actions in order to select the best action(s) for execution in a given situation.

One can consider the above mentioned tasks of approximation of action guards as the task of complex game discovery (see Fig. 1) from data and domain knowledge in cooperation with the domain experts.

The discovery process of complex games, in particular complex vague concepts which are embedded in them, often is based on hierarchical learning supported by domain knowledge [2, 23]. An agent is interacting with the environment for discovering the concepts and the cause-effect relationships relevant for the complex games. Next, these concepts and relationships are used by the agent to judge the results of interactions for efficient initiation of relevant actions. It is also worthwhile mentioning that these games are evolving with time (drifting with time) together with the data and knowledge about the approximated concepts as well as with the relevant strategies for adaptation of games used by agents. Hence, adaptive strategies are required for enabling agents to control their behavior in order to achieve the targets. It is also to be noted that these strategies should be learned from the available uncertain data and domain knowledge.

Let us summarize our considerations on the idea of discovery of games. Decision making under uncertainty involves large number of complex vague concepts. Among them some are concepts related to, e.g., identification of the current situation, discovery of the relevant context relative to which one should consider the actual situation (by considering the past, the possible future, including risks, costs or benefits), discovery of similarity measures of the current situation (or plans) to the observed ones in the past, relevant concepts for measuring the deviation degrees of the predicted situation with the real one. Moreover, for dealing with complex systems there is a need for a language in which adaptive judgement over concepts, relevant for these systems, could be performed. In particular, let us mention the need for adaptive judgement for conflict resolution among the arguments *for* and *against* concerning the satisfiability of these concepts. The reader is referred here to already cited paradigm of CWW, Computing with Words (see, Sect. 1 and e.g., [27–32], http://www.cs.berkeley.edu/~zadeh/presentations.html), the sentences by Pearl from [40] (see Sect. 6.3), as well as to the notion of Perception Based Computing (PBC) (e.g., [28, 30, 32, 41–43]). One of the basic tasks of PBC is hierarchical learning of complex vague concepts used for comprehending the perceived situations. Let us recall some sentences from [41] to explain that PBC is related to BDT:

> Perception is characterized by sensory measurements and ability to apply them to reason about satisfiability of complex vague concepts used, e.g., as guards for actions or invariants to be preserved by agents. Such reasoning is often referred as adaptive judgement. Vague concepts can be approximated on the basis of sensory attributes rather than defined exactly. Approximations usually need to be induced by using hierarchical modeling. [...] Unfortunately, discovery of structures for hierarchical modeling is still a challenge. On the other hand, it is often possible to acquire or approximate them from domain knowledge.

For real-life projects it is hardly possible to expect that the high quality models of the discussed complex vague concepts can only be induced on the basis of automatic methods (see, e.g., [44]) without acquiring the agents' domain knowledge through cooperation with the domain experts.

One natural direction is to construct dialogue systems different from the traditional data mining systems. In the future, it will be then possible for users to formulate hypotheses, which the systems may verify interacting through a dialogue with the users. Such systems will allow us for more efficient discoveries. This view is expressed in [45] in case of biology:

> [...] Tomorrow, I believe, every biologist will use computer to define their research strategy and specific aims, manage their experiments, collect their results, interpret their data, incorporate the findings of others, disseminate their observations, and extend their experimental observations - through exploratory discovery and modeling - in directions completely unanticipated.

One can predict that such systems will be widely used in other domains too.

However, several challenges need to be resolved before such systems get used widely. In particular, they are related to the ontology of (complex vague) concepts, and relations among them on which agents can base for problem solving. Moreover, one should consider a language in which adaptive judgement about satisfiability of these concepts and relations can be performed. A challenge is to transfer the ontology and the language to the system so that the system becomes able to perform the necessary judgements with satisfactory quality. In the following sections, we present some preliminary discussion illustrating how rich such ontology can be, and how complex tasks are to be solved by using judgement.

3   Postulates about Physical World and Agents

In this section, we present some basic postulates concerning agents' operations with the physical world. Agents are perceiving a part of the open physical world, and they are interacting with the perceived world. Concept and relations postulated in this section create the key ontological basis of WisTech [2, 3]. There are several groups of postulates. Some of them are related to the physical character of the agent, c-granules and interaction models, while the others concern the efficiency management of judgements and the realization of the prioritized needs of the agent. The postulates are specifying some basic concepts which are important for interactive computations on complex granules realized by agents for achieving their goals. It is worthwhile mentioning that we specify only a general preliminary framework for applications in real-life (intelligent) systems. There is a need of further work for making this specification more detailed and precise. A further step toward this direction is presented in [2].

3.1 Physical Character of Agents, C-Granule, and Interaction Models

In this section, we present the basic postulates concerning the physical character of agent, c-granule, and interaction model.

Physical World

1. *The physical world* consists of *physical beings*.
2. Physical beings may *interact*.
3. *Interactions* are satisfying some *interaction cause-effect relationships* that proceed from the laws of the physical world.

Agent

1. An agent is a *physical being*.
2. An agent *perceivs* and *records* some interaction effects. Due to uncertainty only some interaction effects (results) may be partially recorded and perceived by agents.
3. Agents are equipped with some private *clocks*.
4. *An agent control* consists of physical beings for the realization of some cause/effect relationships. The agent control may activate realization of cause-effect relationships. This leads to the creation of structures of physical beings which thereby activate interactions. The agent control may try to predict the results of these interactions.
5. *An agent control* has the skills to perceive and record some physical beings and/or their interactions through perceiving and recording respective properties of interaction effects. More precisely, the agent control has the skills to perceive and record some properties of physical beings and the results of their interactions. This is carried out by the control of agent perceiving, recording and verifying the results (effects) of the interactions of the agent control with those physical beings which are in interactions with each other.
6. *An agent control* has some skills for perceiving the physical beings and/or their interactions in a specific space consisting of the agent's potentially perceived physical beings. This space is called the *agent's activity environment*. The physical beings perceived in the agent's activity environment are called *hunks*.

C-Granule

1. The agent control has skills for aggregation, reconfiguration and selection of the hunk configuration aggregates (e.g., corresponding to the agent c-granules) for the activation of complex interactions in environments and for achievement of complex results (effects) (or for the generation of new agent c-granules). In these activities the agent control may use: (i) processes in environments leading to the self-organization of hunk configuration aggregates which are in interaction with each other or (ii) other complex networks of interconnected cause-effect relationships.
2. At any moment t of the agent time, the agent perceives the physical world using c-granules generated by the agent control. By employing c-granules the agent is developing such skills as:

    (a) initialization, storing (registering) and judgement of interaction results over hunks accessible by c-granules,
    (b) aggregation and decomposition of c-granules,
    (c) archiving, retrieval, reconstruction and destruction of c-granules.

3. Any c-granule consists of the following three "architectural layers":

    i. *Soft_suite*—consists of used c-granule configurations of hunks representing properties of the agent's memory states (m-hunks) perceived by the agent control.
    ii. *Link_suite*—consists of configurations of hunks called links. Links are used as transmission carriers of interactions between m-hunks (called the beginnings of links) from the soft suite and interactions among other hunks (in particular, m-hunks) (called the ends of links). In particular, links are used by the c-granule for establishing the current relationships between the beginnings and ends of links. Links also enable the functioning of sensors and actuators accessible by the c-granule.
    iii. *Hard_suite*—consists of hunks which through interactions among themselves as well as with the environment enable the functioning of the soft suite and the link suite of the c-granule. It may happen that interactions initiated by the c-granule may be perturbed by unpredictable interactions of hunks with the environment. This may result in the agent control's expected interactions differing from the real ones.
    iv. *Any c-granule* may be activated by the agent control and/or by another c-granule. An active c-granule may process, modify, and/or produce other c-granules. In this way c-granules can be parts of other more complex c-granules creating (hierarchical) networks of interconnected c-granules.

4. Any c-granule belongs to at least one agent who is the owner of the c-granule. A team of interacting agents may produce common meta c-granules. These meta c-granules belong to a meta-agent composed of agents of the considered team. The control of meta agents obtained in this way treats the meta c-granules as c-granules of the meta-agent. Each agent from the team creating the meta-agent

may be treated by this meta-agent as a c-granule, accessible (fully or partially) by the meta-agent.

Interaction Models

1. Physical world perception by the agent is carried out by means of encompassing c-granules interaction models constructed by the agent and their results.
2. By using *interaction models* the agent is representing: (i) the observed, simulated and predicted models of the physical world states as well as (ii) interactions causing changes of physical world states, called *transition interactions* in models.
3. Interaction models are represented by agent using c-granules. These, in turn, are composed out of other c-granules representing: (i) physical world states, (ii) transition interactions, and (iii) state and interaction properties. Roughly speaking, c-granules are the agent "windows" through which the agent can perceive and record the physical reality fragments and interfere with there ongoing interactions.
4. C-granules modeling states of the physical world, transition interactions and their properties may concern past, present or future of interactions in the physical world. Transition interactions may be initiated by the agent control and/or by the agent activity environment.
5. *Agent interaction plan* is a c-granule with the following skills: (i) a skill defining the agent motion strategies in the agent activity environment carried out by using rules for selection transition interactions in different parts of physical world, (ii) a skill specifying properties of final states of the physical world for a given interaction plan.
6. The agent uses interaction models for the construction of interaction plans and for their realization, adaptation, judgement and efficient learning of construction of new c-granules.
7. Agent control states change over time. An agent may perceive some of these states and relations among them by means of relevant c-granules.
8. An agent has skills for construction of interaction models on the basis of perceivable interactions between the agent control states.
9. An agent can construct of interaction models encompassing properties of perceived:

   (a) agent control states,
   (b) relations among agent control states,
   (c) other interactions causing changes in agent control states.

3.2 Efficiency Management of Task Realization by a Single Agent and Agent Society

In this section we present the basic postulates concerning efficiency management of tasks realization by a single agent and agent society.

Adaptive Judgement Relative to The Agent Needs Hierarchy

1. By using c-granules an agent control is able to perceive some properties of physical control states. One of the properties of these states are *agent needs*. Any
   agent need is represented by a c-granule. Examples of the agent needs concern:
   (i) acquisition of resources necessary for agent functionality, (ii) judgement of
   solutions of problems (e.g., problems related to the construction of an interaction
   plan leading to satisfaction of another need), (iii) judgement if conditions for the
   initialization of an interaction plan are satisfied to a satisfactory degree for making a decision about their initialization, (iv) judgement if results of realization of
   the interaction plan expected by the agent are satisfactory, and if not the agent
   decides which actions should be performed for development and initialization of
   actions responsible for correction.
2. Usually, an agent has c-granules for adaptive judgement of solutions of problems related to the c-granule functionality. They are used for construction, actualization, and aggregation of current variants of problem solutions along with
   judgement on created arguments *for* and *against* selection of the variants in the
   agent activity context. In particular, this judgement may lead to conflict resolution among arguments *for* and *against* for selecting one solution. Changes in the
   agent activity environment may lead to changes caused by c-granules for adaptive
   judgement in arguments *for* and *against*, their aggregations in the form of solution
   variants as well the results of aggregation in the form of adapted judgement of
   resolved issue.
3. For any agent need and any agent time moment t, the c-granule corresponding
   to the need has a skill for adaptive judgement of such attributes of agent needs
   as: (i) timeliness and importance at t of the agent need (especially important
   is judgement of the need importance, relative to importance at t of other agent
   needs), (ii) degree of the need realization at t (i.e., degree in which the agent is
   satisfied and/or unsatisfied at t from the need realization).
4. Agent control has the ability to perceive the *agent needs hierarchy*. This c-granule
   is an aggregation of c-granules of needs (representing the properties of perceived
   needs by agent control), c-granules representing (adaptive) relations among them
   and adaptive judgements over the actualization of components in the agent needs
   hierarchy. Agent control has skills for identifying which of the needs from the
   agent needs hierarchy are currently important. Such needs are aggregated into a
   c-granule called the *current hierarchy of prioritized needs*. This c-granule is a
   part of the agent hierarchy of needs.

Cost/Benefit Analysis and Interaction Plans

1. An agent develops, realizes and adopts plans (represented by c-granules) which
   on the basis of the agent judgement will increase degrees of the agent needs realization (especially the most important needs according to the current prioritized
   hierarchy of needs). In general, the realization of plans causes some agent costs
   (often they are related to decreasing degrees of realization of the agent needs).

The agent expects some benefits after the realization of the interaction plan (often they result in increasing degrees of agent needs realization).

2. The agent has some skills for the estimation of costs and benefits for supporting (i) analysis and selection of interaction plans and (ii) judgement performed by an agent control for the selection of plans for realization. Agent has *interaction plan cost c-granule, interaction plan benefit c-granule*, and *c-granule for comparison of costs and benefits of interaction plans*. The implementation of this approach should make it possible for agents to base judgement on different variants of interaction plans in a framework related to a well known approach in economy called Cost/Benefit Analysis (http://en.wikipedia.org/wiki/Cost-benefit_analysis).

Swot Analysis and Interaction Plan

1. The realization of interaction plans may be disturbed. As a consequence the expected properties of the interaction plan realization may be different from the real ones. A disruption of the interaction plan is *negative* if at least one of the following conditions is satisfied: (i) "ratio" of total benefits to costs of the interaction plan realization (under this disruption) substantially decreases or (ii) the total cost of the interaction plan realization (under this disruption) is not acceptable. A disruption of the interaction plan is *positive* if it is not negative and at the same time it substantially increases the "ratio" of total benefits to costs of interaction plan realization (under this disruption). Any interaction plan realization should be linked to a prediction of likelihood and consequences of negative and/or positive disruptions. Hence, realization of the interaction plans is related to the risk management encompassing the risk assessment and the risk treatment. The basis for analysis and comprehension by agents of their current situation is the SWOT (Strengths, Weaknesses, Opportunities and Threats) analysis [46–49]. According to the definition, a negative disruption causes negative consequences for the interaction plan realization. The larger is the likelihood of the negative consequences, the larger is the risk. Hence, the risk of negative disruption in a given interaction plan realization is an aggregation of the negative consequences of this negative disruption and the disruption likelihood. The risk of the interaction plan is an aggregation over the interaction plan of all risks of the negative disruption of this plan.

Corisks and the Efficiency of Interaction Plan

1. Agent may use the strength for utilizing the (not predicted before) opportunities. Hence, it may occur a positive disruption during interactive plans realization. This leads to the *corisk* concept, analogously to the risk concept. More formally, corisk of a positive disruption of the interaction plan realization is an aggregation of consequences of this positive disruption and its likelihood, under assumption that they are not changing the risk of negative disruptions below acceptable level. The corisk of an interaction plan is an aggregation of all corisks of this interaction plan. Agent judgement concerning the developed interaction plans may take into account the *interaction plan efficiency*, i.e., an aggregation of the total cost, benefit, risk, and co-risk of the interaction plan.

Realization of the Most Important Tasks of the Agent Based on Perception Processes Leading to Comprehension of Perceived Situations by the Agent

1. Perception of situation by the agent is a process leading to construction by agent of a c-granule representing comprehension of the perceived situation from sensory information. This c-granule is called *situation comprehension c-granule*.
2. The situation comprehension c-granule is an aggregation of relevant c-granules resulting from classification of interactions. This c-granule is representing such *computational building blocks* (see cited Sect. 6.4 sentences by Leslie Valiant) for the perceived situation as contexts, SWOT, risks, corisks, prioritized needs or prioritized initialization and realization of interactions in the given situation. The computational building blocks are used by the agent control for concurrent initialization and realization of possibly efficient interaction plans aiming at, in the agent belief, realization of the agent prioritized needs (including the need of better comprehension of perceived situation).
3. The agent has skills for perception and perception evolving. These skills are used for possibly efficient actualization, improvement, and satisfying of the agent hierarchy of needs which is adaptively changing. Agent is performing these tasks by development, realization, verification (judging), and adaptation of interaction plans—following as much as possible the framework of PDCA cycles (Plan-Do-Check-Act) [50, 51]—aiming at construction of the interaction plans which are possibly more and more efficient and are increasingly less risky as well as have possibly more and more high corisk. Agent in searching—as far as possible in PDCA cycles—for 'optimal' c-granules expressing relevant features of processes (such as discovery, learning and satisfying the prioritized and adaptively changing agent needs and relations among them) aims at possibly efficient improving these processes. This'optimization' of c-granules supports the agent in more and more efficient construction, realization, adaptive judgement and adaptation of interaction plans that change over time for a more efficient realization of the agent's priority needs.

Communication Among Agents

1. Communications of agents are realized trough interactions. c-granules are the basic agent constructs for interaction with the environment.
2. Agents are using some specialized c-granules for increasing the efficiency performance. Among such c-granules the *semiotic c-granules* play the important role. A semiotic c-granule is obtained by aggregation of a c-granule g with another c-granule g' called the *context interpretation* of g. Semiotic c-granules are supporting the agent's control in improving the approximation tasks (e.g., identification, specification, and comprehension of the current context of the realization of the agent's activities) as well as construction and realization of (semi)optimal plans (in a given context). A *private agent language* consists of a distinguished family of semiotic c-granules and is closed with respect to some selected operations of construction of new semiotic c-granules.

3. Agents from *agent societies* are constructing, using and developing communication skills among agents from these societies. For these purposes they are constructing, using, and developing *communication languages* relevant for their needs. In communication processes, agents with two roles are fundamental: *sender* and *receiver*. For example, a sender agent is activating some plans of interactions, which leads to producing or distinguishing some hunks, called the *sender artefacts*. Receiver agent is perceiving, judging and sometimes storing co-existing situations, and behavioral patterns of the sender-agent together with created or pointed by the receiver-agents's artefacts. If, for the receiver agent, the co-existence of situations or artefacts is relevant then it is represented as a c-granule, stored, and a special name (and type) to this c-granule is assigned. Agents may change their roles as sender and receiver; this leads to dialogues.

4. In *dialogues* agents may learn from each other and/or adjust properties of interaction plans. This leads to common comprehension of structures and properties of *social c-granules* (in particular, common plans realized by the society). Each agent is producing a c-granule and treating it as an interpretation of the social c-granule. It may happen that different agents may have slightly different comprehension of a given social c-granule. Then agents may try through dialogues to reach a consensus about comprehending this social c-granule (or in some cases to modify it).

Agent Team Cooperation in Problem Solving

If the society of agents is satisfying a collection of conditions following from the specificity of the class of problems to be solved, then this society may be able to undertake and realize constructive cooperation during problem solving processes. For cooperation processes the following aspects are important:

1. The *quality of the common concept ontology* used for initiating and realization of the project for resolving problems from a given class.
2. Relevant selection (especially with respect to initiation and realization) of an *adaptive decomposition* of a given problem into subproblems to be solved by properly prepared agent subteams.
3. *Communication efficiency* among agents, in particular in the scope of prioritized issues, to be solved.

A society of agents may increase efficiency of problem solving using strategies for discovery and improving (semi)optimal social c-granules relevant for concepts (and relations among them) which efficiently support all the required or predicted problems of concern.

In this section, we have presented a preliminary discussion on basic intuition of c-granule usage in the WisTech framework. From this discussion one can observe how complex the concept ontology of WisTech is. In particular, this concerns the concept of c-granule as well as diversity of c-granules. Taking into account the necessity of judgement under uncertainty, this ontology should encompass the concepts necessary for specification, realization, verification and development of such processes, as:

(a) identification of the current situation and characterization of the most important features from the agent needs hierarchical perspective,
(b) discovery of the relevant context of the current situations should be considered (by considering its past, future, risks, costs, benefits as well as likelihoods and vulnerabilities),
(c) similarity of the current situation and plans with the situations and interaction plans analyzed (observed and/or simulated) in the past needs to be recognized,
(d) relevant deviations of expected or predicted interaction results from the real ones need to be identified,
(e) appearance of unexpected possible changes in the recognized situation needs to be counted by the agent.

We have pointed out that in complex real-life projects there is also a need for developing a language used for carrying out the relevant adaptive judgements concerning above mentioned concepts and satisfiability degrees; in particular, conflict resolution among the arguments *for* and *against*, as well as a language for expressing different patterns used for inducing new concepts and properties of observed phenomena, are a few to name. For example, the properties of observed phenomena are useful in further activities and cooperation of societies of agents. This need is very well expressed by Zadeh and Pearl (Sect. 1, [40], and Sect. 6.3).

Let us also note that the undertaken efforts over the last decades for developing AI techniques based on fully automatic learning; representation and processing of concepts are not satisfactory from the point of view of complex real-life projects. Classical examples can be found in research related to reinforcement learning [44] and different variants of natural computing [52]. This follows from the difficulty of coping with complexity and diversity of complex vague concepts which should be efficiently learned (discovered), approximated with satisfactory quality, and efficiently processed with proper judgement. Searching spaces for discovery of satisfactory approximations of such concepts are so huge that existing data mining methods, other AI methods, as well as the current and expected hardware technology do not allow us for effective searching over such spaces in realistic time.

At the end of the previous section, we have mentioned that the dialogue systems, where users will cooperate with computing devices towards solving problems, can be brought into the progress. However, the existing technology is yet not fully satisfactory for developing such systems. We expect, that further development of the ideas presented in this chapter, concerning interactive computations based on c-granules, will help in realization of this goal.

4   Interactive Granular Computing (IGrC)

The essence of the proposed approach is to develop BDT based on IGrC [2, 38, 42, 53–55]. In this sense IGrC creates the basis for BDT. The approach is based on foundations for modeling IGrC relevant for BDT in which computations are progressing through interactions [26]. In IGrC interactive computations are performed on c-granules linking, e.g., information granules [17] with spatiotemporal physical objects, called hunks [2, 56].

Infogranules are widely discussed in the literature. They can be treated as specifications of compound objects which are defined in a hierarchical manner together with descriptions regarding their implementations. Such granules are obtained as the result of information granulation [32]:

> Information granulation can be viewed as a human way of achieving data compression and it plays a key role in implementation of the strategy of divide-and-conquer in human problem-solving.

Infogranules belong to those concepts which play the main role in developing foundations of Artificial Intelligence (AI), data mining, and text mining [17]. They grew up as some generalizations from fuzzy set theory, [30, 32, 33], rough set theory, and interval analysis [17]. In GrC, rough sets, fuzzy sets, and interval analysis are used to deal with vague concepts.

However, the issues related to the interactions of infogranules with the physical world, and their relationship to perception of interactions in the physical world are not well elaborated yet [26, 57]. On the other hand, in [58], it is mentioned that:

> [...] interaction is a critical issue in the understanding of complex systems of any sorts: as such, it has emerged in several well-established scientific areas other than computer science, like biology, physics, social and organizational sciences.

Computations of agents proceed by interaction with the physical world and they have roots in c-granules [2]. Any c-granule consists of three components, namely soft_suit, link_suit and hard_suit. These components make it possible to incorporate abstract objects as infogranules from the soft_suit as well as physical objects from hard_suit. The link_suit of a given c-granule is used as a kind of c-granule interface for handling interaction between soft_suit and hard_suit (see Fig. 2). One can relate this to the statement http://www.en.wikipedia.org/wiki/Embodiment:

> [...] Embodied agent, in artificial intelligence, an intelligent agent that interacts with the environment through a physical body within that environment

Calculi of c-granules are defined by elementary c-granules (e.g., indiscernibility or similarity classes). Then with the help of the calculi it is possible to generate new c-granules from the already defined ones (see Fig. 2, where the presented c-granule produces new output c-granules from the given input c-granules).

The discussed c-granules may represent complex objects. In particular, agents and their societies can be treated as c-granules too. An example of c-granule representing a team of agents is presented in Fig. 3 where some guidelines for implementation of AI projects in the form of a cooperation scheme itself among different agents responsible for relevant cooperation areas is illustrated [2]. This cooperation scheme may be treated as a higher level c-granule. We propose to model a complex system as a society of agents.

Moreover, c-granules create the basis for the construction of the agent's language of communication and the language of evolution. The hierarchy of c-granules is illustrated in Fig. 4.

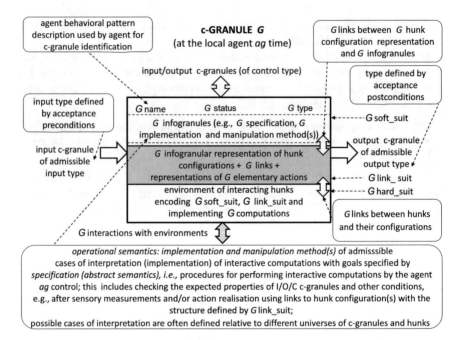

Fig. 2 General structure of a c-granule [36]

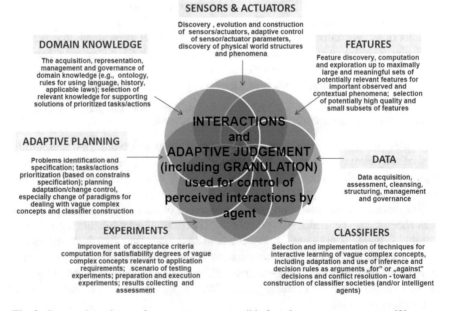

Fig. 3 Cooperation scheme of an agent team responsible for relevant competence area [2]

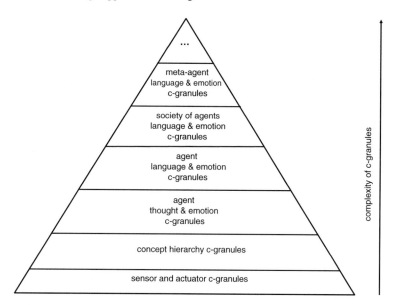

Fig. 4 Hierarchy of c-granules [36]

An agent operates on a local world of c-granules. The control of an agent aims at controlling computations performed on c-granules, from the respective local world of the agent for achieving the target goals. Actions from link_suits of c-granules are used by the agent's control in exploration and/or exploitation of the environment on the way to achieve their targets. C-granules are also used for representation of perception by agents concerning the interactions with the physical world. Due to the limited ability of agent's perception usually only a partial information about the interactions of the physical world may be available to the agents. Hence, in particular the results of performed actions by agents cannot be predicted with certainty. For more details on IGrC based on c-granules the readers are referred to [2].

One of the key issues of the approach related to c-granules presented in [2], is a kind of integration between investigations of physical and mental phenomena. This idea of integration follows from the suggestions presented by many scientists. For illustration, let us consider the following two quotations:

As far as the laws of mathematics refer to reality, they are not certain; and as far as they are certain, they do not refer to reality.

—Albert Einstein [59]

Constructing the physical part of the theory and unifying it with the mathematical part should be considered as one of the main goals of statistical learning theory.

—Vladimir Vapnik ([57], p. 721)

In IGrC, information or decision systems based on the rough set approach play a special role [60–63]. They are used to record information of the processes of interacting configurations of hunks. In order to represent interactive computations (used, e.g., in searching for new features), information systems of a new type, namely interactive information systems, are needed [2, 42, 55].

5 Interactive Computations on Complex Granules Realized by Agents

Figure 5 illustrates the basic components of an agent for interactions. Among them are:

  (i) control (C),
 (ii) internal memory (M),
(iii) interactions realized by the control C between the control granule and memory granule by means of c-granules generated by control C for eliciting interactions

    (a) with the external environment (c-granules with parts: M, link l-2 (l-3) and hunk H-2 (H-3)) and
    (b) with internal parts of the agent other than memory M (c-granule with parts: M, link l-1, and hunk H-1).

In Fig. 6, the basic control cycle of agent is illustrated. At the first stage, starting at time t, the current interactions between control C and memory M are established (at time $t + \delta$). Next, the c-granules relevant for a given moment of time, are established at time $t + \delta + \varepsilon$. The agent is expected to use them for interactions with the external environment and with the agent's internal parts as well. After this the interactions are initiated, their results are recorded in the internal memory (M) of the agent, at time t_1. Once the recording is finalized the agent's control starts a new cycle.

Fig. 5 Basic agent components for interactions

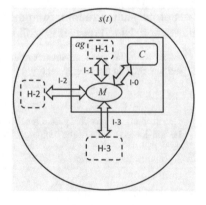

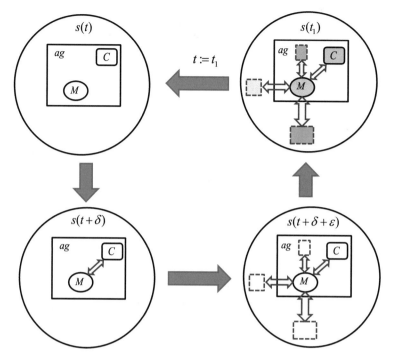

Fig. 6 Basic control cycle of agent

It is worthwhile mentioning that contrary to the existing computation models realized by Turing machine, the results of interactions can be only predicted by the agent's control, but the results of this prediction can be in general different from the results of real interactions between agent and the environment due to uncertainty of the unpredictable environment. In particular, this may be implied by the uncertain information possessed by agent about the environment due to limitations of the available resources, e.g., sensors, which are necessary for building agent's perception strategies.

In Fig. 7, we illustrate how the abstract definition of operation from soft_suit interacts with other suits of c-granule. It is necessary to distinguish two cases. In the case of soft_suit, the results of operation realized by the interactions of the hunks, available in the soft_suit itself, should be consistent with the specifications encoded in the link_suit. However, the result specified in the soft_suit can be treated only as an estimation of the real one which may be different due to the unpredictable interactions in the hard_suit.

The point of view, that the interactive computations on complex granules needs to be based on the process of interactions with the physical world, is important for Natural Computing too. The agent's observation to understand such computations is dependent on the physical world (see [25], p. 268).

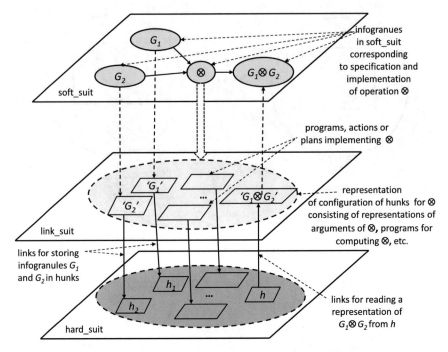

Fig. 7 Explanation of roles of different suits of a c-granule for operation ⊗ [36]

The agent hypotheses about the models of computations can be verified only through interactions running in the physical world. These models should be adaptive to incorporate changes when deviations of the predicted trajectories of computations from the perceived real ones become significant (see Fig. 8).

The issues discussed in this section raise a question about the control over interactive granular computations. In the following sections, we emphasize the importance of the risk management by the agent's control.

Fig. 8 Adaptation of trajectory approximations [64]

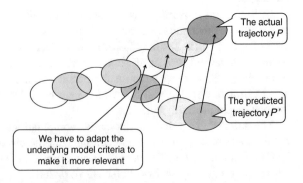

6 BDT and Problem Solving

Certainly, the concept of Big Data will drift with time. The data sets, which are nowadays treated as *big*, may be easily analyzed in the future using new software and/or hardware. One of the most important challenges for the BDT development is to get scalable methods for data analytics [6, 11, 65–69] including (i) scalable techniques for data management, relative to different classes of problems from different domains, as well as (ii) efficient hybridization and integration of relevant techniques relative to the specificity of applications (expressed, e.g., by specification of the class of problems to be solved). Let us note that scalability cannot be achieved without *collective wisdom*. Hence, the important area for the further development of the WisTech program arises.

In this section, a preliminary discussion on some main tasks, which should be supported by BDT, is included. Among them, two are as follows.

1. Filtering Big Data relative to the user's view expressed by the higher-level primitives.
2. Filtering Big Data relative to the user view of specific problem (or class of problems).

The first task is related to the relevant ontology development for the considered domain, as well as to the methods of transferring it to systems for further use. Nowadays, there are many available tools or ontologies (see, e.g., [70]). Methods for approximation of ontology have been also developed (see, e.g., [23, 24]) making it possible to transfer ontology approximation to the systems. Then, the system may use approximated concepts and relations for generating new granules (e.g., new features or patterns) relevant for the approximation of complex vague concepts. However, further work is needed for making these methods scalable for Big Data and domain knowledge.

In the following sections, we give some comments on the issues related to the second task.

6.1 Problem Specification by Users

In this section, we discuss issues related to the specification of problems, may be faced by users for developing systems based on Big Data.

Let us consider the following challenge mentioned in [20]:

> If users are to compose and build complex analytical pipelines over Big Data, it is essential they have appropriate high-level primitives to specify their needs.

Let us observe that the above mentioned high-level primitives are often complex vague concepts, which are semantically "far away" from the raw data. Hence, to make such concepts available by the system it is necessary to develop methods for constructing (inducing) high quality classifiers for such concepts. The problem of specification, given by an user, is defined over such concepts.

The user's task may be to deliver the relevant granules representing complex objects, satisfying the specification to a satisfactory degree [71–73]. Such granules are discovered and/or constructed using the hierarchical approach, where relevant strategies are to search for relevant granules through granulation and degranulation processes. The delivered granules may be treated as computational building blocks for approximation of complex vague concepts representing the user's specification. These approximations represent how the system is comprehending the user's specification. We have already justified that the process of inducing such approximations is challenging.

It is also worthwhile mentioning that approximations of concepts (such as concepts related to comprehending the user higher-level primitives or some expressions over them describing the situation and/or user needs) related to perception are induced by the system with the help of actions.

For example, in the context of reasoning about changes of situation, one should take into account that the predicted actions or/and plans may depend not only on the changes of past situations but also on the performed actions (or plans) in the past. This is strongly related to the idea of perception pointed out in [74]:

> The main idea of this book is that perceiving is a way of acting. It is something we do. Think of a blind person tap-tapping his or her way around a cluttered space, perceiving that space by touch, not all at once, but through time, by skillful probing and movement. This is or ought to be, our paradigm of what perceiving is.

Figrue 9 illustrates this idea.

On the basis of the partial understanding of the user's specification, the system may deliver some proposals for solutions. Next, the user may add some comments on them, which in turn may help in improving or reconstructing the delivered granules. The system should be able to "understand" these comments and search for the granules more relevant to the user's specification. A continuation of such a dialogue between

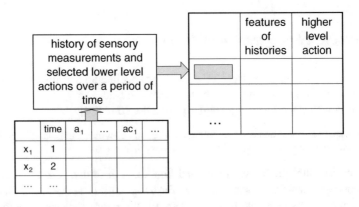

Fig. 9 Action in perception [36]

user and the system should lead to a satisfactory solution corresponding to the user's requirements. Moreover, the whole "dialogue trajectory" should have an acceptable quality. The acceptability criterion could depend on the consumption of time in a dialogue for reaching to a satisfactory solution. This mens that the system should control the schemes of computation for achieving the target goal.

One can treat the above discussed case of problems as a special case of checking satisfiability of complex vague concepts. These concepts can be interpreted as guards for initiation of actions or plans by the agent (see Sect. 2). In the discussed example related to Big Data, these actions may represent users' reactions on the solutions proposed by the system. In this more general case, the granules constructed by the system are interpreted as the degrees (representing arguments *for* and *against*) of satisfiability of complex vague concepts. Let us note that the system should be equipped with strategies for resolving conflicts between these arguments *for* and *against*.

In the next section, we present some approaches which appear to be very useful in searching for relevant calculi of granules and particular granules from the families of granules defined by these calculi.

6.2   Construction and Discovery of Relevant Granules

In GrC, we create calculi of granules by specifying elementary granules (e.g., indiscernibility or similarity classes) and some operations constructing new granules from the already defined ones [34, 73, 75]. In this section, we briefly outline some of the approaches for new granule generation. For a given problem, one should discover a relevant calculi of granules, and deliver a method of searching for relevant granules (in a selected calculi) which could be used as computational building blocks for approximation of vague concepts used in the problem specification. These vague concepts may represent guards of actions or plans performed by an agent. The actions are initiated on the basis of the judgement of satisfiability degrees of these guards in a given situation.

We start from granule aggregation defined by join operation with the constraints over information systems [76]. This approach allows us to generate new granules specifying the granules of different types. Some of these granules are giving rise to new information systems. Next, these systems can be used for generation of new granules such as indiscernibility or similarity classes of granules of a given type, new attributes or features, classifiers, clusters, and different patterns. We also explain that this kind aggregation of granules can be used in modeling self-organization of agents. Through self-organization, new kinds of granules are generated. Next, we discuss how discovery of a relevant hierarchy of the basic logical tools, namely satisfiability relations, can be used for new granule generation. We also discuss interaction of granules realized through dialogues of agents. Such interactions are leading towards generation of new granules relevant for agents. Important classes of granules are related to private and social languages of agents [2]. Strategies of granule generation by self-organization and communication of agents are especially important for complex adaptive systems, where the goal is to obtain relevant emergent behavioral patterns satisfying a given specification to a satisfactory degree [77, 78]. We also emphasize the role of risk management in controlling computations performed by agents over granules. Finally, we discuss a special kind of reasoning called adaptive judgement used by the agent's control for reasoning about granules and computations over them. This reasoning is also based on constructions over relevant granules.

Context, Structural Objects, and Self-organization. One of the important problems in hierarchical learning of complex vague concept approximations, is the discovery of relevant contexts on different levels of hierarchical learning. Contexts can be modeled by aggregation of information (decision) systems based on the join operations with their respective constraints [76] (see Fig. 10). Cartesian product of the universes of aggregated information systems is filtered by constraints. Constraints are specifying the structure of objects on the new hierarchical level obtained by aggregation. The structure is defined by relations over the vectors of attribute values from the aggregated information systems. Constraints can also be treated as specification of types of objects in the aggregated information systems. For more details, the reader is referred to [76].

In the discussed case, there are two groups of (conditional) attributes. The values of attributes from the first group are fixed by the agent's control while the values of attributes from the other group are the results of a function of values of attributes from the first group and interactions with environments. As an instance, let us consider that parameters of sensors or actions, which need to be activated, belong to the first group of attributes. Then the sensory measurements, based on the values from the first group and interactions with environments, constitute the values of the attributes of the second group.

Top-down decomposition strategies of specification create such schemes with the help of which construction of relevant patterns can be discovered. Let us consider an example of decomposition of information systems with type of objects characterized by relation R over tuples of attribute-value vectors into two information systems with object types characterized by relations R_1, R_2. The corresponding join operation

Fig. 10 Join with constraints of information systems $A_1, \ldots, A_k$ to information system A

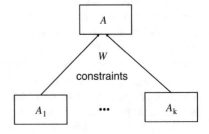

with constraints specifies a construction of R from R_1, R_2. One can ask if such a construction can be modeled using simple *local* interactions only. For example, such local interactions may concern dependencies between values of attributes from the group of attributes defined by the control of neighboring agents, i.e., they can be fixed by neighboring agents. Here, we assume that the values of control attributes from the neighborhood of one agent are perceived by the other agents from the same neighborhood. Searching for relevant contexts under simple constraints can be feasible. However, one should consider that such simplified searching not always can give the relevant aggregated constraints. One can observe here an analogy with the 13th Hilbert problem [79]:

> Can every continuous function of 3 variables be written as a composition of continuous functions of 2 variables?

and the result by Vitushkin [80]:

> There are continuously differentiable functions of 3 variables which are not the superposition of continuously differentiable functions of 2 variables.

The discussed problem is related to self-organization leading from local interactions to global emergent patterns.

The issues of self-organization have been intensively studied for years (e.g., [21, 81–84]). Methods based on self-organization are crucial for dealing with Big Data, and further research is required in this regard.

Let us refer here once again to [21] (p.1088):

> [...] viewing an [...] agent [...] as a complex dynamical system enables us to employ concepts such as self-organization and emergence rather than hierarchical top-down control. [...] autonomous agents display self-organization and emergence at multiple levels: at the level of induction of sensory stimulation, movement generation, exploitation of morphological and material properties, and interaction between individual modules and entire agents.

We propose to use the top-down decompositions for generation of decomposition schemes, along which discovery of agent's self-organization, e.g., aiming at discovery of relevant contexts or object structures for relevant emergent patterns generation, may proceed. These schemes are making the discovery process of self-organization feasible by bottom-up realization using the top-down decomposition schemes acquired from users. However, one should also note that the decomposition schemes generated in the top-down decomposition create only hypotheses; searching for discovery of relevant decompositions requires backtracking. Further development of methods for discovery of self-organization still requires much more work. Here, we would like to mention only an important interaction in this learning process of top-down decompositions with bottom-up self-organization.

Satisfiability and New Granules. Let us observe that the satisfiability relations in the IGrC framework can be treated as tools for constructing new information granules. In fact, for a given satisfiability relation, the semantics of formulas relative to this relation is defined. In this way the candidates for new relevant information granules are obtained. We would like to emphasize on this very important feature

Fig. 11 Interactive
hierarchical structures (*gray
arrows* show interactions
between hierarchical levels
and the environment, *arrows*
at hierarchical levels point
from information (decision)
systems representing partial
specifications of satisfiability
relations to those which are
induced from the theories
consisting of rule sets) [36]

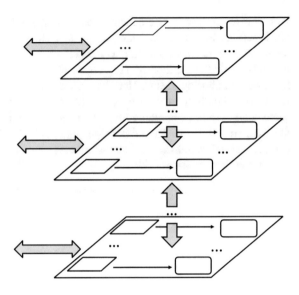

that the relevant satisfiability relation for the considered problems, is not given but it
should be induced (discovered) on the basis of a partial information encoded in the
respective information (decision) systems. For real-life problems, it is often necessary
to discover a hierarchy of satisfiability relations before we obtain the relevant target
level. Information granules constructed at different levels of this hierarchy finally
lead to relevant ones for the approximation of complex vague concepts represented
by complex granules expressed in natural language (see Fig. 11).

Let us discuss some examples of c-granules constructed over a family of satisfiability relations being at the disposal of a given agent. This discussion has some roots
in intuitionism (see, e.g., [85]). Let us consider a remark made by Per Martin-Löf in
[85] about judgement presented in Fig. 12.

In the approach based on c-granules, the judgement for checking values of descriptors (or more compound formulas) pointed by links from simple c-granules is based
on interactions of some physical parts considered over time and/or space (called
hunks) and pointed by links of c-granules. The judgement for the more compound
c-granules is defined by a relevant family of procedures also realized by means of
interactions of physical parts.

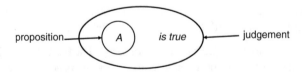

Fig. 12 Judgement of truth in a metalanguage: "when we hold a proposition to be true, then we
make a judgement" [36, 85]

Let us explain the above claims in more detail.

Let us assume that a given agent ag has at the disposal a family of satisfiability relations

$$\{\models_i\}_{i \in I}, \tag{1}$$

where $\models_i \subseteq Tok(i) \times Type(i)$, $Tok(i)$ is a set of tokens and $Type(i)$ is a set of types (using the terminology from [86]). The indices of satisfiability relations are vectors of parameters related to time, space, spatio-temporal features of physical parts represented by hunks, or actions (plans) to be realized in the physical world.

In the discussed example of elementary c-granules, $Tok(i)$ is a set of hunks and, $Type(i)$ is a set of descriptors (elementary infogranules) respectively, pointed by the link represented by $\models_i$. The procedure for computing the value of $h \models_i \alpha$, where h is a hunk and α is an infogranule (e.g., descriptor or formula constructed over descriptors), is based on the interaction of α with the physical world represented by the hunk h.

The agent's control can aggregate some simple c-granules into more compound c-granules, e.g., by selecting some constraints on subsets of I, it is possible to select a relevant sets of simple c-granules, and consider them as a new, more compound c-granule. As constraints, values of descriptors pointed by links of elementary c-granules can also be taken into account, and sets of such more compound c-granules can be aggregated into a new c-granule. Values of new descriptors pointed by links of these more compound granules are computed by new procedures. The computation process again is realized by interaction of the physical parts represented by hunks, which are pointed by links of c-granules, included in the considered more compound c-granule. Moreover, a procedure for computing values of more compound descriptors from values of descriptors included in the elementary c-granules (of the considered more compound c-granule), is used. It is to be noted that this procedure is also realized in the physical world with the help of relevant interactions.

In hierarchical modeling aiming at inducing relevant c-granules (e.g., for approximation of complex vague concepts), one can consider so far constructed c-granules as tokens. For example, they can be used to define structured objects representing corresponding hunks, and using new satisfiability relations (from a given family) they can be linked to the relevant higher order descriptors together with the appropriate procedures (realized by interactions of hunks) for computing values of these descriptors. This approach generalizes hierarchical modeling developed for infogranules (see, e.g., [23, 24]) in the context of hierarchical modeling of c-granules, which is important for many real-life projects.

We have assumed before that the agent ag is equipped with a family of satisfiability relations. However, in real-life cases the situation is more complicated. The agent ag should have strategies for discovery of new relevant satisfiability relations on the way of searching for target goals (solutions of problems). This is related to issue of the adaptive judgement, relevant to the agent's performance of computations based

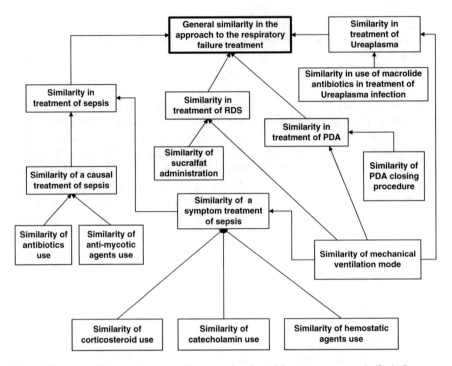

Fig. 13 Fragment of the ontology used for approximation of the vague concept *similarity* between plans of the treatment of new born infants with respiratory failure [23, 36]

on configurations of c-granules. In the framework of granular computing, based on c-granules, satisfiability relations are tools for constructing new c-granules. In fact, for a given satisfiability relation, the semantics of descriptors (and more compound formulas) relative to this relation can be defined.

Figure 13 presents a fragment of domain ontology used for approximation of the vague concept, namely the *similarity between plans of the therapy administered for the cases of respiratory failure*. Approximations of concepts and relations from the ontology are used for inducing the model of similarity relation between the treatment plans, in particular delivered by medical expert, and predicted by the decision support system. For details, the reader is referred to [23, 24].

Comments on Dialogues of Agents in BDT. In this section, we present some preliminary comments on dialogues among agents. Dialogues of agents from a given team can lead to a common understanding of the problems of concern and help, to get a cooperative problem solving strategy by the team. The issues related to reasoning based on dialogues are not trivial, especially when one would like to propose a treatment incorporating the possibility of combining different dominating paradigms of reasoning in logic. This point of view was well expressed by Johan van Benthem in [87] (see Foreword, p.viii):

> I see two main paradigms from Antiquity that come together in the modern study of argumentation: Platos Dialogues as the paradigm of intelligent interaction, and Euclids Elements as the model of rigour. Of course, some people also think that formal mathematical proof is itself the ultimate ideal of reasoning - but you may want to change your mind about reasonings peak experiences' when you see top mathematicians argue interactively at a seminar.

Dialogues enable the agents to (efficiently) search for solutions. Very often a query, formulated in BDT by an agent, involves vague concepts from natural language, e.g., one can consider queries given by an user to a dialogue based search engine. Agents are expecting to receive c-granules satisfying their specifications to some satisfactory degrees. The meaning of *satisfiability to a degree* should be learned on the basis of dialogues among agents embedded in the systems based on BDT. Satisfiability to a degree gives some flexibility in searching for solutions. The solutions do not need to be *exact*. This may make the process of searching for constructions of such c-granules feasible. It is worthwhile mentioning that such constructions should be robust relative to the deviations of components. The interested reader may find more details on these issues in (e.g., [71, 73, 88]), where the development is based on the rough mereological approach.

By using dialogues agents may try to recognize the meaning of c-granules received from other agents. They can do this by learning approximations of received c-granules in their own languages. A given agent may acquire the ontology of the concepts used by another agent. However, usually a given agent can only acquire an approximation of concept-ontology possessed by another agent. This idea of shared knowledge among agents may be very useful in solving problems by any individual agent (see e.g., [23, 24]). Let us note that the ontology approximation may also be used in efficient searching for relevant contexts of queries received by agents from other agents.

One of the challenges for adaptive judgement, performed by a given agent ag, is the task of learning of *approximation of derivations* performed by another agent ag', assuming that an approximated concept-ontology of ag' is already available to ag. The agent ag may approximate, to a satisfactory degree, the derivations performed by ag' with the help of the constructions of solutions delivered by ag'.

6.3 Risk Management by Agents in BDT

In Sect. 3, we have formulated some postulates concerning the issues of efficiency management by agents. In this section, we add some comments on the risk management. Let us note that practical judgement is involved in efficient management [2].

Risk may be understood as interaction (of agents) with uncertainty (of environment). Perception of risk is a subjective judgement, which people make about the severity and/or probability of a risk. This may vary from one person to another. Any human endeavor carries some risk, but some are much riskier than others (The Stanford Encyclopedia of Philosophy: http://plato.stanford.edu/archives/spr2014/entries/risk/).

Since the very beginning, all human activities were done at risk of failure. The recent years have shown the low quality of risk management in areas such as finance, economics, and many others. In this context, improvement in the risk management has a particular importance for the further development of complex systems. The importance of risk management illustrates the following example from the financial sector. Many of financial risk management experts consider Basel II rules (see http://en.wikipedia.org/wiki/Basel_Committee_on_Banking_Supervision) as a causal factor in the credit bubble prior to the 2007-8 collapse. Namely, in Basel II one of the principal factors of financial risk management was

outsourced to companies that were not subject to supervision, credit rating agencies.

Of course, now we do have a new "improved" version of Basel II, called Basel III. However, according to an OECD (see http://en.wikipedia.org/wiki/Basel_III) *the medium-term impact of Basel III implementation on GDP growth is negative and estimated in the range of* −0.05 % *to* −0.15 % *per year* (see also [89]).

On the basis of experience in many areas, we have now many valuable studies on different approaches to risk management. Currently, the dominant terminology is determined by the standards of ISO 31K [39]. However, the logic of inferences in risk management is dominated by the statistical paradigms, especially by Bayesian data analysis initiated about 300 years ago by Bayes, and regression data analysis initiated about 200 years ago by Legendre and Gauss. They initiated many detailed methodologies specific for different fields. A classic example is the risk management methodology in the banking sector, based on the recommendations of Basel II standards for mathematical models of risk management [90]. The current dominant statistical approach is not satisfactory because it does not give effective tools for inferences about the vague concepts and relations between them (see the afore-mentioned sentences by Valiant cited in Sect. 6.4).

A particularly important example of a vague concept relation in the risk management is the relation of a cause-effect dependencies between various events. It should be noted that the concept of risk in ISO 31K is defined as *the effect of uncertainty on objectives*. Thus, by definition, the vagueness is also an essential part of the risk concept.

To paraphrase the motto of this study by Judea Pearl, we can say that traditional statistical approach to risk management inference *is strong in devising ways of describing data and inferring distributional parameters from sample*. However, in practice risk management inference requires two additional ingredients:

- *a science-friendly language for articulating risk management knowledge*, and
- *a mathematical machinery for processing that knowledge, combining it with data and drawing new risk management conclusions about a phenomenon*.

One can observe that this is a slightly modified version of the opinion of Judea Pearl [40].

Adding both the above mentioned components is an extremely difficult task, and relates to the core of AI research, as very accurately specified by the Turing test. In the context of our applications, the idea of Turing test boils down to the fact that on the basis of a "conversation" with a hidden risk management expert and a hidden machine one will not be able to distinguish who is the man and who is the machine.

We propose to extend the statistical paradigm by adding the two above discussed components for designing the high quality risk management systems in BDT.

For the risk management in BDT one of the most important task is to develop strategies for inducing approximations of the vague complex concepts involved in the domain of concern of the risk management. Let us note that the approximations are providing methods for checking their satisfiability (to a degree). A typical example of such vague concept is the statement of the form: "now we do have very risky situation". Among such concepts, the complex vague concepts representing the role of guards, on which the activation of actions performed by agents are based, are of special importance.

These vague complex concepts are represented by the agent's hierarchy of needs. In the risk management, one should consider a variety of complex vague concepts and relations between them, as well as the reasoning schemes related to the bow-tie diagram (see Fig. 14).

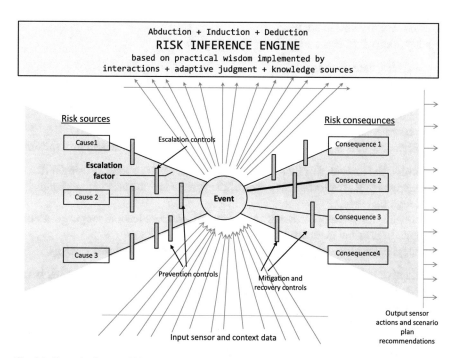

Fig. 14 Bow-tie diagram [37]

6.4 Adaptive Judgement

The reasoning, which makes it possible to derive relevant granules used in the processes of solving a set of given problems, is called an *adaptive judgement*. The *Intuitive judgement* and the *rational judgement* are distinguished as different kinds of judgement in [91]. Among the tasks for adaptive judgement following are the ones supporting reasoning towards,

- inducing relevant classifiers, e.g.,

  - searching for relevant approximation spaces,
  - discovery of new features,
  - selection of relevant features (attributes),
  - rule induction,
  - discovery of inclusion measures,
  - strategies for conflict resolution,
  - adaptation of measures based on the minimum description length principle,

- prediction of changes,
- initiation of relevant actions or plans,
- discovery of relevant contexts,
- adaptation of different sorts of strategies e.g., for

  - existing data models,
  - quality measure over computations realized by agents,
  - objects structures,
  - knowledge representation and interaction with knowledge bases,
  - ontology acquisition and approximation,
  - hierarchy of needs, or for identifying problems to be solved according to priority,

- learning the measures of inclusion between granules from sources using different languages (e.g., the formal language of the system and the user natural language) through dialogue,
- strategies for development and evolution of communication language among agents in distributed environments, and
- strategies for efficiency management in distributed computational systems.

Adaptive judgement on interactive computations is a mixture of reasoning based on deduction, abduction, and induction. In particular, case based or analogy based reasoning, reasoning which make use of experience, reasoning based on observed changes in the environment, or reasoning with application of meta-heuristics from natural computing (see Fig. 15).

The meaning of practical judgement goes beyond typical tools for reasoning based on deduction or induction [92]:

> Practical judgement is not algebraic calculation. Prior to any deductive or inductive reckoning, the judge is involved in selecting objects and relationships for attention and assessing

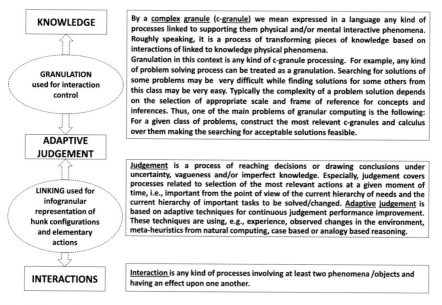

| KNOWLEDGE | By a complex granule (c-granule) we mean expressed in a language any kind of processes linked to supporting them physical and/or mental interactive phenomena. Roughly speaking, it is a process of transforming pieces of knowledge based on interactions of linked to knowledge physical phenomena. |

Granulation in this context is any kind of c-granule processing. For example, any kind of problem solving process can be treated as a granulation. Searching for solutions of some problems may be very difficult while finding solutions for some others from this class may be very easy. Typically the complexity of a problem solution depends on the selection of appropriate scale and frame of reference for concepts and inferences. Thus, one of the main problems of granular computing is the following: For a given class of problems, construct the most relevant c-granules and calculus over them making the searching for acceptable solutions feasible.

GRANULATION used for interaction control

ADAPTIVE JUDGEMENT

Judgement is a process of reaching decisions or drawing conclusions under uncertainty, vagueness and/or imperfect knowledge. Especially, judgement covers processes related to selection of the most relevant actions at a given moment of time, i.e., important from the point of view of the current hierarchy of needs and the current hierarchy of important tasks to be solved/changed. Adaptive judgement is based on adaptive techniques for continuous judgement performance improvement. These techniques are using, e.g., experience, observed changes in the environment, meta-heuristics from natural computing, case based or analogy based reasoning.

LINKING used for infogranular representation of hunk configurations and elementary actions

INTERACTIONS

Interaction is any kind of processes involving at least two phenomena /objects and having an effect upon one another.

Fig. 15 Interactions, adaptive judgement and granulation

their interactions. Identifying things of importance from a potentially endless pool of candidates, assessing their relative significance, and evaluating their relationships is well beyond the jurisdiction of reason.

For example, a particular question for the agent's control concerns discovering strategies for models of dynamic changes of the agent's attention. This may be related to discovery of changes in a relevant context necessary for the judgement.

We would like to stress that still much work should be done to develop approximate reasoning methods about complex vague concepts for the progress of the development of BDT, in particular for the efficiency management in BDT systems. This idea was very well expressed by Leslie Valiant (see, e.g., http://en.wikipedia.org/wiki/Vagueness, http://people.seas.harvard.edu/~valiant/researchinterests.htm):

A fundamental question for artificial intelligence is to characterize the computational building blocks that are necessary for cognition. A specific challenge is to build on the success of machine learning so as to cover broader issues in intelligence. [...] This requires, in particular a reconciliation between two contradictory characteristics – the apparent logical nature of reasoning and the statistical nature of learning.

The views by Zadeh and Pearl, which are already cited in this chapter (see Sect. 1, and Sect. 6.3), are also of relevance here.

7   Conclusions

The approach for modeling interactive computations based on c-granules is presented, and its importance for the efficiency management of controlling computations in Big Data Technology is outlined. It is worthwhile mentioning that in modeling and/or discovering granules, tools from different areas are used. Among these areas some are, machine learning, data mining, multi-agent systems, complex adaptive systems, logic, cognitive science, neuroscience, and soft computing. Granular Computing is aiming at developing a unified methodology for modeling and controlling computations over complex objects, called granules, as well as for reasoning about such objects and computations over them. In particular, such a methodology is of great importance for Big Data Technology.

The discussed concepts such as interactive computation and adaptive judgement are among the basic ingredients in the field of Wisdom Technology. Let us mention here the WisTech meta-equation:

$$\text{WISDOM} = \tag{2}$$
$$\text{INTERACTIONS} +$$
$$\text{ADAPTIVE JUDGEMENT} +$$
$$\text{KNOWLEDGE}.$$

The presented approach has a potential for being used for developing computing models in different areas, such as natural computing (including, e.g., computing models for meta-heuristics or computing models for complex processes in molecular biology), computing in distributed environments under uncertainty realized by multi-agent systems (including, e.g., in social computing), modeling of computations for feature extraction (constructive induction) used for approximation of complex vague concepts, hierarchical learning, discovery of planning strategies or strategies for coalition formation by agents as well as for approximate reasoning about interactive computations based on such computing models.

In our research, we plan to further develop the foundations of interactive computations based on c-granules. The approach will be used for development of modeling and analysis of computations in Natural Computing [52], Wisdom Web of Things [13], or Cyber-Physical Systems [12]. Foundations of interactive computations using c-granules create the basis on which methods for BDT can be developed.

References

1. Berman, J.J.: Principles of Big Data. Sharing, and Analyzing Complex Information. Elsevier, Amsterdam, Preparing (2013)
2. Jankowski, A.: Complex Systems Engineering: Conclusions from Practical Experience. Springer, Heidelberg (2015). (in preparation)

3. Jankowski, A., Skowron, A.: A WisTech paradigm for intelligent systems. Trans. Rough Sets VI: J. Subline 94–132
4. Arthur, L.: Big Data Marketing. Wiley, Hoboken (2013)
5. Chu, W.W. (ed.): Data Mining and Knowledge Discovery for Big Data Methodologies. Challenges and Opportunities. Springer, Berlin (2014)
6. Kudyba, S. (ed.): Big Data, Mining, and Analytics: Components of Strategic Decision Making. CRC Press Taylor & Francis, Boca Raton (2014)
7. Mayer-Schönberger, V., Cukier, K.: Big Data: A Revolution That Will Transform How We Live, Work, and Think. John Murray Pub, London (2013)
8. O'Reilly Media, I.T.: Big Data Now: 2012 Edition. O'Reilly Media, Inc., Sebastopol (2012)
9. Pollak, B. (ed.): Ultra-Large-Scale Systems. Carnegie Mellon University, Pittsburgh, PA, The Software Challenge of the Future. Software Engineering Institute (2006)
10. Schmarzo, B.: Big Data: Understanding How Data Powers Big Business. Wiley, Indianapolis (2013)
11. Zikopoulos, P.C., Eaton, C., deRoos, D., Deutsch, T., Lapis, G.: Understanding Big Data. Analytics from Enterprise Class Hadoop and Streaming Data. McGraw-Hill, New York (2012)
12. Lamnabhi-Lagarrigue, F., Di Benedetto, M.D., Schoitsch, E.: Introduction to the special theme cyber-physical systems. Ercim News **94**, 6–7 (2014)
13. Zhong, N., Ma, J.H., Huang, R., Liu, J., Yao, Y., Zhang, Y.X., Chen, J.: Research challenges and perspectives on wisdom web of things (W2T). J. Supercomput. **64**, 862–882 (2013)
14. Cyber-physical and ultra-large scale systems (2013), http://resources.sei.cmu.edu/library/asset-view.cfm?assetid=85282
15. Zadeh, L.A.: Toward a theory of fuzzy information granulation and its centrality in human reasoning and fuzzy logic. Fuzzy Sets Syst. **90**, 111–127 (1997)
16. Bargiela, A., Pedrycz, W. (eds.): Granular Computing: An Introduction. Kluwer Academic Publishers (2003)
17. Pedrycz, W., Skowron, S., Kreinovich, V. (eds.): Handbook of Granular Computing. Wiley, Hoboken (2008)
18. Pedrycz, W.: Granular Computing Analysis and Design of Intelligent Systems. CRC Press, Taylor & Francis, Boca Raton (2013)
19. Skowron, A., Pal, S.K., Nguyen, H.S. (eds.): Special issue on rough sets and fuzzy sets in natural computing. Theor. Comput. Sci. **412**(42), (2011)
20. Jagadish, H., Gehrke, J., Labrinidis, A., Papakonstantinou, Y., Patel, J.M., Ramakrishnan, R., Shahabi, C.: Big data and its technical challenges. Commun. ACM **57**, 86–94 (2014)
21. Pfeifer, R., Lungarella, M., Iida, F.: Self-organization, embodiment, and biologically inspired robotic. Science **318**, 1088–1093 (2007)
22. Amershi, S., Cakmak, M., Knox, W.B., Kulesza, T.: Power to the people: the role of humans in interactive machine learning. AI Mag. **35**, 105–120 (Winter 2014)
23. Bazan, J.: Hierarchical classifiers for complex spatio-temporal concepts. Trans. Rough Sets IX: J. Subline LNCS **5390**, 474–750 (2008)
24. Nguyen, S.H., Bazan, J., Skowron, A., Nguyen, H.S.: Layered learning for concept synthesis. Trans. Rough Sets I: J. Subline LNCS **3100**, 187–208 (2004)
25. Deutsch, D., Ekert, A., Lupacchini, R.: Machines, logic and quantum physics. Bull. Symbolic Logic **6**, 265–283 (2000)
26. Goldin, D., Smolka, S., Wegner, P. (eds.): Interactive Computation: The New Paradigm. Springer (2006)
27. Mendel, J.M., Zadeh, L.A., Trillas, E., Yager, R., Lawry, J., Hagras, H., Guadarrama, S.: What computing with words means to me. IEEE Comput. Intell. Mag. 20–26 (February 2010)
28. Zadeh, A.: Computing with Words: Principal Concepts and Ideas, Studies in Fuzziness and Soft Computing, vol. 277. Springer, Heidelberg (2012)
29. Zadeh, L.A.: Fuzzy logic = computing with words. IEEE Trans. Fuzzy Syst. **4**, 103–111 (1996)
30. Zadeh, L.A.: From computing with numbers to computing with words—from manipulation of measurements to manipulation of perceptions. IEEE Trans. Circuits Syst. **45**, 105–119 (1999)
31. Zadeh, L.A.: Foreword. In: Pal et al. [48], pp. IX–XI

32. Zadeh, L.A.: A new direction in AI: toward a computational theory of perceptions. AI Mag. **22**(1), 73–84 (2001)
33. Zadeh, L.A.: Fuzzy sets and information granularity. In: Advances in Fuzzy Set Theory and Applications, pp. 3–18. North-Holland, Amsterdam (1979)
34. Skowron, A., Stepaniuk, J.: Information granules and rough-neural computing. In: Pal et al. [48], pp. 43–84
35. Jankowski, A., Skowron, A., Swiniarski, R.W.: Interactive computations: toward risk management in interactive intelligent systems. In: Maji, P., Ghosh, A., Murty, M.N., Ghosh, K., Pal, S.K. (eds.) Pattern Recognition and Machine Intelligence—5th International Conference, PReMI 2013, Kolkata, India, December 10–14, 2013. Proceedings. Lecture Notes in Computer Science, vol. 8251, pp. 1–12. Springer (2013)
36. Jankowski, A., Skowron, A., Swiniarski, R.W.: Interactive complex granules. Fundamenta Informaticae **133**, 181–196 (2014)
37. Jankowski, A., Skowron, A., Swiniarski, R.W.: Perspectives on uncertainty and risk in rough sets and interactive rough-granular computing. Fundamenta Informaticae **129**, 69–84 (2014)
38. Skowron, A., Jankowski, A., Wasilewski, P.: Risk management and interactive computational systems. J. Adv. Math. Appl. **1**, 61–73 (2012)
39. ISO 31000 standard, http://webstore.ansi.org/
40. Pearl, J.: Causal inference in statistics: an overview. Stat. Surv. **3**, 96–146 (2009)
41. Skowron, A., Wasilewski, P.: An introduction to perception based computing. In: Kim, T.H., Lee, Y.H., Kang, B.H., Ślęzak, D. (eds.) Proceedings of FGIT 2010. Lectures Notes in Computer Science, vol. 6485, pp. 12–25. Springer, Heidelberg (2010)
42. Skowron, A., Wasilewski, P.: Interactive information systems: toward perception based computing. Theor. Comput. Sci. **454**, 240–260 (2012)
43. Zadeh, L.A.: Computing with words and perceptions a paradigm shift. In: Proceedings of the IEEE International Conference on Information Reuse and Integration (IRI 2009), Las Vegas, Nevada, USA. pp. viii–x. IEEE Systems, Man, and Cybernetics Society (2009)
44. Sutton, R.S., Barto, A.G.: Reinforcement Learning: An Introduction. The MIT Press (1998)
45. Bower, J.M., Bolouri, H. (eds.): Computational Modeling of Genetic and Biochemical Networks. MIT Press (2001)
46. Press, Harvard Business School: SWOT Analysis I: Looking Outside for Threats and Opportunities. Harvard Business School Publishing Corporation, Boston (2006)
47. Press, Harvard Business School: SWOT Analysis II: Looking Inside for Strengths and Weaknesses. Harvard Business School Publishing Corporation, Boston (2006)
48. Osterwalder, A., Pigneur, Y.: Business Model Generation: A Handbook for Visionaries, Game Changers, and Challengers. Wiley, Hoboken (2010)
49. Pahl, N., Richter, A.: Swot Analysis. Methodology and a Practical Approach. GRIN Verlag GmbH, Münich, Idea (2009)
50. Imai, M., Kaizen, G.: A Commonsense Approach to a Continuous Improvement Strategy, 2nd edn. McGraw-Hill Professional, New York (2012)
51. Sobek II, D.K., Smalley, A.: Understanding A3 Thinking: A Critical Component of Toyota's PDCA Management System. Productivity Press, Boca Raton (2008)
52. Rozenberg, G., Bäck, T., Kok, J. (eds.): Handbook of Natural Computing. Springer (2012)
53. Jankowski, A., Skowron, A.: Wisdom technology: a rough-granular approach. In: Marciniak, M., Mykowiecka, A. (eds.) Bolc Festschrift. Lectures Notes in Computer Science, vol. 5070, pp. 3–41. Springer, Heidelberg (2009)
54. Skowron, A., Stepaniuk, J., Swiniarski, R.: Modeling rough granular computing based on approximation spaces. Inf. Sci. **184**, 20–43 (2012)
55. Skowron, A., Wasilewski, P.: Information systems in modeling interactive computations on granules. Theor. Comput. Sci. **412**(42), 5939–5959 (2011)
56. Heller, M.: The Ontology of Physical Objects. Cambridge University Press, Four Dimensional Hunks of Matter. Cambridge Studies in Philosophy (1990)
57. Vapnik, V.: Statistical Learning Theory. Wiley, New York (1998)

58. Omicini, A., Ricci, A., Viroli, M.: The multidisciplinary patterns of interaction from sciences to computer science. In: Goldin et al. [18], pp. 395–414
59. Einstein, A.: Geometrie und Erfahrung (Geometry and Experience). Julius Springer, Berlin (1921)
60. Pawlak, Z., Skowron, A.: Rudiments of rough sets. Inf. Sci. **177**(1), 3–27 (2007)
61. Pawlak, Z.: Rough sets. Int. J. Comput. Inf. Sci. **11**, 341–356 (1982)
62. Pawlak, Z.: Rough Sets: Theoretical Aspects of Reasoning about Data, System Theory, Knowledge Engineering and Problem Solving, vol. 9. Kluwer Academic Publishers, Dordrecht (1991)
63. Stepaniuk, J.: Rough-Granular Computing in Knowledge Discovery and Data Mining. Springer, Heidelberg (2008)
64. Skowron, A., Stepaniuk, J., Jankowski, A., Bazan, J.G., Swiniarski, R.: Rough set based reasoning about changes. Fundamenta Informaticae **119**(3–4), 421–437 (2012)
65. Abbott, D.: Applied Predictive Analytics: Principles and Techniques for the Professional Data Analyst. Wiley, Indianapolis (2014)
66. Bartlett, R.: A Practitioner's Guide To Business Analytics: Using Data Analysis Tools to Improve Your Organization's Decision Making and Strategy. McGraw-Hill, New York (2013)
67. Provost, F., Fawcett, T.: Data Science for Business: What You Need to Know About Data Mining and Data-analytic Thinking. O'Reilly Media, Sebastopol (2013)
68. Marr, B.: Big Data: Using SMART Big Data. Analytics and Metrics to Make Better Decisions and Improve Performance. Wiley, Hoboken (2015)
69. Siegel, E.: Predictive Analytics: The Power to Predict Who Will Click, Buy, Lie, or Die. Wiley, Hoboken (2013)
70. Staab, S., Studer, R. (eds.): Handbook on Ontologies. International Handbooks on Information Systems. Springer, Heidelberg (2004)
71. Polkowski, L., Skowron, A.: Rough mereology: a new paradigm for approximate reasoning. Int. J. Approximate Reasoning **15**(4), 333–365 (1996)
72. Polkowski, L., Skowron, A.: Towards adaptive calculus of granules. In: Zadeh, L.A., Kacprzyk, J. (eds.) Computing with Words in Information/Intelligent Systems, pp. 201–227. Physica-Verlag, Heidelberg (1999)
73. Polkowski, L., Skowron, A.: Rough mereological calculi of granules: a rough set approach to computation. Comput. Intell. Int. J. **17**(3), 472–492 (2001)
74. Noë, A.: Action in Perception. MIT Press (2004)
75. Skowron, A., Stepaniuk, J., Peters, J., Swiniarski, R.: Calculi of approximation spaces. Fundamenta Informaticae **72**, 363–378 (2006)
76. Skowron, A., Stepaniuk, J.: Hierarchical modelling in searching for complex patterns: constrained sums of information systems. J. Exp. Theor. Artif. Intell. **17**, 83–102 (2005)
77. Desai, A.: Adaptive complex enterprises. Commun. ACM **45**, 32–35 (2005)
78. Liu, J.: Autonomous Agents and Multi-Agent Systems: Explorations in Learning, Self-organization and Adaptive Computation. World Scientific Publishing (2001)
79. Hilbert, D.: Mathematische probleme. Nachr. Akad. Wiss. Göttingen, pp. 253–297 (1900), (Gesammelte Abhandlungen,. Bd. 3, Springer, Berlin, 1935, pp. 290–329)
80. Vitushkin, A.G.: On Hilbert's thirteenth problem. Dokl. Acad. Nauk. SSSR **156**, 1003–1006 (1954)
81. Estep, M.: Self-organizing Natural Intelligence: Issues of Knowing, Meaning, and Complexity. Springer, Heidelberg (2014)
82. Holland, J.: Signals and Boundaries Building Blocks for Complex Adaptive Systems. MIT Press, Cambridge (2014)
83. Jarrah, K., Guan, L., Kyan, M., Muneesawang, P.: Unsupervised Learning: A Dynamic Approach. IEEE Press Series on Computational Intelligence, Wiley-IEEE Press, Hoboken (2014)
84. Nolfi, S., Fioreano, D.: Evolutionary Robotics: The Biology, Intelligence, and Technology of Self-organizing Machines. MIT Press, Cambridge (2000)
85. Martin-Löf, P.: Intuitionistic Type Theory (Notes by Giovanni Sambin of a Series of Lectures Given in Padua, June 1980). Bibliopolis, Napoli (1984)

86. Barwise, J., Seligman, J.: Information Flow: The Logic of Distributed Systems. Cambridge University Press (1997)
87. Rahwan, I., Simari, G.R.: Argumentation in Artificial Intelligence. Springer, Berlin (2009)
88. Polkowski, L., Skowron, A.: Rough mereological approach to knowledge-based distributed AI. In: Lee, J.K., Liebowitz, J., Chae, J.M. (eds.) Critical Technology, Proc. Third World Congress on Expert Systems, February 5–9, Soeul, Korea, pp. 774–781. Cognizant Communication Corporation, New York (1996)
89. Slovik, P., Cournède: Macroeconomic Impact of Basel III, Working Papers, vol. 844. OECD Economics Publishing, OECD Economics Department (2011), http://www.oecd.org/eco/ Workingpapers
90. Shevchenko, P. (ed.): Modelling Operational Risk Using Bayesian Inference. Springer (2011)
91. Kahneman, D.: Maps of bounded rationality: psychology for behavioral economics. Am. Econ. Rev. **93**, 1449–1475 (2002)
92. Thiele, L.P.: The Heart of Judgment: Practical Wisdom, Neuroscience, and Narrative. Cambridge University Press, Cambridge (2010)
93. Pal, S.K., Polkowski, L., Skowron, A. (eds.): Rough-Neural Computing: Techniques for Computing with Words. Cognitive Technologies. Springer, Heidelberg (2004)

An Overview of Concept Drift Applications

Indrė Žliobaitė, Mykola Pechenizkiy and João Gama

We dedicate this chapter to Dr. Alexey Tsymbal who passed away suddenly and unexpectedly in November 2014 at age of 39. Alexey contributed to the progress of data mining and medical informatics on several topics, including notable work on handling concept drift.

Abstract In most challenging data analysis applications, data evolve over time and must be analyzed in near real time. Patterns and relations in such data often evolve over time, thus, models built for analyzing such data quickly become obsolete over time. In machine learning and data mining this phenomenon is referred to as concept drift. The objective is to deploy models that would diagnose themselves and adapt to changing data over time. This chapter provides an application oriented view towards concept drift research, with a focus on supervised learning tasks. First we overview and categorize application tasks for which the problem of concept drift is particularly relevant. Then we construct a reference framework for positioning application tasks within a spectrum of problems related to concept drift. Finally, we discuss some promising research directions from the application perspective, and present recommendations for application driven concept drift research and development.

Keywords Evolving data · Adaptation · Concept drift · Data streams

I. Žliobaitė (✉)
Department of Computer Science, Aalto University, Espoo, Finland
e-mail: indre.zliobaite@aalto.fi

I. Žliobaitė
Helsinki Institute for Information Technology, Espoo, Finland

I. Žliobaitė
University of Helsinki, Helsinki, Finland

M. Pechenizkiy
Eindhoven University of Technology, Eindhoven, The Netherlands
e-mail: m.pechenizkiy@tue.nl

J. Gama
LIAAD, INESC TEC and University of Porto, Porto, Portugal
e-mail: jgama@fep.up.pt

© Springer International Publishing Switzerland 2016
N. Japkowicz and J. Stefanowski (eds.), *Big Data Analysis: New Algorithms for a New Society*, Studies in Big Data 16, DOI 10.1007/978-3-319-26989-4_4

1 Introduction

Realism of the perfect world assumptions in machine learning has been challenged years ago [31]. One of these challenges relates to an observation that in the real world the data tends to change over time. As a result, predictions of the models trained in the past may become less accurate as time passes or opportunities to improve the accuracy might be missed. Thus, learning models need to have mechanisms for continuous diagnostics of performance, and be able to adapt to changes in data over time.

In machine learning, data mining and predictive analytics unexpected changes in underlying data distribution over time are referred to as concept drift [27, 58, 71, 73]. In pattern recognition the phenomenon is known as covariate shift or dataset shift [58]. In signal processing the phenomenon is known as non-stationarity [36]. Changes in underlying data occur due to changing personal interests, changes in population, adversary activities or they can be attributed to a complex nature of the environment.

The traditional supervised learning assumes that the training and the application data come from the same distribution, as illustrated in Fig. 1a. In real life the predictions need to be made online, often in real time. An online setting brings additional challenges, since it may be expected for the data distribution to change over time. Thus, at any point in time the testing data may be coming from a different distribution than the training data has come, as illustrated in Fig. 1.

The problem of concept drift is of increasing importance as more and more data is organized in the form of data streams rather than static databases, and it is unrealistic to expect that data distributions stay stable over a long period of time. It is not surprising that the problem of concept drift has been studied in several research communities including but not limited to pattern mining, machine learning and data mining, data streams, information retrieval, and recommender systems. Different approaches for detecting and handling concept drift have been proposed in research literature, and many of them have already proven their potential in a wide range of application domains.

One of the most illustrative cases, is learning against an adversary (e.g. spam filters, intrusion detection). A predictive model aims at identifying patterns characteristic of the adversary activity, while the adversary is aware that adaptive learning is used,

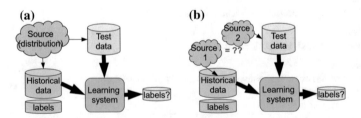

Fig. 1 Stationary supervised learning (**a**) and learning under concept drift (**b**)

and tries to change the behavior. Another context is learning in the presence of hidden variables. User modelling is one of the most popular learning tasks, where the learning system constructs a model of the user intentions, which of course are not observable and may change time to time. Drift also occurs in monitoring tasks and predictive maintenance. Learning the behaviour of a system (e.g. the quality of products in industrial process) where degradation or corrosion of mechanical pieces occur over time.

Concept drift is used as a generic term to describe computational problems with changes over time. These changes may be of countless different types and there are different types of applications that call for different adaptation techniques. Thus, a solution "one-size-fits-all" is hardly possible and not desirable for handling concept drift. On the other hand, real application tasks being seemingly different from each other may share common properties and may have similar needs for adaptation. In order to transfer adaptive techniques from application to application we need to have means to characterize application tasks in a systematic manner.

The main aim and contribution of this chapter is to present tools for describing application tasks with concept drift in a systematic way, to position the existing application driven work using these tools, and define promising directions for future research. To keep the focus on applications we leave a detailed discussion of concept drift handling methods out of the scope of this paper, a reader is referred to existing reviews of the methods and techniques [27, 40, 58, 71]. Our study focuses on describing the *research tasks* driven by application needs.

The chapter is organized as follows. In Sect. 2 we discuss knowledge discovery process in the context of learning from streaming data and handling concept drift. Section 3 presents a reference framework of concept drift tasks and applications. This framework is intended to serve as a tool for describing an application oriented task in a systematic way. In Sect. 4 we survey application oriented published work on adaptive learning, focusing on task formulations, while leaving the techniques out of the scope of this study. Section 5 gives our recommendations towards promising and urgent future research directions from the concept drift application perspective, and concludes the study.

2 Knowledge Discovery Process and Industry Standards

In the era of big data, many data mining projects shift their emphasis towards evolving nature of the data that requires to study the automation of feedback loops more thoroughly. In the standard data mining and machine learning settings the majority of algorithmic techniques have been researched and developed under the assumption of identical and independent data distribution (IID). In big data applications data arrives in a stream, and patterns in the data are expected to evolve over time, therefore, it is not practical, and often is not feasible to involve a data mining expert to monitor the performance of the models and to retrain the models every time they become outdated.

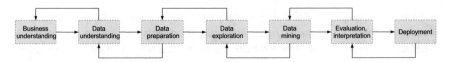

Fig. 2 Knowledge discovery process: from problem understanding to deployment. *Arrows* indicate the most important and frequent dependencies between the phases

Therefore, interest towards automating development and update of predictive models in the streaming data settings has been increasing.

CRISP-DM model [11] describes the classical data mining process, where the life cycle of a data mining project spans over six phases: business understanding, data understanding, data preparation, modeling, evaluation and deployment. Reinartz's framework [65] follows CRISP-DM with some modifications, making modeling steps more explicit. The high-level process steps are summarized in Fig. 2.

Business understanding phase aims at formulating business questions, and translating them into data mining goals. Data understanding phase aims at analyzing and documenting the available data and knowledge sources in the business according to the formulated goals, and providing initial characterization of data. Data preparation phase starts from target data selection that is often related to the problem of building and maintaining useful data warehouses. After selection, the target data is preprocessed in order to reduce the level of noise, pre-process the missing information, reduce data, and remove obviously redundant features. Next, data exploration phase aims at providing the first insight into the data, evaluate the initial hypotheses, usually, by means of descriptive statistics and visualization techniques. Data mining phase covers selection and application of data mining techniques, initialization and further calibration of their parameters to optimal values. Evaluation phase typically considrs offline evaluation on historical data. In predictive modeling, one would typically analyze simulated performance of the data mining system with respect to some suitable measures of accuracy (such as precision, recall, or AUC, among others), or utility (for instance, expressed as cost-sensitive classification). Finally, the most promising predictive model is deployed in operational settings, and the performance is regularly followed up.

CRISP-DM model assumes that most of the data mining processes, including data cleaning, feature engineering, algorithm and parameter selection, and final evaluation, performed offline. If anything goes wrong with the deployed model, a data mining expert would analyze the problem, and try to fix it revisiting one or more steps in the process, and retraining the model.

In the streaming settings, it is common to expect changes in data and model applicability. Therefore, monitoring of model performance and model update or relearning becomes a natural and core part of the data mining process. Figure 3 presents our view towards adaptive data mining process. The main difference with the standard process is that now data preparation, mining, and evaluation steps are automated, there is no manual data exploration, and there is automated monitoring of performance, including change detection and alert services, after deployment.

Fig. 3 Towards CRISP for Adaptive Data Mining

Different strategies for updating learning models have been developed. Two main strategies can be distinguished. Learning models may evolve continuously, for instance, models can be periodically retrained using a sliding window of a fixed size over the past data (e.g. FLORA1 [73]). Alternatively, learning models may use trigger mechanisms, to initiate a model update. Typically, statistical change detection tests are used as triggers (e.g. [26]). Incoming data is continuously monitored, if changes are suspected, the trigger issues an alert, and adaptive actions are taken. When a change is signalled, the old training data is dropped and the model is updated using the latest data.

Learning systems can use single models or ensembles of models. Single model algorithms employ only one model for decision making at a time. Once the model is updated, the old one is permanently discarded. Ensembles, on the other hand, maintain some memory of different concepts. The prediction decisions are made either fusing the votes casted by different models or nominating the most suitable model for the time being from the pool of existing models.

Ensembles can be evolving or have trigger mechanisms as well. Evolving ensembles build and validate new models as new data arrives, the rule for model combination is dynamically updated based on the performance (e.g. [55]). Ensembles with triggers proactively assign the most relevant models for decision maxing based on the context (e.g. [72]).

Table 1 summarizes the taxonomy of adaptive learning strategies.

An important aspect with respect to evaluation of performance of adaptive learning models relates to data collection. An adaptive system collects data, which is biased with respect to adaptations performed. For example, consider a recommender system, where so called "rich-gets-richer" phenomenon boosts the popularity of already popular items. In such situations relying on learning and evaluation of models on offline data is particularly dangerous, since within-system data does not give an unbiased view towards outside world. Consequently, it is important to develop techniques allowing for online evaluation and online adaptation.

Table 1 Adaptive learning strategies

	With triggers	Evolving
Single model	Detectors	Forgetting
Multiple models	Contextual	Dynamic ensembles

Overall, we are not aware of fully automated and functioning adaptive learning system. It could be that well functioning fragments of such systems already exist in industry, especially in big data (web sized) data analysis, where manual attendance to all the running models is simply infeasible. In academia, except for some isolated cases (e.g. [9]), there has been little attention towards automating data mining process for big data, and we anticipate seeing more of such research efforts in the future.

In the following section we first categorize different big data applications where handling of concept drift is important and then refer to different data mining techniques that are suitable for data preprocessing, predictive modeling and evaluation in the streaming settings.

3   Categorization of Concept Drift Tasks and Applications

We start this section by describing relations between concept drift tasks and applications. We analyze application tasks in three steps:

(a) properties of tasks,
(b) landscape of applications,
(c) links between tasks and applications.

The following subsections describe each component.

3.1   Characterization of Application Tasks

Real application tasks, where concept drift is expected, can be mapped into three dimensions: (i) a type of the learning task, (ii) environment from which data comes, and (iii) online operational settings.

3.1.1   Data and Task

Different types of tasks may be required depending on the intended application (even using the same data source): regression, ranking, classification, novelty detection, clustering, itemset mining.

Prediction makes assertions about the future, or about unknown characteristics of the present. Predictions is probably the most common use of data mining, it covers regression and classification tasks. Regression is typically considered in demand planning, resource scheduling optimization, user modelling, and, generally, in applications, in which the main objective is to anticipate future behavior of customers. Ranking is a special form of prediction, where partial ordering of alternative choices is required. Classification is a typical task in diagnosis and decision support, for example, antibiotic resistance prediction, e-mail spam classification, or news categorization. Ranking is a common task in recommendation, information retrieval, credit scoring and preference learning systems domains. Regression, ranking and classification are supervised learning tasks, where models are trained on examples, where the ground truth is available.

Novelty detection is a common task in fault, fraud detection applications, or identifying abnormal behavior. Faults in machines, frauds in credit cards transactions, intrusion detection in computer networks, emergent topics in text news, requires some sort of outlier or anomaly detection, which is a basic form of novelty detection. Novelty detection is a semi-supervised, or unsupervised learning task. Typically, normal examples are available, but abnormal examples are unknown.

Clustering produces a grouping of people or objects, and is a popular task, for instance, in marketing. Itemset mining aims at finding items that commonly appear together, the task is relevant, for instance, in analyzing shopping baskets in retail. Patterns may evolve in those groups, new groups may appear or disappear due to changes in the data generating process. Clustering and itemset mining belong to unsupervised learning tasks, the ground truth is not known.

Orthogonally to different learning tasks, input data may have different forms. Data can be single or multi-relational, sequential, time series, general graph or particular complex structure, bags of instances or a mix. Instances can be noisy or highly accurate. Relational data can be of low or high dimensionality, have a few or lots of missing value, be almost complete or very sparse, have binary, categorical, ordered or numerical attributes.

Moreover, input data can be organized in different ways in terms of its accessibility. Data can come as a stream of instances or batches, or it can arrive in time-stamped batches. Data re-access can be allowed, or a single pass over the data may need to be strictly enforced. There might be randomly or systematically missing values in the incoming data.

3.1.2 Characteristics of Changes

When designing adaptive learning systems one needs to consider, what is the source of drift in data, as different adaptive learning algorithms may be better suited for handling different types of changes. Data may change due to evolution in individual preferences (a person used to like accordion and jazz music earlier, but does not like it any more), a population change (in time of a crisis everybody tend to get lower salaries), adversary actions (new actions are tried to overcome the security system aiming to commit credit card frauds), or complexity of the environment (in automated vehicle navigation the environment is so complex that it is not feasible to take into account all possibilities of landscape deterministically, thus the environment is assumed to be changing).

In addition to types of drifts, it is important to consider, in which patterns changes are expected to occur in the future. Patterns of changes can be categorized according to the transition speed from one concept to another into sudden, or gradual. A drift can include a combination of multiple changes, for instance incremental drift features small steps of sudden changes, resulting in a trend. In terms of reoccurrence drifts can be categorized into novel, or reoccurring concepts.

Finally, it is advisable to consider, to what extent future changes may be predictable in a particular application. Concept drift can be completely unpredictable (e.g. evolution of the financial markets), somewhat predictable or identifiable (e.g. an upcoming financial crisis may be anticipated using a signal from external early warning systems), or the environment might be well identifiable due to seasonality, or reoccurring contexts (e.g. an increase sales of ice-cream in summer).

3.1.3 Operational Settings

Determine availability of the ground truth in an online operation, such as, arrival of true labels in classification, or true target values in regression tasks. Labels may become known immediately in the next time step after casting the prediction (e.g. food sales prediction). Labels may arrive within a fixed or variable time lag (in credit scoring typically the horizon of bankruptcy prediction is fixed, for instance, to one year, thus true labels become known after one year has passed). Alternatively, the setting may allow to obtain labels on demand (e.g. in spam categorization we can ask the user the true status of a given message).

Requirements for the speed of decision making need to be considered when selecting, which algorithms to deploy. In some applications prediction decisions may be required immediately (fraud detection), the sooner the better, while for other analytical decisions timing may be more flexible (e.g. credit scoring decision may reasonably take one–two weeks).

The cost of errors is an aspect to consider when selecting an evaluation metric for monitoring of performance. In traditional supervised learning different types of errors (e.g. false positives, false negatives) may resolve to different losses. In some applications prediction accuracy may be the main performance metric (e.g. in online mass flow prediction), in other applications accurate and timely identification of changes as well as accurate prediction are important (e.g. in demand prediction). In the online setting, discrepancies in time may as well have associated error costs (for instance, too early prediction of a peak in food sales would still allow to sell the extra products later, but too late prediction would lead to throwing away the excess products).

Finally, the ground truth labels may be objective based on clearly defined and accepted rules (e.g. bankrupt or not bankrupt company) or subjective, based on a personal opinion (e.g. interesting or not interesting article). Alternatively, the true labels may not be available at all being impossible or too costly to measure or define in a direct way.

Table 2 summarizes the identified properties of the concept drift application tasks. The identified properties are relevant for describing the type of task, the associated environment and the operational settings of an application under consideration. This information is essential to determine the characteristics that the adaptive learning

Table 2 Summary of properties of concept drift applications

Data and task	*Task*: prediction, classification, detection, clustering, itemset mining
	Input data type: time series, relational, graph, bags or mix
	Incoming data: stream, batches, collection iterations on demand
	Complexity: volume; multiple scans; dimensionality
	Missing values: unlikely, random, systematic
Characteristics of changes	*Change source*: adversary activities, changes of preferences, population change, complex environment
	Change type: sudden, incremental, gradual, reoccurring
	Change expectation: unpredictable, predictable, identifiable (meta)
	Change visibility: direct/indirect; visual inspection, ground truth, external source, result of statistical hypothesis testing
Operational settings	*Label availability*: real time, on demand, fixed lag, variable lag
	Decision speed: real time, analytical
	Costs of mistakes: balanced, unbalanced
	True labels: objective, subjective

system needs to possess, the properties that must be prioritized when designing such a system and the evaluation criteria of the system performance.

3.2 A Landscape of Concept Drift Application Areas

Now as we have identified the properties that characterize concept drift application tasks, our next goal is to categorize application areas, and present typical applications for each category.

We recall application domains, where data mining already plays an important role, or it has a high potential to be deployed. For surveying and summarizing the application domains we combine the taxonomies from the ACM classification[1] and KDnuggets polls.[2]

Table 3 presents our categorization of applications within the identified industries. We group different application areas into three application blocks:

(a) monitoring and control,
(b) information management, and
(c) analytics and diagnostics.

For a compact representation each industry (rows) is assigned a group of applications that share common supervised learning tasks. As it can be seen from the table, for

---

[1] http://www.acm.org/about/class/ccs98-html.

[2] http://www.kdnuggets.com/polls/2010/analytics-data-mining-industries-applications.html.

Table 3 Categorization of applications by type and industry

Indust.	Appl.		
	Monitoring and control	Information management	Analytics and diagnostics
Security, police	Fraud detection, insider trading detection, adversary actions detection	Next crime place prediction	Crime volume prediction
Finance, banking, telecom, insurance, marketing, retail, advertising	Monitoring and management of customer segments, bankruptcy prediction	Product or service recommendation, including complimentary, user intent or information need prediction	Demand prediction, response rate prediction, budget planning
Production industry	Controlling output quality	–	Predict bottlenecks
Education (e-learning, e-health), media, entertainment	Gaming the system, drop out prediction	Music, VOD, movie, news, learning object personalized search and recommendations	Player-centered game design, learner-centered education

each of the industries or groups of industries, more than one application type can be relevant.

The *monitoring and control* block mostly relates to the detection tasks, where an abnormal behavior needs to be signaled. It includes such tasks as detection of adversary activities on the web, computer networks, telecommunications, financial transactions. In most of these tasks the normal behavior is modeled and the goal is to alarm when an abnormal behavior is observed.

The *information management* applications address personalized learning, they include (web) search, recommender systems, categorization and organization of textual information, customer profiling for marketing, personal mail categorization and spam filtering.

The *analytics and diagnostics* block includes predictive analytics and diagnostics tasks, such as evaluation of creditworthiness, demand prediction, drug resistance prediction.

After identifying three blocks of application areas, we now assign the most likely properties to the respective application areas based on our subjective judgement. Table 4 presents the assignment of the properties.

We acknowledge that contradictory examples within each area are always possible to find, yet we believe that the identified properties are the most common for given areas.

It should be noted also that this summary is aimed to cover the majority of cases that would be traditionally associated with applications of machine learning, data mining, and pattern recognition, in which the term concept drift was originally coined

Table 4 Mapping between properties and application areas

	Monitoring and control	Information management	Analytics and diagnostics
Task			
Task	Detection, prediction	Prediction ranking	Prediction classification
Input data	Sequential	Relational transactional	Time series sequential relational
Incoming	Stream	Batches	Stream iterations
Volume	High	Moderate	Moderate
Multiple scans	No/yes	Yes	Yes
Missing values	Random	Unlikely	Systematic
Environment			
Change source	Adversary complex	Preferences contextual	Population
Change type	Sudden	Gradual incremental	Incremental reoccurring
Expectations	Unpredictable	Unpredictable predictable	Identifiable unpredictable
Operational settings			
Label speed ground labels	Fixed lag objective	On demand subjective	Real time objective

and studied most. More recent examples of big data applications in web information retrieval and recommender systems also fit well to our categorization. However, the wider adoption of the big data perspective in other research areas and application domains may bring new interesting aspects. Thus, e.g. handling concept drift has been recognized as an important problem in process mining research dealing with the different kinds of analysis of (business) processes by extracting information from event logs recorded by an information system [8, 10].

In the following section we overview application oriented studies on learning from evolving data and through considered examples illustrate peculiarities of handling concept drift under different application settings.

4   An Overview of Application Oriented Studies on Learning from Evolving Data

Following the categorization of applications, we distinguish three main groups of application tasks: monitoring and control, information management, and diagnostics. Besides having different goals, the groups also differ in data types. Monitoring and control applications typically use streaming sensory data as inputs, concept drift typically happens fast and suddenly. Information management applications work with time-stamped documents, concept drift happens slower than in the previous case, changes can be sudden or gradual. Diagnostics applications typically use relational data tables, where observations are time-stamped. Concept drift, also known as population drift, typically happens slowly. Changes are typically incremental, or evolving. Sudden shifts are not very typical in these applications.

In this section we briefly characterize each group, overview application studies that fall within each group and touch upon the issue of concept drift, and present three studies in more detail, illustrating how the prediction task is formulated, and how concept drift is handled. We discuss research challenges, and highlight interesting aspects of these application tasks from concept drift handling perspective.

We do not claim that this is an exhaustive list of concept drift applications. Our goal is to include examples from a wide range of application tasks.

4.1 Monitoring and Control

The first group of concept drift application tasks aim at real-time monitoring or control of some automated activity, for example, operation of a chemical plant. Input data typically consists of streaming sensory readings, and the target is often related to describing the quality of the activity or process. The goal of such monitoring could be to oversee operation of the system (without interfering, unless something goes wrong), to control the system, or to detect abnormal behaviour (possibly due to adversary actions). Concept drift typically happens fast (in the order of seconds or minutes), and changes are sudden. Table 5 summarizes example studies related to handling concept drift in monitoring and control applications.

4.1.1 Monitoring for Management

Monitoring for management tasks are often found in production industry and transportation domains. Concept drift is typically observed due to complexity of the process, or human (operators) factors. So many factors are affecting the process, that it is not possible to take all of them in the predictive model. When some of those factors, that have been fixed for a while, suddenly change—a concept drift is observed. For example, production quality in a chemical plant may be different depending on the supplier of raw materials. If we make a model when one supplier is used, such a model may not be as accurate when the supplier changes, and some adaptation may be required.

In transportation, traffic control centers use data driven traffic management systems for predicting traffic conditions [13], such as car density in a particular area, or anticipating traffic accidents. Public transportation travel time prediction [57] is used for scheduling and human resource (drivers) planning purposes. In remote sensing relevant application tasks include place recognition [52], activity recognition [51], interactive road segmentation [77]. In production industry relevant tasks

Table 5 Summary of monitoring and control studies

Goal	Domain	Application task	References
Monitoring for management	Transportation	Traffic management	[13, 57]
	Remote sensing	Place, activity recognition	[51, 52, 77]
	Production industry	Production quality control	[41, 61]
	Telecom. network	Telecommunication monitoring	[60]
Automated control	Mobile systems	Controlling robots, vehicles	[48, 63, 70]
	Smart home	Intelligent appliances	[2, 64]
	Virtual reality	Computer games, flight simulators	[12, 34]
Anomaly detection	Computer security	Intrusion detection	[50]
	Telecommunications	Intrusion detection, fraud	[37, 54]
	Finance	Fraud, insider trading	[7, 19]

include monitoring the output quality, for example, in chemical production [41], or the process itself, for example, boilers producing heat [61]. Monitoring models in production industry are called *soft sensors* [40]. In *service monitoring* detection of defects or faults in telecommunication network [60] present relevant tasks.

4.1.2 Automated Control

In automated control applications the problem of concept drift is often referred to as the dynamically changing environment. The objects learn how to interact with the environment and since the environment is too complex to take all the playing factors into a predictive model, therefore predictive models need to be adaptive.

Examples of application domains in automated control include: mobile systems and robotics, smart homes, and virtual reality. Ubiquitous knowledge discovery deals with distributed and mobile systems, operating in a complex, dynamic and unstable environment. The word 'ubiquitous' means distributed at a time. Relevant tasks include navigation systems [70], soccer playing robots [48], vehicle monitoring, household management systems, music mining are examples. Intelligent systems, or smart home systems [64] aim to develop intelligent household appliances [2]. Virtual reality includes application tasks in computer game design [12], where adversary actions of the players (cheating) or improving skills of a player, may be cause concept drift. Virtual reality is also used in flight simulators, where skills and strategies change from user to user [34].

4.1.3 Anomaly Detection

Anomaly detection is often tackled as one class classification task, where the properties of a normal behavior are well defined, while the properties of abnormal behovior may be changing. Concept drift happens due to changes in behavior, characteristics of legitimate users, or new creative adversary actions.

Anomaly detection is very relevant for computer security domain, in particular, network intrusion detection [50]. In telecommunications fraud prevention [37] or mobile masquerade detection [54] are the relevant tasks. In finance data mining techniques are employed to monitor streams of financial transactions (credit cards, internet banking) to alert for possible frauds or insider trading [3, 30, 68].

4.1.4 Credit Scoring

In retail banking, credit risk assessment often relies in credit scoring models developed with supervised learning methods used to evaluate a person's credit worthiness. The output of these models is a score that translates a probability of a customer becoming a defaulter, usually in a fixed future period, so-called scoring or PD models. Nowadays, these models are at the core of the banking business, because they are imperative in credit decision-making, in price settlement, and to determine the cost of capital. Moreover, central banks and international regulation have dramatically evolved to a structure where the use of these models is implicit, to achieve sound standards for credit risk valuation in the banking system.

Developing and implementing a credit scoring model can be time and resources consuming—easily ranging from 9 to 18 months, from data extraction until deployment. Hence, it is not rare that banks use an unchanged credit scoring model for several years (a 5 year period is commonly exceeded). Bearing in mind that models are built using a sample file frequently comprising 2 or more years of historical data, in the best case scenario, data used in the models are shifted 3 years away from the point they will be used. An 8 years shift is frequently exceeded. Should conditions remain unchanged, then this would not significantly affect the accuracy of the model, otherwise, its performance can greatly deteriorate over time. The recent financial crisis came to confirm that financial environment greatly fluctuates, in an unexpected manner, posing renewed attention regarding scorecards built upon frames that are by far outdated. By 2007–2008, many financial institutions were using stale scorecards built with historical data of the early-decade. The degradation of stationary credit scoring models is an issue with empirical evidence in the literature [14, 32], however research is still lacking application oriented solutions.

4.1.5 Example Study: Online Mass Flow Estimation

Industrial boilers are used for heating buildings in winter times. Some boilers operate on biofuel, which is a mix tree branches, peat and plants; the mix is not necessarily

Fig. 4 An example of boiler data

uniform and the proportions may vary. The authors of the first example study [61] consider the problem of online mass flow estimation in boiler operation. During burning phase the mass of fuel inside the boiler container decreases, and as new fuel is added to the container, while at the same time the burning process continues, the fuel feeding phase starts that is reflected by a rapid mass increase.

Input data comes from physical sensors with a negligible lag. The task is to estimate the current mass flow (similarly to fuel consumption indicators in passenger cars), and detect the points of phase switch in real time.

There are three main sources of drifts in the signal (an exsample is depicted in Fig. 4). First, fuel feeding is manual and non-standardized process, which is not necessarily smooth and may have short interruptions. Second, rotation of the feeding screw adds noise to the measured signal. Finally, there is a low amplitude rather periodic noise, which is caused by the mechanical rotation of the system parts, the magnitude of this noise depends on operational setting.

The main focus is on constructing a learning system that can deal with two types of change points: an abrupt change to feeding and slower but still abrupt switch to burning, and asymmetric outliers, which in online settings can be easily mixed with the changes to feeding. These change points need to be identified in real time, and they should not be mixed with noise. When these regime switch points are known, a new predictive model can be incrementally started after each feed to reflect the most recent fuel characteristics.

The optimization criteria for change detection is to minimize the detection delay (from the actual change point to detection), and minimize the number of false alarms, when an outlier is singled as a change. All true change points have to be detected, no misses are allowed. In addition, the final performance indicator is the mean square error (MSE) of the mass flow estimation. It is critical for algorithm design to understand, how different types of errors in detection affect the overall accuracy of classification. Such sensitivity analysis can be performed by varying the detection thresholds.

Evaluation of the performance of the algorithms is challenging, since there is no ground truth available. The authors construct an approximation to the ground truth, and use it for the evaluation purposes in online settings (only in the experiments, but not in real operational setting). Absence of ground truth is a common problem in monitoring applications, since, if it was easily available, there would be no need for the predictive model, which is being designed.

Table 6 Summary of information management studies

Goal	Domain	Application task	References
Personal assistance	Text processing	News categorization, document classification	[4, 49, 59, 74]
		Spam filtering	[17, 21]
	Web mining	Web personalization	[15, 16, 67]
		Library, media search	[22, 35]
Marketing	Customer segmentation	Customer segmentation, profiling	[5, 13, 47, 66]
	Recommender systems	Movie recommendations	[18, 45]
Management	Project management	Software project mgmt.	[20]
	Archiving	Article, mail organization	[43, 76]

4.2   Information Management

These tasks aim at organizing, and personalizing information. Typically, data is comes as time stamped entities, for instance, web documents, and the goal is to character-ize each entity. Information management application tasks can be further split into personal assistance, marketing, and management tasks. Concept drift happens not so fast (in the order or days or weeks), changes could be sudden or gradual. Table 6 summarizes example studies related to handling concept drift for information man-agement.

4.2.1   Personal Assistance

Personal assistance applications aim at user modeling. The goal is to personalize information flow, the process is often referred to as information filtering. A rich technical presentation on user modeling can be found in [28]. One of the primary applications of user modeling is representation of queries, news, blog entries with respect to *current* user interests. Changing user interests over time is the main cause of concept drift.

Large part of personal assistance applications are related to handling *textual data*, example tasks include news story classification [4, 74], or document categorization [49, 59]. In web search, detecting changes in user satisfaction has been recognized to be important [42]. Personal assistance tasks relate to other types of data, such as networked multimedia, music, video, as well as digital libraries [35]. Large body of applications relate to web personalization and dynamics [15, 16, 67], where interim system data (logs) is mined.

4.2.2 Marketing

Customer profiling applications use aggregated data from many users. The goal is to segment customers based on their interests and needs. Concept drift happens due to changing individual interests and behavior over time.

Relevant tasks include direct marketing, based on product preferences, for example cars [13], or service usage, for example telecommunications [5], identifying and analyzing shopping baskets [66], social network analysis for customer segmentation [47], recommender systems [45].

4.2.3 Management

A number of studies aim at adaptive organization or categorization of web documents, e-mails, news articles [43, 76]. Concept drift happens due to evolving nature of the content. In business software project management, careful planning may be inaccurate if concept drift is not taken into account [20].

4.2.4 Example Study: Movie Recommendation

Interest of data mining community in recommender systems domain has been boosted by Netflix competition.[3] One of the lessons learnt from it was that taking temporal dynamics is important for building accurate models. Handling concept drift has another set of peculiarities here. Both items and users are changing over time. Item-side effects include first of all changing product perception and popularity. The popularity of some movies is expected to follow seasonal patterns. User-side effects include changing tastes and preferences of customers, some of which may be short-term or contextual and therefore likely reoccurring (mood, activity, company, etc.), changing perception of rating scale, possible change of rater within household and alike problems.

As suggested in [45] popular windowing and instance weighing approaches for handling concept drift are not the best choice simply because in collaborative filtering the relations between ratings is of the main importance for predictive modeling.

In this application labels are soft, data comes in batches, and the rating matrix is high-dimensional and extremely sparse containing only about 1 % of non-zero elements (that makes the application of most machine learning predictors inapplicable and boost the development of advanced collaborative filtering approaches).

---

[3] www.netflixprize.com.

Table 7 Summary of diagnostics studies

Goal	Domain	Application task	References
Forecasting	Banking	Bankruptcy prediction	[38, 69]
	Economics	Forecasting	[29, 33, 44]
Medicine	Drug research	Antibyiotic resistance	[39, 53, 72]
		Drug discovery	[23]
	Clinical research	Disease monitoring	[6, 24, 46]
Security	Biometrics	Authentication	[62, 75]

4.3 Analytics and Diagnostics

Analytics and diagnostics tasks aim at characterizing health, well-being, or a state of humans, economies, or entities. Data typically comes as time stamped relational data. Concept drift often happens due to population drift, and changes are typically slow (in the order of months or years) and incremental.

Analytics and diagnostics tasks can be further split into forecasting, medicine, or security applications. Table 7 summarizes example studies related to handling concept drift in diagnostics.

Changes happen due to changing environment, such as economic situation, which includes a large number of influencing factors.

4.3.1 Forecasting

Forecasting applications typically relate to analytics tasks in economics and finance, such as macroeconomic forecasting, demand prediction, travel time predictions, event prediction (e.g. crime maps, epidemic outbreaks). Changes over time often happen due to population drift, which typically happens much slower, than, for instance, changes in personal preferences in information management applications, or adversary actions in monitoring applications.

In finance relevant tasks include bankruptcy prediction or individual credit scoring [38, 69], in economics concept drift appears in making macroeconomic forecasts [29], predicting phases of a business cycle [44], or stock price prediction [33].

4.3.2 Security

In biometric authentication [62, 75] concept drift can be caused by changing physiological factors, e.g. growing a beard.

4.3.3 Medicine

Medicine applications, such as antybiotic resistance prediction, or predicting epidemic outbreaks or nosocomial infections, may be subject to concept drift due to adaptive nature of microorganisms [39, 53, 72]. Clinical studies and systems need adaptivity mechanisms to changes caused by human demographics [24, 46].

4.3.4 Example Study: Predicting Antibiotic Resistance

Antibiotic resistance is an important problem and it is an especially difficult problem with nosocomial infections in hospitals because pathogens attack critically ill patients who are more vulnerable to infections than the general population and therefore require more antibiotics.

Prediction model is based on information about patients, hospitalization, pathogens and antibiotic themselves. The data arrives in batches, the labels become available with a variable lag depending on the size of the hospital and intensiveness of the patients flow. The size of the data is relatively small both in number of instances and the number of features to be considered.

The peculiarity of concept drift is that it may happen for various reasons particularly because pathogens may develop resistance and share this information with peers in different ways. Consequently, the type and severity of changes may depend on the location in the instance space. Furthermore, the drift is expected to be local and reflect e.g. a pathway in the hospital where the resistance was taking place and spread around. This calls for the direct or indirect identification of the regions or subgroups in which concept drift is occurring. Handling concept drift with dynamic integration of classifiers that takes this peculiarity into account was shown to be effective [72].

5 Discussion and Conclusions

The main lesson in this study is related to the evolving characteristic of data and the implications in data analysis. Nowadays, digital data collection is easy and cheap. Data analytics in applications where data is collected over time, must take into account the evolving nature of data.

The problem of concept drift has been recognized in different application domains. Interest in different research communities has been reinforced by several recent competitions including e.g. controlling driverless cars at the DARPA challenge, credit risk assessment competition at PAKDD'09), and Netflix movie recommendation.

However, concept drift research field is still in an early stage. The research problems, although motivated by a belief that handling concept drift is highly important for practical data mining applications, have been formulated and addressed often in artificial and somewhat isolated settings. Approaches for handling concept drift are rather diverse and have been developed from two sides—theory-oriented and applications-oriented. Recent studies however do highlight the peculiarities of particular applications and give intuition and/or empirical evidence why traditional general-purpose concept drift handling techniques are not expected to perform well and suggest tailored or more focused techniques suitable for a particular application type.

In this work we categorized the applications, where handling concept drift is known or expected to be an important component of any learning system. We identified three major types of applications, identified key properties of the corresponding settings, and provided a discussion emphasizing the most important application oriented aspects. Summarizing those we can speculate that the concept drift research area is likely to refocus further from studying general methods to detect and handle concept drift to designing more specific, application oriented approaches that address various issues like delayed labeling, label availability, cost-benefit trade-off of the model update and other issues peculiar to a particular type of applications.

Most of the work on concept drift assumes that the changes happen in hidden context that is not observable to the adaptive learning system. Hence, concept drift is considered to be unpredictable and its detection and handling is mostly reactive. However, there are various application settings in which concept drift is expected to reappear along the time line and across different objects in the modeled domain. Seasonal effects with vague periodicity for a certain subgroup of object would be common e.g. in food demand prediction [78]. Availability of external contextual information or extraction of hidden contexts from the predictive features may help to better handle recurrent concept drift, e.g. with use of a meta-learning approach [25]. Temporal relationships mining can be used to identify related drifts, e.g. in the distributed or peer-to-peer settings in which concept drift in one peer may precede another drift in related peer(s) [1]. Thus, we can expect that for many applications more accurate, more proactive and more transparent change detection mechanisms may become possible.

Moving from adaptive algorithms towards adaptive systems that would automate full knowledge discovery process and scaling these solutions to meet the computational challenges of big data applications is another important step for bringing research closer to practice. Developing open-source tools like SAMOA [56] certainly facilitates this.

Domain experts play an important role in acceptance of big data solutions. They often want to go away from non interpretable black-box models and to develop trust in underlying techniques, e.g. to be certain that a control system is really going to react to changes when they happen and to understand how these changes are detected and what adaptation would happen. Therefore we anticipate that there will be also a change in the focus from change detection to change *description*, from *when a change happen* to *how and why it happened* as such research would be helpful in improving utility, usability and trust in adaptive learning systems being developed for many of the big data applications.

Acknowledgments This work was partially supported by European Commission through the project MAESTRA (Grant number ICT-2013-612944).

References

1. Ang, H.H., Gopalkrishnan V., Zliobaite I., Pechenizkiy M., Hoi S.C.H.: Predictive handling of asynchronous concept drifts in distributed environments. IEEE Trans. Knowl. Data Eng. **25**, 2343–2355 (2013)
2. Anguita, D.: Smart adaptive systems: state of the art and future directions of research. In: Proceedings of the 1st European Sympposium on Intelligent Technologies, Hybrid Systems and Smart Adaptive Systems, EUNITE (2001)
3. Becker, R.A., Volinsky, C., Wilks, A.R.: Fraud detection in telecommunications: History and lessons learned. Technometrics **52**(1), 20–33 (2010)
4. Billsus, D., Pazzani, M.: A hybrid user model for news story classification. In: Proceedings of the 7th International Conference on User Modeling, UM, pp. 99–108 (1999)
5. Black, M., Hickey, R.: Classification of customer call data in the presence of concept drift and noise. In: Proceedings of the 1st International Conference on Computing in an Imperfect World, pp. 74–87 (2002)
6. Black, M., Hickey, R.: Detecting and adapting to concept drift in bioinformatics, pp. 161–168. In Proc. of Knowledge Exploration in Life Science Informatics, International Symposium (2004)
7. Bolton, R., Hand, D.: Statistical fraud detection: A review. Stat. Sci. **17**(3), 235–255 (2002)
8. Bose, R.P.J.C., van der Aalst W.M.P., Zliobaite, I., Pechenizkiy, M. Dealing with concept drift in process mining. IEEE Trans. Neur. Net. Lear. Syst. accepted (2013)
9. Budka, M., Eastwood, M., Gabrys, B., Kadlec, P., Martin-Salvador, M., Schwan, S., Tsakonas, A., Zliobaite, I.: From sensor readings to predictions: on the process of developing practical soft sensors. In: Procedings of the 13th International Symposium on Intelligent Data Analysis, pp. 49–60 (2014)
10. Carmona, J., Gavaldà, R.: Online techniques for dealing with concept drift in process mining. In: Proceedings of the 11th International Symposium on Intelligent Data Analysis, pp. 90–102 (2012)
11. Chapman, P., Clinton, J., Kerber, R., Khabaza, T., Reinartz, T., Shearer, C., Wirth, R.: CRISP-DM 1.0 step-by-step data mining guide. Technical report, The CRISP-DM consortium (2000)
12. Charles, D., Kerr, A., McNeill, M., McAlister, M. Black, M., Kucklich, J., Moore, A., Stringer, K.: Player-centred game design: player modelling and adaptive digital games. In: Proceedings of the Digital Games Research Conference, pp. 285–298 (2005)
13. Crespo, F., Weber, R.: A methodology for dynamic data mining based on fuzzy clustering. Fuzzy Sets and Syst. **150**, 267–284 (2005)
14. Crook, J., Hamilton, R., Thomas, L.C.: The degradation of the scorecard over the business cycle. IMA J. Manage. Math. **4**, 111–123 (1992)
15. da Silva, A., Lechevallier, Y., Rossi, F., de Carvalho, F.: Construction and analysis of evolving data summaries: an application on web usage data. In: Proceedings of the 7th International Conference on Intelligent Systems Design and Applications, pp. 377–380 (2007)
16. De Bra, P., Aerts, A., Berden, B., de Lange, B., Rousseau, B., Santic, T., Smits, D., Stash, N.: AHA! the adaptive hypermedia architecture. In: Proceedings of the 14th ACM Conference on Hypertext and hypermedia, pp. 81–84 (2003)
17. Delany, S., Cunningham, P., Tsymbal, A.: A comparison of ensemble and case-base maintenance techniques for handling concept drift in spam filtering. In: Proceedings of Florida Artificial Intelligence Research Society Conference, pp. 340–345 (2006)
18. Ding, Y., Li, X.: Time weight collaborative filtering. In: Proceedings of the 14th ACM International Conference on Information and Knowledge Management, pp. 485–492 (2005)

19. Donoho, S.: Early detection of insider trading in option markets. In: Proceedings of the 10th ACM SIGKDD International Conference on Knowledge Discovery and Data Mining, pp. 420–429 (2004)
20. Ekanayake, J., Tappolet, J., Gall, H.C., Bernstein, A.: Tracking concept drift of software projects using defect prediction quality. In: Proceedings of the 6th IEEE International Working Conference on Mining Software Repositories, pp. 51–60 (2009)
21. Fdez-Riverola, F., Iglesias, E., Diaz, F., Mendez, J., Corchado, J.: Applying lazy learning algorithms to tackle concept drift in spam filtering. Expert Syst. Appl. **33**(1), 36–48 (2007)
22. Flasch, O., Kaspari, A., Morik, K., Wurst, M.: Aspect-based tagging for collaborative media organization. In: Proceedings of Workshop on Web Mining, From Web to Social Web: Discovering and Deploying User and Content Profiles, pp. 122–141 (2007)
23. Forman, G.: Incremental machine learning to reduce biochemistry lab costs in the search for drug discovery. In: Proceedings of the 2nd Workshop on Data Mining in Bioinformatics, pp. 33–36 (2002)
24. Gago, P., Silva, A., Santos, M.: Adaptive decision support for intensive care. In: Proceedings of 13th Portuguese Conference on Artificial Intelligence, pp. 415–425 (2007)
25. Gama, J., Kosina, P.: Learning about the learning process. In: Proceedings of the 10th International Conference on Advances in intelligent data analysis, IDA, pp. 162–172, Germany, Springer (2011)
26. Gama, J., Medas, P., Castillo, G., Rodrigues, P.: Learning with drift detection. In: Proceedings of the 17th Brazilian Symposium on Artificial Intelligence, pp. 286–295 (2004)
27. Gama, J., Zliobaite, I., Bifet, A., Pechenizkiy, M., Bouchachia, A.: A survey on concept drift adaptation. ACM Comput. Surv. **46**(4), 44:1–44:37 (2014)
28. Gauch, S. Speretta, M., Chandramouli, A., Micarelli, A.: User profiles for personalized information access. In: Brusilovsky, P., Kobsa, A., Nejdl, W. (eds.) The Adaptive Web, pp. 54–89. Springer (2007)
29. Giacomini, R., Rossi, B.: Detecting and predicting forecast breakdowns. Working Paper 638, ECB (2006)
30. Hand, D.J.: Fraud detection in telecommunications and banking: discussion of Becker, Volinsky, and Wilks (2010); Sudjianto et al. Technometrics **52**(1), 34–38 (2010)
31. Hand, D.: Classifier technology and the illusion of progress. Stat. Sci. **21**(1), 1–14 (2006)
32. Hand, D.J., Adams, N.M.: Selection bias in credit scorecard evaluation. JORS **65**(3), 408–415 (2014)
33. Harries, M., Horn, K.: Detecting concept drift in financial time series prediction using symbolic machine learning. In: In Proceedings of the 8th Australian Joint Conference on Artificial Intelligence, pp. 91–98 (1995)
34. Harries, M., Sammut, C., Horn, K.: Extracting hidden context. Mach. Learn. **32**(2), 101–126 (1998)
35. Hasan, M., Nantajeewarawat, E.: Towards intelligent and adaptive digital library services. In: Proceedings of the 11th International Conference on Asian Digital Libraries, pp. 104–113 (2008)
36. Haykin, S., Li, L.: Nonlinear adaptive prediction of nonstationary signals. IEEE Trans. Sig. Process. **43**(2), 526–535 (1995)
37. Hilas, C.: Designing an expert system for fraud detection in private telecommunications networks. Expert Syst. Appl. **36**(9), 11559–11569 (2009)
38. Horta, R., de Lima, B., Borges, C.: Data pre-processing of bankruptcy prediction models using data mining techniques (2009)
39. Jermaine, C.: Data mining for multiple antibiotic resistance. Online (2008)
40. Kadlec, P., Grbic, R., Gabrys, B.: Review of adaptation mechanisms for data-driven soft sensors. Comput. Chem. Eng. **35**, 1–24 (2011)
41. Kadlec, P., Gabrys, B.: Local learning-based adaptive soft sensor for catalyst activation prediction. AIChE J. **57**(5), 1288–1301 (2011)
42. Kiseleva, J., Crestan, E., Brigo, R., Dittel, R.: Modelling and detecting changes in user satisfaction. In: Proceedings of the 23rd ACM International Conference on Information and Knowledge Management, pp. 1449–1458 (2014)

43. Kleinberg, J.: Bursty and hierarchical structure in streams. In: Proceedings of the 8th ACM SIGKDD International Conference on Knowledge Discovery and Data Mining, pp. 91–101. ACM (2002)
44. Klinkenberg, R.: Meta-learning, model selection and example selection in machine learning domains with concept drift. In: Proceedings of annual workshop of the Special Interest Group on Machine Learning, Knowledge Discovery, and Data Mining, pp. 64–171 (2005)
45. Koren, Y.: Collaborative filtering with temporal dynamics. Commun. ACM **53**(4), 89–97 (2010)
46. Kukar, M.: Drifting concepts as hidden factors in clinical studies. In: Proceedings of the 9th Conference on Artificial Intelligence in Medicine in Europe, pp. 355–364 (2003)
47. Lathia, N., Hailes, S., Capra, L.: kNN CF: a temporal social network. In: Proceedings of the ACM Conference on Recommender Systems, pp. 227–234 (2008)
48. Lattner, A., Miene, A., Visser, U., Herzog, O.: Sequential pattern mining for situation and behavior prediction in simulated robotic soccer. In: Proceedings of Robot Soccer World Cup IX, pp. 118–129 (2006)
49. Lebanon, G., Zhao, Y.: Local likelihood modeling of temporal text streams. In: Proceedings of the 25th International Conference on Machine Learning, pp. 552–559 (2008)
50. Lee, W., Stolfo, S.J., Mok, K.W.: Adaptive intrusion detection: A data mining approach. Artif. Intell. Rev. **14**(6), 533–567 (2000)
51. Liao, L., Patterson, D., Fox, D., Kautz, H.: Learning and inferring transportation routines. Artif. Intell. **171**(5–6), 311–331 (2007)
52. Luo, J., Pronobis, A., Caputo, B., Jensfelt, P.: Incremental learning for place recognition in dynamic environments. In: Proceedings of the IEEE/RSJ International Conference on Intelligent Robots and Systems, pp. 721–728 (2007)
53. Martin, M.T., Knudsen, T.B., Judson, R.S., Kavlock, R.J., Dix, D.J.: Economic benefits of using adaptive predictive models of reproductive toxicity in the context of a tiered testing program. Syst. Biol. Reprod. Med. **58**, 3–9 (2012)
54. Mazhelis, O., Puuronen, S.: Comparing classifier combining techniques for mobile-masquerader detection. In: Proceedings of the The 2nd International Conference on Availability, Reliability and Security, pp. 465–472 (2007)
55. Minku, L.L., White, A.P., Yao, X.: The impact of diversity on online ensemble learning in the presence of concept drift. IEEE Trans. Knowl. Data Eng. **22**(5), 730–742 (2010)
56. Morales, G.D.F., A, Bifet.: SAMOA: Scalable advanced massive online analysis. J. Mach. Learn. Res. **16**, 149–153 (2015)
57. Moreira, J.: Travel time prediction for the planning of mass transit companies: a machine learning approach. PhD thesis, University of Porto (2008)
58. Moreno-Torres, J.G., Raeder, T., Alaiz-Rodríguez, R., Chawla, N.V., Herrera, F.: A unifying view on dataset shift in classification. Pattern Recogn. **45**(1), 521–530 (2012)
59. Mourao, F., Rocha, L., Araujo, R., Couto, T., Goncalves, M., Meira, W.: Understanding temporal aspects in document classification. In: Proceedings of the International Conference on Web Search and Web Data Mining, pp. 159–170 (2008)
60. Pawling, A., Chawla, N., Madey, G.: Anomaly detection in a mobile communication network. Comput. Math. Organ. Theory **13**(4), 407–422 (2007)
61. Pechenizkiy, M., Bakker, J., Zliobaite, I., Ivannikov, A., Karkkainen, T.: Online mass flow prediction in CFB boilers with explicit detection of sudden concept drift. SIGKDD Explor. **11**(2), 109–116 (2009)
62. Poh, N., Wong, R., Kittler, J., Roli, F.: Challenges and research directions for adaptive biometric recognition systems. In: Proceedings of the 3rd International Conference on Advances in Biometrics, pp. 753–764 (2009)
63. Procopio, M., Mulligan, J., Grudic, G.: Learning terrain segmentation with classifier ensembles for autonomous robot navigation in unstructured environments. J. Field Robot. **26**(2), 145–175 (2009)
64. Rashidi, P., Cook, D.: Keeping the resident in the loop: Adapting the smart home to the user. IEEE Trans. Syst. Man Cybern. Part A Syst. Hum **39**(5), 949–959 (2009)

65. Reinartz, T.P.: Focusing solutions for data mining: analytical studies and experimental results in real-world domains. In: Lecture Notes in Computer Science, vol. 1623. Springer (1999)
66. Rozsypal, A., Kubat, M.: Association mining in time-varying domains. Intell. Data Anal. **9**(3), 273–288 (2005)
67. Scanlan, J., Hartnett, J., Williams. R.: DynamicWEB: adapting to concept drift and object drift in cobweb. In: Proceedings of the 21st Australasian Joint Conference on Artificial Intelligence, pp. 454–460 (2008)
68. Sudjianto, A., Nair, S., Yuan, M., Zhang, A., Kern, D., Cela-Diaz, F.: Statistical methods for fighting financial crimes. Technometrics **52**(1), 5–19 (2010)
69. Sung, T., Chang, N., Lee, G.: Dynamics of modeling in data mining: interpretive approach to bankruptcy prediction. J. Manage. Inf. Syst. **16**(1), 63–85 (1999)
70. Thrun, S., Montemerlo, M., Dahlkamp, H., Stavens, D., Aron, A., Diebel, J., Fong, P., Gale, J., Halpenny, M., Hoffmann, G., Lau, K., Oakley, C., Palatucci, M., Pratt, V., Stang, P., Strohband, S., Dupont, C., Jendrossek, L.-E., Koelen, C., Markey, C., Rummel, C., van Niekerk, J., Jensen, E., Alessandrini, P., Bradski, G., Davies, B., Ettinger, S., Kaehler, A., Nefian, A., Mahoney, P.: Winning the darpa grand challenge. J. Field Robot. **23**(9), 661–692 (2006)
71. Tsymbal, A.: The problem of concept drift: definitions and related work. Technical report, Department of Computer Science, Trinity College Dublin, Ireland (2004)
72. Tsymbal, A., Pechenizkiy, M., Cunningham, P., Puuronen, S.: Dynamic integration of classifiers for handling concept drift. Inf. Fusion **9**(1), 56–68 (2008)
73. Widmer, G., Kubat, M.: Learning in the presence of concept drift and hidden contexts. Mach. Learn. **23**(1), 69–101 (1996)
74. Widyantoro, D., Yen, J.: Relevant data expansion for learning concept drift from sparsely labeled data. IEEE Trans. Knowl. Data Eng. **17**(3), 401–412 (2005)
75. Yampolskiy, R., Govindaraju, V.: Direct and indirect human computer interaction based biometrics. J. Comput. **2**(10), 76–88 (2007)
76. Yang, Y., Wu, X., Zhu, X.: Mining in anticipation for concept change: Proactive-reactive prediction in data streams. Data Min. Knowl. Discov. **13**(3), 261–289 (2006)
77. Zhou, J., Cheng, L., Bischof, W.: Prediction and change detection in sequential data for interactive applications. In: Proceedings of the 23rd AAAI Conference on Artificial Intelligence, pp. 805–810 (2008)
78. Zliobaite, I., Bakker, J., Pechenizkiy, M.: Beating the baseline prediction in food sales: How intelligent an intelligent predictor is? Expert Syst. Appl. **31**(1), 806–815 (2012)

Analysis of Text-Enriched Heterogeneous Information Networks

Jan Kralj, Anita Valmarska, Miha Grčar, Marko Robnik-Šikonja
and Nada Lavrač

Abstract This chapter addresses the analysis of information networks, focusing on heterogeneous information networks with more than one type of nodes and arcs. After an overview of tasks and approaches to mining heterogeneous information networks, the presentation focuses on text-enriched heterogeneous information networks whose distinguishing property is that certain nodes are enriched with text information. A particular approach to mining text-enriched heterogeneous information networks is presented that combines text mining and network mining approaches. The approach decomposes a heterogeneous network into separate homogeneous networks, followed by concatenating the structural context vectors calculated from separate homogeneous networks with the bag-of-words vectors obtained from textual information contained in certain network nodes. The approach is show-cased on the analysis of two real-life text-enriched heterogeneous citation networks.

1 Introduction

The field of *network analysis* has its roots in two research fields: mathematical graph theory and social sciences. Network analysis started as an independent research discipline in the late seventies [42] and early eighties [5], when sociologists became increasingly aware that the study of social relations—and not only individual attributes—is necessary for in-depth analysis of human societies. Since

J. Kralj (✉) · A. Valmarska · N. Lavrač
Jožef Stefan Institute, Jamova 39, 1000 Ljubljana, Slovenia
e-mail: jan.kralj@ijs.si

J. Kralj · A. Valmarska · N. Lavrač
Jožef Stefan International Postgraduate School, Jamova 39,
1000 Ljubljana, Slovenia

M. Robnik-Šikonja
Faculty of Computer and Information Science, Večna pot 113,
1000 Ljubljana, Slovenia

N. Lavrač
University of Nova Gorica, Vipavska 13, 5000 Nova Gorica, Slovenia

© Springer International Publishing Switzerland 2016
N. Japkowicz and J. Stefanowski (eds.), *Big Data Analysis: New Algorithms for a New Society*, Studies in Big Data 16, DOI 10.1007/978-3-319-26989-4_5

this early research, network analysis has grown substantially: the field now covers not only social networks but also general networks originating from any (scientific) discipline.

In recent years, analysis of *heterogeneous information networks* [34] has gained momentum. In contrast to standard *homogeneous* information networks, heterogeneous information networks describe heterogeneous types of entities and different types of relations. Moreover, in *enriched heterogeneous information networks*, nodes of certain type contain additional information, for example in the form of experimental results or documents. After an overview of tasks and approaches to mining heterogeneous information networks, we focus on *text-enriched heterogeneous information networks*. We present a particular approach to mining text-enriched heterogeneous information networks, together with its application in two complex real-life domains. In the first example, video lectures from the VideoLectures.NET website, forming a network of lectures, authors and viewers, are enriched with their abstracts. The results show that using both structural context vectors and bag-of-words vectors improves category prediction compared to using only one type of vectors. In the second example, scientific publications forming a network of publications and authors, are enriched with their abstracts. The results show that increasing the network size and combining text and network structure information improves the accuracy of paper categorization.

The chapter is structured as follows. Section 2 introduces the concepts of homogeneous and heterogeneous information networks and presents examples of such networks. Section 3 presents data analysis tasks applicable in homogeneous or heterogeneous networks. Section 4 presents an approach to the analysis of text-enriched information networks. Sections 5 and 6 present the applications of the described methodology in two real-life domains: a network of video lectures and their authors and a citation network of psychology papers, respectively. The chapter concludes with a summary and opportunities for further work.

2 Information Networks

This section introduces the area of *information network analysis*, illustrated with some real-world examples of information networks.

Standard data sets used in data mining and machine learning are usually available in a tabular form, where a data instance (corresponding to a row in the data table) is characterized by its properties described in terms of the values of a selected set of attributes (each corresponding to a table column). In contrast, the motivation for information network mining is due to the fact that information may exists both at the instance level and in the way how the instances interact.

Intuitively, an information network is a network composed of entities (for example, web pages) that are in some way connected to other entities (one page may contain links to other pages). In mathematical terms, such structures are represented by graphs.

Definition 1 A *graph* $G = (V, E)$ is a mathematical object, composed of a set of vertices V and a set of edges E connecting the vertices. Set of edges E is the union of two sets, $E = E_U \cup E_D$, where set E_U contains undirected edges $\{x, y\}$ and set E_D contains directed edges (x, y) between pairs of vertices x, y.

- If all edges present in E are undirected, we call the graph *undirected*. If all edges are directed, the graph is *directed*. Graphs containing both directed and undirected edges are sometimes referred to as *mixed* graphs.
- A graph with no loops (edges connecting a node to itself) and no multiple edges (meaning that a pair of nodes is connected by at most one edge) is called a *simple* graph.

Graphs are a convenient way to represent relations between different entities, but do not contain any real data themselves. An *information network* is a graph in which each vertex has certain properties. Networks are a richer way of representing data than using either graphs or tables, but can lack the power to represent truly complex interactions between entities of different types. To this end, we define the concept of a heterogeneous information network.

Definition 2 A *heterogeneous information network* is a tuple $(V, E, \mathcal{A}, \mathcal{E}, \tau, \phi)$, where $G = (V, E)$ is a directed graph, $\mathcal{A}$ a set of object types, $\mathcal{R}$ a set of edge types and $\tau : V \to \mathcal{A}$ and $\phi : E \to \mathcal{R}$ are functions satisfying the conditions: if edges $e_1 = (x_1, y_1)$ and $e_2 = (x_2, y_2)$ belong to the same edge type ($\phi(e_1) = \phi(e_2)$), then their start points and their end points belong to the same vertex type ($\tau(x_1) = \tau(x_2)$ and $\tau(y_1) = \tau(y_2)$).

Remark 1 In many information networks, vertices of two types a_1 and a_2 are only connected by edges of one type. In this case, the edge type is uniquely defined by the type of its starting and ending vertex type and is not explicitly stated. It is common to view the elements of the set $\mathcal{A}$ as disjoint sets of vertices from V, instead of abstract types. This gives rise to the style of writing, where for type $t_1 \in \mathcal{A}$, and vertex $v \in V$, we write $v \in t_1$ instead of the usual $\tau(v) = t_1$ to denote the fact that v is of type t_1.

Remark 2 A heterogeneous information network may also be represented in a relational, RDF-like form as a set of triplets. Edge types, in this representation, would be represented as relations. In this representation, network schemas of heterogeneous information networks correspond to RDF schemas. The constraints put on edge types in heterogeneous information networks (i.e. that all edges of a certain type start in nodes of the same type) can be encoded using `rdfs:domain` and `rdfs:range` properties.

Sun and Han [34] note that sets $\mathcal{A}$ and $\mathcal{E}$, along with the restrictions imposed by the definition of a heterogeneous information network, can be seen as a network as well, with edges connecting two vertex types if there exists an edge type whose edges connect vertices of the two vertex types. The authors call this 'meta-level' description of a network a *network schema*.

Definition 3 For a heterogeneous information network $G = (V, E, \mathcal{A}, \mathcal{E}, \tau, \phi)$, a *network schema of* G, denoted T_G, is a directed graph with vertices $\mathcal{A}$ and edges $\overline{\mathcal{E}}$, where edge type $t \in \mathcal{E}$ whose edges connect vertices of type $t_1 \in \mathcal{A}$ to vertices of type $t_2 \in \mathcal{A}$, defines an edge in $\overline{\mathcal{E}}$ from type t_1 to t_2.

Given such a broad definition of heterogeneous information networks, a large amount of human knowledge can be expressed in the form of networks. Some examples are listed below.

Example 1 Bibliographic information networks or *citation networks*, such as the DBLP network examined by [34] or the network examined by [13] are networks connecting authors of scientific papers with their papers. Thus, in their elementary form, they contain at least two types of entities (authors (A) and papers (P)), and at least one type of edges, connecting authors to the papers they have (co)authored. On top of this, the network may also include several other entity types, including journals and conferences (which can be merged into one type, *venue*), institutions and so on. Along with the entity types, the list of edge types is also expanded: papers are, for example, written by authors, published at venues and contain certain terms. Papers may cite other papers, meaning that a paper in the network can be connected to entities of all other entity types in the network.

Example 2 Online social networks model the structure of popular online social platforms such as Twitter and Facebook. In the case of Twitter the network entity types are *user, tweet, hashtag*, and *term*. The connections between the types are as follows: users *follow* other users and *post* tweets, tweets *reply* to other tweets and *contain* both terms and hashtags.

Example 3 Biological networks represent a starting point for a large number of different heterogeneous information networks and can contain entity types such as species, genes, Gene Ontology [9] annotations, proteins, metabolites and so on. The types of links between such mixed entities are diverse. For example, genes can belong to species, encode proteins, be annotated by a GO annotation, and so on.

3 Analysis of Information Networks

We present some analytic tasks which can be applied to information networks. First, we present general tasks that can be applied to homogeneous networks, followed by approaches to mining heterogeneous networks.

3.1 Tasks in Homogeneous Information Network Analysis

The field of information network analysis covers a wide variety of tasks. Some of them are listed below.

3.1.1 Classification

Classification of network data is a natural generalization of classification tasks encountered in a typical machine learning setting. The problem formulation is simple: given a network and class labels for some of the entities in the network, predict the class labels for the rest of the entities in the network. Another name for this problem is *label propagation*. A common approach used for this task is the algorithm proposed by [43]. The approach finds a probability distribution f of a vertex v_i being labeled with label 1 (as opposed to 0). The classification problem in this case is a binary classification problem. The method finds f by minimizing the function

$$f^T(I - D^{-1/2}MD^{-1/2})f + \mu||f - y||^2, \tag{1}$$

where M is the adjacency (or weight) matrix of the network and D is a diagonal matrix defined as $d_{ii} = \deg(v_i) = \sum_j m_{ij}$. The two summands in Equation (1) represent two demands that have to be fulfilled: (*i*) the label distribution for vertices which are connected (especially the ones connected with strong edges) must be similar, and (*ii*) the distribution must be close to the original distribution of the data already labeled. Parameter μ determines the strength of two influences on the result: a large value of μ results in a labeling that closely matches the known labels while a small value of μ strongly penalizes connections between differently labeled vertices. This approach was tried by [38], where the method was used to discover new genes, associated with a disease.

3.1.2 Link Prediction

While classification tasks try to discover new knowledge about network entities, link prediction focuses on unknown connections between the entities. The assumption is that not all network edges are known. The task of link prediction is to predict new edges that are missing or likely to appear in the future. A common approach to link prediction is assigning a score $s(u, v)$ to each pair of vertices u and v which models a probability of the vertices being connected. Approaches used include calculating the score as a product of vertex degrees [2, 27], or using the number of common neighbors of two vertices, $|N_u \cap N_v|$ [28]. The latter approach can be modified to the Jaccard coefficient of N_u and N_v, which is defined as $\frac{|N_u \cap N_v|}{|N_u \cup N_v|}$ with values in $[0,1]$. This normalization prevents high degree vertices to overshadow low degree vertex pairs, which may have a large share of common neighbors. The Adamic/Adar measure, used in [1], further increases the impact of low degree vertices by calculating the distance as follows: $\sum_{n \in N_u \cap N_v} \frac{1}{\log(|N_n|)}$.

3.1.3 Community Detection

There is a general consensus on what a network community is, however, there is no strict definition of the term. The idea is well summarized in the definition by [41]: a community is a group of network nodes, with dense links within the group and sparse links between the groups. An extensive overview of community detection methods is presented in [31].

3.1.4 Ranking

The objective of ranking in information networks is to asses the relevance of a given object either globally (with regard to the whole graph) or locally (relative to some object in the graph). A well known ranking method is PageRank [30], which was used in the Google search engine. The idea of PageRank—frequently abbreviated as PR—is simple: for a given network with the adjacency matrix M, the score of the ith vertex is equal to the ith component of the dominant eigenvector of M'^{T}, where M' is the matrix M with rows normalized so that they sum to 1. This is motivated by two different views.

The first is the random walker approach: a random walker starts walking from a random vertex v of the network and in each step walks to one of the neighboring vertices with a probability proportional to the weight of the edge traversed. The PageRank of a vertex is then the expected proportion of time the walker spends in the vertex, or, equivalently, the probability that the walker is in the particular vertex after a long time. The second view of PageRank is the view of score propagation. The PageRank of a vertex is its score, which it passes to the neighboring vertices. A vertex v_i with a score $PR(i)$ transfers its score to all its neighbors. Each neighbor receives a share of the score proportional to the strength of the edge between it and v_i. This view explains the PageRank with a principle that in order for a vertex to be highly ranked, it must be pointed to by many highly ranked vertices.

Other methods for ranking include Personalized-PageRank [30]—frequently abbreviated as $P\text{-}PR$ that calculates the vertex score locally to a given network vertex, SimRank [16], diffusion kernels [23], hubs and authorities [21] and spreading activation [10].

3.2 Tasks in Heterogeneous Information Network Analysis

Most data mining tasks in homogeneous information networks can be applied to heterogeneous networks by simply ignoring the heterogeneous structure. This, however, decreases the amount of information available in subsequent steps and can therefore decrease the performance of algorithms [11]. Approaches that take the heterogeneous network structure into account are therefore preferable.

3.2.1 Authority Ranking

Sun and Han [34] introduce *authority ranking* to rank the vertices of a bipartite network, where vertices are comprised of a set of authors $X = \{x_1, \ldots, x_m\}$ and a set of papers $Y = \{y_1, \ldots, y_n\}$. There are two edge types: links from papers to authors and links from authors to papers. The adjacency matrix of the network can therefore be written as

$$M = \begin{bmatrix} 0 & M_{XY} \\ M_{YX} & 0 \end{bmatrix}$$

where M_{YX} contains weights of edges pointing from authors to papers and M_{XY} contains weights of edges pointing from papers to authors.

The concept of authority ranking is a generalization of PageRank for bipartite networks, defining two functions r_X (ranking the set X) and r_Y (ranking the set Y) to rank papers and authors separately. The functions are defined as follows:

$$r_X(x_i) = \sum_{j=1}^{n} r_{ij}^{XY} r_Y(x_j) \tag{2}$$

$$r_Y(y_j) = \sum_{i=1}^{m} r_{ji}^{YX}(j) r_X(y_i) \tag{3}$$

where r_{ij}^{XY} is the weight of the edge between vertices i and j. The weights the matrix R, obtained from the matrix M by normalizing the row sums to 1, as in the PageRank approach. The above equations can be rewritten as an eigenproblem for a block matrix, since vectors r_X and r_Y satisfy $r_X = R_{XY} r_Y$ and $r_Y = R_{YX} r_X$ or, in matrix form:

$$\begin{bmatrix} r_X \\ r_Y \end{bmatrix} = \begin{bmatrix} 0 & R_{XY} \\ R_{YX} & 0 \end{bmatrix} \begin{bmatrix} r_X \\ r_y \end{bmatrix}.$$

Similarly, [36] define authority ranking on a star heterogeneous network with a central type Z, where instead of propagating authority directly from a node of type X to a node of type Y, authority is propagated indirectly through a node of type Z, yielding equations $r_X = R_{XY} R_{ZY} r_Y$ for all pairs of types X and Y.

3.2.2 Ranking Based Clustering

While both ranking and clustering can be performed on heterogeneous information networks, applying only one of the two may sometimes lead to results which are not truly informative as there is a high risk of apples-to-pears comparisons being made. For example, simply ranking authors in a bibliographic network may lead to a comparison of scientists in completely different fields of work which may not be comparable. Sun and Han [34] propose joining the two seemingly orthogonal approaches to information network analysis (ranking and clustering) into one. They propose two algorithms: RankClus [35] and NetClus [36], both of which cluster entities of a certain type (for example, authors) into clusters and rank the entities within clusters. Algorithm RankClus is tailored for bipartite information networks, while NetClus can be applied to networks with a star network schema.

The RankClus algorithm starts with a starting clustering of elements, which it then iteratively improves. The ranking of objects within each type is used to define ranking functions $r_{Y|X_i}$, which rank elements of type Y only taking into account elements of type X, belonging to cluster X_i. For the next step, the algorithm considers the ranking $r_{Y|X_i}$ as values proportional to probabilities that objects from Y belong to cluster X_i. This is justified by the fact that when the clustering is discovered, the elements of Y will only have a high rank within the cluster they belong to. Using this view the algorithm constructs a mixture model (using the EM algorithm [3]) to evaluate the probabilities of *links* belonging to each of the clusters. Using this knowledge, new clusters of type X are constructed and the process is repeated until convergence. The NetClus algorithm shares its idea with the RankClus. Instead of applying probabilities to links, as in RankClus, the role of links in NetClus is replaced by objects belonging to the central type in the star network.

3.2.3 Classification Through Label Propagation

The problem of classification is generalized from homogeneous to heterogeneous networks: given a network and class labels for some of the entities in the network, predict the labels of the remaining entities in the network. In [15] the idea of label propagation used by [43] is expanded to include multiple parameters μ_{ij} in place of a single parameter μ appearing in Eq. (1). A similar approach is taken by [34]. Ji et al. [18] propose the GNetMine algorithm which uses the idea of knowledge propagation through a heterogeneous information network to find probability estimates for labels of the unlabeled data. A strong point of this approach is that it has no limitations on the network schema, meaning it can be applied to both highly complex heterogeneous and homogeneous networks.

3.2.4 Ranking Based Classification

Building on the idea of GNetMine, [34] propose a classification algorithm that relies on within-class ranking functions to achieve better classification results. The idea is that nodes, connected to *high ranked* entities belonging to class c, most likely belong to the same class. This idea is implemented in the RankClass framework for classification in heterogeneous information networks.

Ranking and classification in RankClass are interlinked, since only elements *within* each class are ranked rather than the whole set. The methodology consists of two steps which are applied successively until the convergence. In the ranking step, the network elements are ranked according to the authority ranking principle. Then, given the rankings of elements, the EM algorithm calculates new estimates of probabilities that elements belongs to a certain class. Edges connecting elements likely to belong to the same class are increased and within class rankings are recalculated.

3.2.5 Multi-relational Link Prediction

Expanding the ideas of link prediction for homogeneous information networks, [11] propose a link prediction algorithm for each pair of object types in the network. The score is higher if the two objects are likely to be linked. Two objects o_1 and o_2 of types t_1 and t_2 have a high score if there exist many common neighbors of o_1 and o_2, which are neighbors to connected objects of types t_1 and t_2 (for example, if two authors often attend the same conferences, and it is common for authors at a conference to be paper co-authors, it is probable that the two authors are going to become co-authors of a paper).

3.2.6 Semantic Link Association Prediction

Chen et al. [6] constructed a heterogeneous network consisting of 295,897 nodes and 727,997 edges from 17 publicly available data sources about drug target interaction, including semantically annotated knowledge sources in the form of ontologies. The constructed heterogeneous network contains 10 node types and 12 edge types. Two most important node types are target nodes, representing individual genes, and chemical compound nodes. These two node types are connected by two edge types: a chemical compound can bind to a certain target gene or can change the expression of the gene. In addition to these two link types, target nodes are linked to nodes representing Gene Ontology [9] concepts, KEGG [20] pathways, tissues and diseases. Chemical compound nodes are linked to nodes representing chemical ontology concepts, chemical substructures, medical side effects and diseases. The authors developed a statistic model called Semantic Link Association Prediction (SLAP) to measure associations between network elements. Scores are calculated for drug-target pairs for each possible meta path between the two. The scores are normalized for each meta path, with the sum giving an actual association score between the elements. Element pairs with significant scores (smaller p-values) are then discovered.

3.3 Data-Enriched Network Analysis

The methods described in Sects. 3.1 and 3.2 rely solely on the network structure to extract information. However, as an information network includes both the network structure and the data itself, it is sensible to include the data attached to each node into the network analysis process as well.

3.3.1 Network Guided Forest

Dutkowski and Ideker [12] present a method based on decision trees to analyze a protein-protein interaction network. They analyze gene expression data from several studies on human cancers. The data consists of gene expression levels, obtained through microarray experiments and contains a series of expression levels, one for each gene of each sample. The proposed method, named Network-Guided Forests (NGF), constructs a forest of trees to classify an example into the appropriate class according to the expression levels of the examined genes. The final result is obtained through the aggregation of all results. The NGF method is similar to the random forest method [4] as it constructs several decision trees and each decision tree classifies examples according to their gene expression. The difference is that in NGF the construction of trees is guided by the underlying network of protein-protein interactions, which helps to find the best gene to split the data in each tree node. This approach is interesting from a conceptual point of view, as it is composed of both network analysis methods and standard statistical and data mining algorithms. It can be viewed as either mining data enriched with a network component, or analysis of networks enriched with experimental data.

3.3.2 Two-Step Clustering

Hofree et al. [14] also combine network analysis and data mining. They analyze a complex network of gene-gene interactions to analyze cancer patient data. The data consists of a large binary matrix F with values indicating if a given gene is mutated for a patient. They propose a two-step patient clustering method. First, a network propagation approach based on [43] (see Sect. 3.1) is applied to the network which transforms the original binary matrix into a matrix with values in [0,1]. In the second step, authors use non-negative matrix factorization [25] to find candidate features to use in clustering.

3.3.3 Network Extraction Using Text Mining

While human readable documents may contain a lot of information, this information is not conveniently structured for data analysis. As information networks are a better way of representing knowledge, several methods and applications converted databases of scientific articles into large (and usually heterogeneous) information networks. One of the first attempts is described by [17], where a network of human genes is constructed from titles and abstracts of over 10 million MEDLINE records. Kok and Domingos [22] use relational clustering to cluster both vertices and edges and construct a semantic network from the text. In [7], a NLP-based text mining approach called Chilibot was introduced. The methodology can construct networks about biological entities using articles collected from PubMed. [37] used a semantic network extracted from PubMed articles with protein–protein and regulatory inter-actions from experimental databases to discover clusters of tightly connected genes.

4 Mining Text-Enriched Heterogeneous Information Networks

This section introduces the methodology of mining text-enriched information net-works first described by [13]. The methodology uses both text mining and network analysis of text-enriched heterogeneous information networks (such as the citation network of scientific papers) to construct feature vectors which describe both, the location of nodes in the network and internal structure of nodes.

4.1 Data Structure

The data in a text-enriched heterogeneous information network is a fusion of two diffedrent data types: heterogeneous information networks and texts. Our data thus comprises of a heterogeneous information network with different node and edge types, where nodes of one designated type are text documents. An example of a heterogeneous citation network in which the text documents are papers is shown in Fig. 1 and its network schema is presented in Fig. 2.

Remark 3 In a directed heterogeneous network, an edge from vertex v to vertex w (for example, an author *writes* a paper) implicitly defines an 'inverse' edge going from vertex w to vertex v (a paper is *written by* an author).

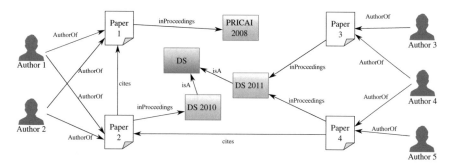

Fig. 1 An example of a citation network (from [13])

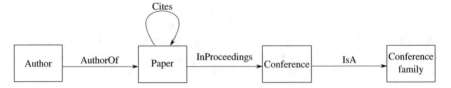

Fig. 2 The network schema of the citation network, shown in Fig. 1

4.2 Network Decomposition

The first step of the methodology focuses on the network structure. The original heterogeneous information network is decomposed into a set of homogeneous networks. Each homogeneous network is constructed from a circular walk in the original network schema. If a sequence of node types $t_1, t_2, \ldots, t_n$ forms a circular walk (meaning that $t_1 = t_n$) in the network schema, then two nodes n and m are connected in the decomposed network if there exists a walk $n_1, n_2, \ldots n_n$ such that $n_1 = n$, $n_n = m$ and each node n_i in the walk is of type t_i.

Take for example the network shown in Fig. 1. From it (using the implicitly defined inverse edges, described in Remark 3), we construct three homogeneous networks of papers, shown in Fig. 3:

- The first network (Fig. 3a) is constructed using the walk Paper *HasAuthor* Author *AuthorOf* Paper, i.e., two papers are connected if they share a common author.
- The second network (Fig. 3b) is constructed using the walk Paper *inProceedings_of* Conference *isA* Conference family *hasConference* Conference *containsPaper* Paper and two papers are connected if they appeared at the same conference family.
- The third network (Fig. 3c) is constructed using the walk Paper *cites* Paper and two papers are connected if one paper cites another.

This step of the methodology is the only one which cannot be made fully automatic. For each heterogeneous network, different meta-paths can be considered, and

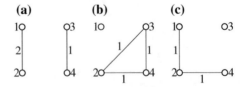

Fig. 3 The decomposition of the network from Fig. 1 according to the **a** paper-author-paper, **b** paper-conference family-paper and **c** paper-paper meta-paths. The nodes in the decompositions correspond to the papers in the original network. The weights, assigned to the edges in this example, are obtained by simply counting the number of paths in the original network which correspond to a link in the decomposed network. The weight in the Paper-Author-Paper decomposition corresponds to the number of authors, shared by two papers

expert judgment is required to asses their importance. Usually meta-paths of heterogeneous networks have a real-world meaning, so field experts may provide insights into paths importance.

4.3 Feature Vector Construction

In the second step of the methodology, a set of feature vectors is calculated for each node in the original heterogeneous network: a bag-of-words vector constructed from the text document, and feature vectors constructed from all homogeneous networks.

In the bag-of-words (BOW) construction, each text is processed using standard natural language processing techniques. Typically the following steps are performed: preprocessing using a tokenizer, stop-word removal, stemming, construction of N-grams of certain length, and removal of infrequent words from the vocabulary. The resulting vectors are normalized according to the Euclidean metric.

For each homogeneous networks, obtained through network decomposition, the personalized PageRank $P\text{-}PR$ algorithm [30] is used to construct feature vectors for each node in the network.

Personalized page rank of node v $P\text{-}PR_v$ in a network is a vector, which for each other node w of the network, tells how simple it is to randomly walk from v to w. It is defined as the stationary distribution of the position of a random walker which starts its walk in node v and at either selects one of the outgoing connections or travels to his starting location. The probability (denoted p) of continuing the walk is a parameter of the algorithm and is usually set to 0.85. The resulting PageRank vectors are normalized according to the Euclidean norm to make them compatible with the BOW vector calculated for the same document.

The PageRank vector is calculated iteratively. In the first step, the rank of node v is set to 1 and the other ranks are set to 0 to construct r^0, the 0th estimation of the PageRank vector. Then, at each step, the rank is spread along the connections of the network using the formula

$$r^{(k+1)} = p(A^T r^{(k)}) + (1 - p)r^{(0)}. \tag{4}$$

In Eq. (4), $r^{(k)}$ is the estimation of the PageRank vector after k iterations, and A is the coincidence matrix of the network, normalized so that the elements in each row sum to 1. If all elements in a given row of the coincidence matrix are zero (i.e., if a vertex has no outgoing connections), all values in that row are set to $\frac{1}{n}$, where n is the number of vertices (this simulates the behaviour of the walker when jumping from a node with no outgoing connections to any other node in the network).

Remark 4 Continuing from Remark 2, if heterogeneous information networks are viewed as RDF-graphs, we can consider the feature vector construction as a further enrichment of the RDF-graph. Bag-of-words vectors can be represented as a set of triplets using the hasTerm relation, as seen in [26], and $P\text{-}PR$ vectors may be represented as a set of triplets using a `distanceFrom` relation with a numeric property. In this way, the PageRank vector for node v would be encoded in all relations of the type `distanceFrom` which start in v.

4.4 Data Fusion

The result of running both the text mining procedure and the $P\text{-}PR$ algorithm is a set of vectors $\{v_0, v_1, \ldots, v_n\}$ for each node v, where v_0 is the BOW vector, and where for each i ($1 \leq i \leq n$, where n is the number of network decompositions), v_i is the personalized PageRank vector of node v in the ith homogeneous network. In the final step of the methodology, these vectors are combined to create a final feature vector. Using positive weights $\alpha_0, \alpha_1, \ldots, \alpha_n$, which sum to 1, a unified vector is constructed describing the node v. The vector is constructed as

$$v = \sqrt{\alpha_0}b \oplus \sqrt{\alpha_1}v_1 \oplus \cdots \oplus \sqrt{\alpha_n}v_n,$$

where the symbol $\oplus$ represents the concatenation of two vectors. The values of weights α_i can be determined automatically.

A simple way to automatically set weights is to use an optimization algorithm such as the multiple kernel learning (MKL), presented in [32], in which the feature vectors are viewed as linear kernels. For each i, the vector v_i corresponds to a linear mapping $\overline{v_i} : x \mapsto x \cdot v_i$. The concatenated vector v then represents the linear mapping

$$[x_0, x_1, \ldots, x_n] \mapsto \alpha_0 x_0 \cdot v_0 + \alpha_1 x_1 \cdot v_1 + \cdots + \alpha_n \cdot v_n.$$

Another possibility is to determine the optimal weights using a general purpose optimization algorithm, e.g., differential evolution [33].

4.5 Scalability Issues

While the calculation of bag-of-words vectors can be done in a single pass over the data, the calculation of $P\text{-}PR$ vectors has to be adapted when the number of basic nodes becomes too large. The iterative process converges to a stationary distribution of the rank after several steps. In our experiments, the required number of steps ranged from 50 to 100, and since each step requires a matrix-vector multiplication, the calculation of a single $P\text{-}PR$ vector may take several seconds for a large network, making the calculation of tens of thousands of $P\text{-}PR$ vectors computationally difficult. Here, we present some ideas to handle the rising computational complexity of large networks.

To reduce the size of the network on which PageRank vectors are calculated we calculated $P\text{-}PR_v$ by performing the PageRank algorithm on a subnetwork of the original network, composed of nodes that have a path leading from v to them in the original network [24]. The P-PR value for all other nodes is set to 0. We can also limit the size of the graph on which the P-PR method is applied by calculating only PageRank values of nodes in a local neighborhood of a given node, setting PageRank values for nodes that are too far from the start node to 0. This in some cases decreases the computation time, but the decrease will not occur in many real world networks, especially *small world networks* [40], in which the shortest path between any two nodes may be very short.

Alternatively, a community detection method [31] can be used as a preprocessing step in the calculation of P-PR vectors. Once the communities in the graph are discovered, one can calculate P-PR_v by only calculating its values on a subgraph containing all the nodes of the same community as v and links between them. We can treat the remaining communities as non-existent by setting the PageRank value of their nodes to 0, or treat them as a single entity by replacing the entire community with one node v. For a node w, the weight of the edge between v and w can be calculated as the sum, average, or maximum of all weights leading from v to the community.

5 VideoLectures.NET Categorization Case Study

The network propositionalization approach, described in Sect. 4, was applied to a network of 3,520 lectures from the VideoLectures.NET website. The aim of the experiment was to develop a method that can assist in categorization of lectures, hosted on the site. This functionality was required due to the rapid growth of the number of hosted lectures (150–200 lectures are added each month) as well as due to the fact that the categorization taxonomy is fine-grained, making manual categorization difficult.

5.1 Data Set

Of the 3,520 lectures 1,156 lectures were manually categorized into 129 categories (one lecture may belong to more than one category) by the curators of the website. The data included 2,706 lecture authors, events at which the lectures were filmed and 62,070 user clicks. From this data we constructed a heterogeneous network containing lectures, authors, events and portal users as nodes.

Each lecture contained a title and possibly an abstract which were used to create the BOW vector for each lecture. The heterogeneous network was decomposed into three homogeneous networks: the *lecture-event-lecture* network, the *lecture-author-lecture* and the *lecture-viewer-lecture* network, in which links between two lectures were weighed in proportion to the number of viewers that viewed both lectures.

5.2   Experiment Description

In the first set of experiments, a pure text mining approach was used to classify the lectures. The lectures were processed using a standard text mining approach using both TF and TF-IDF weighing. The n-gram length, the minimum term frequency and the cut-off percentage were varied to provide several benchmark performance measures. For each parameter setting the centroid classifier was used on the resulting vectors to predict the categories of individual video lecture.

In the second set of experiments, the vectors obtained through text mining were used to train two classifiers: the k-nearest neighbors classifier and the SVM classifier. For the k-NN classifier, k was set to 20, and for SVM, the SVM-Multiclass [19] was used with the termination criterion set to 0.1 and the trade-off between error and margin set to 5,000. In addition to the text mining vectors, the SVM and k-NN classifiers were also applied to diffusion kernels (DK) [23] calculated on the three homogeneous graphs.

The third set of experiments used the methodology proposed in Sect. 4. The method was deployed on each of the three homogeneous graphs from Sect. 5.1. For each homogeneous graph, the three classifiers from the first two sets (the centroid classifier, the SVM classifier and the k-NN classifier) were applied to the resulting feature vectors. Next the feature vectors were combined as described in Sect. 4.4. The feature vectors were combined (a) using equal weights for all feature vectors, or (b) using a stochastic optimizer called differential evolution (DE) [33].

5.3   Evaluation and Results

In the experiments described in Sect. 5.2 the performance of classifiers was evaluated by matching predictions to the pre-categorized classes. Classification accuracy was measured on the top 1, 3, 5 and 10 categories, proposed by the classifier. For each experiment, a 10-fold cross validation was performed. The results are given in Table 1.

The results of the first set of experiments show that using a TF-IDF weighing improves the accuracy of the centroid classifier compared to using TF weights. Varying the minimum frequency, n-gram length and cut-off values resulted in smaller improvements to the performance. The most efficient setting was using $2-$grams and the minimum term frequency of 1, so this setting was used in all BOW constructions in the successive experiments.

The results of the second set of experiments show that the text mining approach performs relatively well and outperforms both the classifier based on the same-event network and the classifier based on the same-author network. The same-author graph contains the least relevant information for the categorization task. The most relevant information is contained in the viewed-together graph. It is noteworthy that the choice of the classification algorithm is less important than the data source from which the similarities between objects are inferred.

Table 1 Accuracies of the algorithms when classifying video lectures

Setting	Top 1	Top 3	Top 5	Top 10
First set (text mining)	Acc. (%)			
TF, $n = 1$, min-freq $= 1$, cut-off $= 0$	53.97	69.46	74.48	81.74
TF-IDF, $n = 1$, min-freq $= 1$, cut-off $= 0$	58.99	75.34	79.50	85.55
TF-IDF, $n = 2$, min-freq $= 1$, cut-off $= 0$	59.60	75.34	80.27	85.20
TF-IDF, $n = 3$, min-freq $= 1$, cut-off $= 0$	59.42	75.77	80.10	85.20
TF-IDF, $n = 2$, min-freq $= 2$, cut-off $= 0$	59.51	76.21	80.79	85.46
TF-IDF, $n = 2$, min-freq $= 3$, cut-off $= 0$	58.13	75.86	80.62	85.20
TF-IDF, $n = 2$, min-freq $= 2$, cut-off $= 0.1$	58.99	75.34	79.15	84.25
Second set (Text mining + DK)	Accuracy (%)			
Text mining + SVM	59.16	73.09	78.28	82.96
Text mining + k-NN	58.47	72.74	78.28	83.91
Text mining + centroid	59.51	76.21	80.79	85.46
DK on viewed-together + SVM	70.75	86.93	90.92	93.68
DK on viewed-together + k-NN	72.74	87.80	90.83	93.94
DK on same-event + SVM	32.00	49.04	54.67	58.65
DK on same-event + k-NN	31.92	47.66	53.37	61.07
DK on same-author + SVM	18.94	27.51	31.22	36.24
DK on same-author + k-NN	19.81	31.74	36.24	43.59
Third set (enriched networks)	Accuracy (%)			
viewed-together + SVM	70.41	85.46	89.71	93.60
viewed-together + k-NN	70.75	84.60	89.36	93.34
viewed-together + centroid	74.91	89.01	92.13	95.33
same-evend + SVM	31.74	50.17	55.97	59.95
same-evend + k-NN	32.34	50.43	55.96	64.79
same-evend + centroid	27.59	46.62	53.63	65.05
same-author + SVM	15.83	24.22	27.33	33.04
same-author + k-NN	15.48	23.70	27.94	32.52
same-author + centroid	14.79	25.52	31.74	42.73
combined-equal weights + centroid	65.73	83.21	87.97	93.42
combined-DE calculated weights + centroid	78.11	91.43	94.03	95.85

The results of the third set of experiments showcase the performance of the methodology presented in this section. Just as in the second set of experiments, the results show that the choice of the classification algorithm results in only minor changes in the classification accuracy compared to the choice of the network decomposition method. The final two rows of the results show that setting equal weights to all feature vectors is far from optimal, as it decreases the accuracy to *below* that of the best individual feature vector. Using differential evolution, on the other hand, improves the performance, as this classifier, using optimized weights and all feature vectors, consistently outperforms other classifiers.

6 Psychology Publications Categorization Case Study

We also applied the methodology, presented in Sect. 4, on almost one million scientific publications from the field of psychology. Like the video lectures, the publications belonged to at least one category from a large set of possible categories. The size of the constructed network allowed us to measure how classifier performance increases as we increase the size of the network on which it is trained. Our motivation was to construct a classifier capable of predicting all categories of a publication with more probable categories listed first. Such a classifier may be used to assist in the manual classification of new psychology articles.

6.1 Data Collection

The first step in the construction of a network is data collection. Because there is no central database containing publications in the field of psychology, we decided to crawl the Wikipedia pages connected with psychology.

We collected the information about psychology publications from the reference section of the articles connected to the category Psychology on English Wikipedia. Due to citation formatting inconsistencies, we extracted only the references containing their DOI (Digital Object Identifier).

We examined the hierarchical tree of Wikipedia categories, belonging to the category Psychology. Categories in lower levels of the hierarchy reveal articles that are connected to psychology, but are also strongly connected to other disciplines. Examples include pages from the categories Religion, Evolution, Biology, etc. We decided to stop our collection at level 5. The decision was based on the difference between the number of visited categories and the number of yet uncollected articles at depths 4, 5 and 6. Our final collection therefore includes all Wikipedia subcategories and pages reached from the top level Psychology category in 5 or less steps.

Once we collected the set of DOIs connected with psychology on Wikipedia, we needed a suitable free citation tool that includes academic publications from the field of psychology, allows a crawling script and offers DOI search. Microsoft Academic Search (MAS) satisfied these conditions and was selected as our citation tool. We queried the MAS for each of the collected DOIs. If a publication was found on MAS, we collected the information about the title, authors, year of publication, the journal, ID of the publication, IDs of the authors, etc. Whenever possible we also extracted the publication's abstract. Additionally, we collected the same information for all the publications that cite the queried publications.

6.2 Data Set

The result of our data collection process was a network consisting of 953,628 publications of which 63,862 'core publications' were obtained directly from Wikipedia pages. Other publications were cited by the core publications. The publications were linked by 1,539,563 citation links and had 1,589,144 authors. The core publications are labeled with the Wikipedia page referencing it. The remaining publications are labeled with the labels of the core papers citing them. Each publication may be labeled by several different articles. The publications were linked by 1,539,563 citation links and had 1,589,144 authors. We collected 93,977 abstracts of the publications, of which 4551 belong to the core publications.

The goal of our experiment was to examine the accuracy of a classifier predicting the labels of publications. To do that, we first decreased the number of labels. Originally, the publications were labeled with Wikipedia pages, resulting in 71,606 different labels. The Wikipedia pages were replaced with the Wikipedia categories listing them, however, this still left us with 3,173 labels, of which many, especially the categories visited in the final step of crawling, were rather obscure. Because of this, we decided to only allow the categories visited at levels 0, 1 and 2 to represent labels of publications. The categories at level 3, 4, 5 and 6 were transformed into publication labels by climbing up the category hierarchy to the level 2 categories that link to them. The result is a data set in which every publication is labeled with one or more Wikipedia categories it is associated to.

The heterogeneous network was decomposed into three homogeneous networks: the *paper-author-paper* (PAP) network, the *paper-cites-paper* (PP) network and a symmetric copy of the PP network in which directed edges are replaced by undirected edges (PPS).

Remark 5 It is not fair to use all homogeneous networks for the prediction of publication categories. Because the non-core publications were labeled with labels of core publications that cited them, using the citation graphs (the PP and the PPS graph) would yield too optimistic error estimation, because it would use the very structure that was used to label the publications.

Remark 6 We use both the directed and undirected citation network because both contain information about the publications, but may have very different effects on the PageRank calculation. In the directed network, a publication will share its rank with all publications it is citing, while in the undirected case, it will also share its rank with publications it is cited by. Because the resulting vectors may contain different information about the publication, we decided to calculate both and evaluate their performance.

6.3 Experiment Description

The settings used to obtain feature vectors is the same as in Sect. 5. As in [13], n-grams of size up to 2 and a minimum term frequency of 0 was used to calculate the BOW vectors. For the calculation of P-PR vectors the damping factor was set to 0.85, as this is the standard setting also used in [30]. Where more than one feature vector was calculated, the vectors were concatenated using weights optimized using the differential evolution optimization algorithm [33]. In all experiments the calculated featured vectors were used with a centroid classifier using the cosine similarity distance. This classifier first calculates the centroid vectors of each class (or category) by summing and normalizing vectors belonging to indices of that class. For a new instance with feature vector w, it calculates the cosine similarity distance

$$d(c_i, w) = 1 - c_i \cdot w,$$

which represents the proximity of the instance to class i. We classify the instance into the class for which the distance is the smallest. We also examine the 'top n' classifier, where the classifier predicts that the instance belongs to one of the n classes with the smallest distances. Just like in [13], we consider a classifier successful if it correctly predicts at least one class with which the instance is labeled.

We used the centroid classifier for two reasons. First, the experiments presented in Sect. 5, show that it performs just as well as the SVM and the k-nearest neighbor classifier, and second, because for large networks, calculating all P-PR vectors is computationally too demanding. As shown in [13], the centroids of classes can be calculated in one iteration of the PageRank algorithm.

In the first set of experiments we use the publications for which an abstracts were available. Because most of the 93,977 selected publications are not core publications, we construct only two feature vectors for each publication: a bag-of-words (BOW) vector and a P-PR vector obtained from the PAP network. We examine how the predictive power of the classifier increases as we use more publications. We used 10,000, 20,000, 30,000, 40,000, 50,000, 70,000 and 93,977 publications.

In the second set of experiments we use only the core publications for which abstracts are available. While this is the smallest data set, it allows us to use all feature vectors the methodology provides: the BOW vectors and the P-PR obtained from all three networks (PP, PPS and PAP).

In the third set of experiments all collected papers are used. Because the papers were labeled using citations, the PP and PPS networks are not used. Since abstracts are not available for most of the papers, only the P-PR vectors obtained from the PAP network are used in the classification.

6.4 Evaluation and Results

With each of the experiments described in Sect. 6.3 we predict the labels of publications. Classification accuracy is measured on the top 1, 3, 5 and 10 labels, proposed by the classifier. For each experiment the data set is split into a training, validation, and test set. Centroids of classes are calculated using the training set and concatenated according to the weights optimized using the validation set. The accuracy of the algorithm (the percentage of papers for which the label is correctly predicted) is estimated using the test set. The results are given in Table 2 and Fig. 4.

The results of the first set of experiments are shown in Fig. 4. The performance of the classifier using BOW vectors does not increase with more data, while the classifier using PAP vectors is steadily improving as more and more publications are added. The classifier using both BOW and PAP vectors consistently outperforms both individual classifiers, showing the utility of combining structural information of the network and the content of the publications. As the performance of the PAP classifier increases, the gap between the BOW classifier and the classifier using both

Table 2 Accuracies of the algorithms classifying publications from the field of psychology

Setting	Top 1	Top 3	Top 5	Top 10
First set	Accuracy (%)			
BOW + PAP	55.5	75.8	85.6	93.5
PAP	35.6	53.7	66.0	78.3
BOW	49.9	72.6	82.8	92.0
Second set	Accuracy (%)			
All	78.6	92.4	94.1	97.4
all but BOW	47.7	62.2	71.7	83.0
all but PAP	45.4	57.9	60.4	96.9
all but PP	44.7	74.3	81.7	93.0
all but PPS	59.4	75.9	80.7	94.4
BOW + PAP	78.7	93.0	95.4	97.5
BOW + PP	79.8	93.0	95.5	97.4
BOW + PPS	79.6	93.0	95.5	97.5
PAP + PP	44.5	58.9	69.4	82.3
PAP + PPS	47.5	61.9	70.6	82.0
PP + PPS	44.4	58.4	68.3	78.9
BOW	78.3	92.9	95.6	97.5
PP	40.7	56.9	67.1	77.7
PPS	44.9	59.6	67.9	80.8
PAP	27.5	45.4	58.2	74.7
Third set	Accuracy (%)			
PAP	38.8	59.3	71.0	81.4

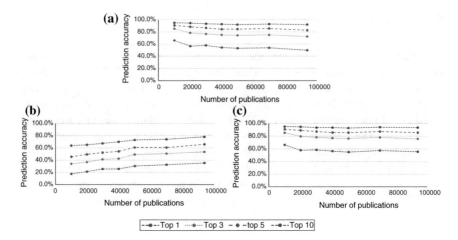

Fig. 4 The classification accuracy of classifiers using different amounts of publications to predict labels. **a** The centroid classifier using BOW. **b** The centroid classifier using PAP. **c** The centroid classifier using both BOW and PAP

vectors also increases. The accuracies obtained with all 93,977 publications are also shown in the first part of Table 2.

The results obtained in the second set of experiments are shown in the second part of Table 2. Because more information was extracted from the network, this is the most comprehensive overview of the methodology. The results show that using a symmetric citation network (PPS), i.e. spreading the PageRank in both directions of a citation yields better results than using a non-symmetric citation network (PP). Combining both the PP and PPS vectors does not improve the performance of the classifier, which means that the vectors, obtained from the PP network, carry no information that is not already contained in the PPS network. The same is not true for other vectors. The results consistently show that including more vectors into the classification increases the prediction accuracy: using both BOW and PAP is better than simply using BOW, but adding PP increases the performance even further.

The performance of the PAP classifier on the full network (calculated in experiment 3 and shown in the last row of Table 2) is higher than PAP results for all other networks, demonstrating that increasing the network size does help the classification. However, the performance is still lower than that of the BOW classifier on smaller networks. It appears that authors in the field of Psychology are not strictly limited to one field of research, making predictions using co-authorship information difficult.

7 Conclusion and Further Work

While network analysis is an established field of research, analysis of heterogeneous networks is a much newer research area. Methods taking the heterogeneous nature of the networks into account show improved performance, as shown by, e.g., [11]. Some methods like RankClus and others presented in [34] are capable of solving tasks that cannot even be defined on homogeneous information networks (like clustering two disjoint sets of entities). Another important novelty is merging network analysis with the analysis of data, either in the form of text documents or results obtained from various past experiments presented in [12–14].

This chapter presents a methodology for mining text-enriched heterogeneous information networks which combines the information from heterogeneous networks with textual data. Compared to the methods described in Sect. 3, the presented methodology combines aspects of network analysis with aspects of text mining. The methodology is applied to text-enriched heterogeneous networks and does not present an alternative, but rather an expanded way of data analysis, compared to these methods. Thus, many other network analysis techniques, especially those that focus on discovering information about *nodes* in the network, can be modified to use it. The presented methodology is comparable to the methods described in Sect. 3.3, in which data enriched networks are analyzed with methods that consider both the network structure and the data enriching the nodes. However, unlike those methods, our methodology deals with textual information enriching network nodes and thus requires a combination of network analysis and text mining. While there are many applications in which text analysis is combined with network mining, they usually apply text analysis to extract knowledge in the form of a network (see Sect. 3.3) and then apply network analysis methods to further analyze it. Unlike these approaches, the approach presented in this Chapter combines two separate knowledge sources and joins them into a single representation.

We show-case the performance of the methodology on two data sets. The results from the VideoLectures.NET data show that using the methodology increases classification accuracy compared to using only texts or only structural information about the instances. The results from the psychology papers experiment show that the relational information hidden in the network structure is beneficial to classification and that its usefulness increases for larger networks.

In our experiments on publications from the field of psychology, we only used a part of information collected about psychology publications. In future, we plan to examine how to incorporate temporal information into the described methodology. We have already collected the year of publication which allows us to observe the dynamics of categories. This additional information may also be used to improve the classification accuracy.

Our approach uses network enrichment with text data and heterogeneous network decomposition and then combines the produced vectors into a single score. Alternative is to use Cartesian product of multiple vector spaces to form a tensor representation as presented by [29]. The tensor space grows exponentially with the number of dimensions but recently several decompositions have been proposed which allow processing in a big data setting [8, 39]. The suggested decompositions allow multi-relational learning, which is a path we want to test in our future work.

In further work we plan to use a combination of network analysis and data mining on a problem of biological networks enriched with experimental data and texts. An experimental data-enriched heterogeneous network centered around genes can be constructed in which network information will be enriched with papers mentioning the genes.

Acknowledgments The presented work was partially supported by the European Commission through the Human Brain Project (Grant number 604102) and by the Slovenian Research Agency project "Development and applications of new semantic data mining methods in life sciences" (Grant number J2-5478).

References

1. Adamic, L.A., Adar, E.: Friends and neighbors on the web. Soc. Netw. **25**(3), 211–230 (2003)
2. Barabási, A.L., Jeong, H., Néda, Z., Ravasz, E., Schubert, A., Vicsek, T.: Evolution of the social network of scientific collaborations. Phys. A: Stat. Mech. Appl. **311**(3–4), 590–614 (2002)
3. Bilmes, J.: A gentle tutorial of the EM algorithm and its application to parameter estimation for Gaussian mixture and hidden Markov models. Technical Report TR-97-021, ICSI (1997)
4. Breiman, L.: Random forests. Mach. Learn. **45**, 5–32 (2001)
5. Burt, R., Minor, M.: Applied Network Analysis: a Methodological Introduction. Sage Publications
6. Chen, B., Ding, Y., Wild, D.J.: Assessing drug target association using semantic linked data. PLoS Comput. Biol. **8**(7), (2012)
7. Chen, H., Sharp, B.M.: Content-rich biological network constructed by mining pubmed abstracts. BMC Bioinf. **5**, 147 (2004)
8. Cichocki, A.: Era of big data processing: a new approach via tensor networks and tensor decompositions (2014)
9. Consortium. Gene ontology: tool for the unification of biology. The gene ontology consortium. Nat. Genet. **25**(1), 25–29 (2000)
10. Crestani, F.: Application of spreading activation techniques in information retrieval. Artif. Intell. Rev. **11**(6), 453–482 (1997)
11. Davis, D., Lichtenwalter, R., Chawla, N.V.: Multi-relational link prediction in heterogeneous information networks. In: Proceedings of the 2011 International Conference on Advances in Social Networks Analysis and Mining, pp. 281–288 (2011)
12. Dutkowski, J., Ideker, T.: Protein networks as logic functions in development and cancer. PLoS Comput. Biol. **7**(9), (2011)
13. Grcar, M., Trdin, N., and Lavrac, N. A methodology for mining document-enriched heterogeneous information networks. *The Computer Journal*, **56**(3), 321–335 (2013)
14. Hofree, M., Shen, J.P., Carter, H., Gross, A., Ideker, T.: Network-based stratification of tumor mutations. Nat. Meth. **10**(11), 1108–1115 (2013)
15. Hwang, T., Kuang, R.: A heterogeneous label propagation algorithm for disease gene discovery. In: Proceedings of SIAM International Conference on Data Mining, pp. 583–594 (2010)
16. Jeh, G., Widom, J.: SimRank: a measure of structural-context similarity. In: Proceedings of the 8th ACM SIGKDD International Conference on Knowledge Discovery and Data Mining, pp. 538–543 (2002). ACM
17. Jenssen, T.-K., Laegreid, A., Komorowski, J., Hovig, E.: A literature network of human genes for high-throughput analysis of gene expression. Nat. Genet. **28**(1), 21–28 (2001)
18. Ji, M., Sun, Y., Danilevsky, M., Han, J., Gao, J.: Graph regularized transductive classification on heterogeneous information networks. In: Proceedings of the 25th European Conference on Machine Learning and Principles and Practice of Knowledge Discovery in Databases, pp. 570–586 (2010)
19. Joachims, T., Finley, T., Yu, C.-N.J.: Cutting-plane training of structural SVMs. Mach. Learn. **77**(1), 27–59 (2009)
20. Kanehisa, M., Goto, S.: KEGG: Kyoto encyclopedia of genes and genomes. Nucleic Acids Res. **28**(1), 27–30 (2000)

21. Kleinberg, J.M.: Authoritative sources in a hyperlinked environment. J. ACM **46**(5), 604–632 (1999)
22. Kok, S., Domingos, P.: Extracting semantic networks from text via relational clustering. In: Proceedings of the 2008 European Conference on Machine Learning and Knowledge Discovery in Databases—Part I, ECML PKDD '08, pp. 624–639. Springer, Heidelberg (2008)
23. Kondor, R.I., Lafferty, J.D.: Diffusion kernels on graphs and other discrete input spaces. In: Proceedings of the 19th International Conference on Machine Learning, pp. 315–322 (2002)
24. Kralj, J., Valmarska, A., Robnik Šikonja, M., Lavrač, N.: Mining text enriched heterogeneous citation networks. In: Proceedings of the 19th Pacific-Asia Conference on Knowledge Discovery and Data Mining (2015)
25. Lee, D.D., Seung, H.S.: Learning the parts of objects by non-negative matrix factorization. Nature **401**(6755), 788–791 (1999)
26. Lytras, M., Sheth, A.: Progressive Concepts for Semantic Web Evolution: Applications and Developments. IGI Global (2010)
27. Newman, M.: Clustering and preferential attachment in growing networks. Phys. Rev. E **64**(2), 025102 (2001a)
28. Newman, M.E.J.: The structure of scientific collaboration networks. Proc. Natl Acad. Sci. USA **98**(2), 404–409 (2001b)
29. Nickel, M.: Tensor Factorization for Relational Learning. PhD thesis, Ludwig–Maximilians–Universitaet Muenchen (2013)
30. Page, L., Brin, S., Motwani, R., Winograd, T.: The PageRank citation ranking: Bringing Order to the web. Technical report, Stanford InfoLab (1999)
31. Plantie, , M., Crampes, M.: Survey on social community detection. In: Ramzan, N., Zwol, R., Lee, J.-S., Cluver, K., Hua, X.-S. (eds) Social Media Retrieval, Computer Communications and Networks, pp. 65–85. Springer, London (2013)
32. Rakotomamonjy, A., Bach, F., Canu, S., Grandvalet, Y.: SimpleMKL. J. Mach. Learn. Res. **9**, 2491–2521 (2008)
33. Storn, R., Price, K.: Differential evolution; a simple and efficient heuristic for global optimization over continuous spaces. J. Glob. Optim. **11**(4), 341–359 (1997)
34. Sun, Y., Han, J.: Mining Heterogeneous Information Networks: Principles and Methodologies. Morgan and Claypool Publishers (2012)
35. Sun, Y., Han, J., Zhao, P., Yin, Z., Cheng, H., Wu, T.: RankClus: integrating clustering with ranking for heterogeneous information network analysis. In: Proceedings of the International Conference on Extending Data Base Technology, pp. 565–576 (2009a)
36. Sun, Y., Yu, Y., Han, J.: Ranking-based clustering of heterogeneous information networks with star network schema. In: Proceedings of the 15th ACM SIGKDD International Conference on Knowledge Discovery and Data Mining, pp. 797–806 (2009b)
37. Van Landeghem, S., De Bodt, S., Drebert, Z.J., Inze, D., Van de Peer, Y.: The potential of text mining in data integration and network biology for plant research: a case study on arabidopsis. Plant Cell **25**(3), 794–807 (2013)
38. Vanunu, O., Magger, O., Ruppin, E., Shlomi, T., Sharan, R.: Associating genes and protein complexes with disease via network propagation. PLoS Comput. Biol. **6**(1), (2010)
39. Vervliet, N., Debals, O., Sorber, L., De Lathauwer, L.: Breaking the curse of dimensionality using decompositions of incomplete tensors: tensor-based scientific computing in big data analysis. Sign. Process. Mag. IEEE **31**(5), 71–79 (2014)
40. Watts, D.J., Strogatz, S.H.: Collective dynamics of 'small-world' networks. Nature **393**(6684), 440–442 (1998)
41. Yang, B., Liu, D., Liu, J.: Discovering communities from social networks: methodologies and applications. In: Handbook of Social Network Technologies and Applications, pp. 331–346. Springer, Heidelberg (2010)
42. Zachary, W.: An information flow model for conflict and fission in small groups. J. Anthropol. Res. **33**, 452–473 (1977)
43. Zhou, D., Bousquet, O., Lal, T.N., Weston, J., Schölkopf, B.: Learning with local and global consistency. Adv. Neural Inf. Process. Syst. **16**(16), 321–328 (2004)

Implementing Big Data Analytics Projects in Business

Françoise Fogelman-Soulié and Wenhuan Lu

Abstract Big Data analytics present both opportunities and challenges for companies. It is important that, before embarking on a Big Data project, companies understand the value offered by Big Data and the processes needed to extract it. This chapter discusses why companies should progressively increase their data volumes and the process to follow for implementing a Big Data project. We present a variety of architectures, from in-memory servers to *Hadoop*, to handle Big Data. We introduce the concept of *Data Lake* and discuss its benefits for companies and the research still required to fully deploy it. We illustrate some of the points discussed in the chapter through the presentation of various architectures available for running Big Data initiatives, and discuss the expected evolution of hardware and software tools in the near future.

1 Introduction

Big Data has become somewhat of a buzz word in recent years with countless news items and scientific articles appearing in the general or scientific press. In 2007, the IDC (International Data Corporation) analysts [13] reported the explosion of the *Digital Universe*, creating the necessity to use new *measurement units* for big data: the usual Megabyte and Terabyte would soon have to be replaced by Exabyte or Zettabyte (1,000,000 or 1,000,000,000 Terabytes), with the Digital Universe estimated at 161 Exabytes in 2006. This trend was made possible by the development of Internet, digital devices (phones, cameras, sensors in the Internet of Things) and the sharp decrease in prices for storage, computing power, memory, and network bandwidth.

In 2011, Mc Kinsey [26] said that a 40 % growth was to be expected in the amount of data generated each year. In the same report, Mc Kinsey showed that all economic

F. Fogelman-Soulié (✉) · W. Lu
School of Computer Software, Tianjin University, Beiyangyuan Campus, 135 Ya Guan Road, Jinan District, Tianjin 300350, China
e-mail: soulie.fr@gmail.com

W. Lu
e-mail: wenhuan@tju.edu.cn

© Springer International Publishing Switzerland 2016   141
N. Japkowicz and J. Stefanowski (eds.), *Big Data Analysis: New Algorithms for a New Society*, Studies in Big Data 16, DOI 10.1007/978-3-319-26989-4_6

sectors could profit from big data: for example, the US health care sector could expect $300 billion potential annual value and the US retail sector could expect 60 % increase in operating margins. Reaping such additional value requires new tools and talents, and this has created the new field of *data science*.

Following these reports and other publications, companies have been embarking on Big Data initiatives, but finding many daunting issues on their way.

In this chapter, we want to describe, in as simple and pragmatic a way as possible, what the difficulties are for companies wanting to run Big Data projects. In Sect. 2, we define Big Data; in Sect. 3, we describe the various stages in a Big Data project process and illustrate these in an example from credit-card fraud on Internet; in Sect. 4, we show how companies should store their Big Data in a Data Lake if they want to implement many Big Data projects; in Sect. 5, we introduce the various elements in a Big Data platform and some of the most widely used analytics packages.

2 Big Data Value for Companies

In 2001, Doug Laney (from Meta Group, now Gartner) published a report [24] in which he showed how the rise of e-commerce, in particular, was producing an explosion in data volumes resulting in growing data management challenges. He introduced three important dimensions: volume, velocity and variety (which have come to be known since as the *3Vs*) and discussed possible solutions to handle them.

- **Volume:** in e-commerce, at the time, lower costs of e-commerce channels started to allow collecting increasing data volumes while, at the same time, enterprises were realizing that such data represented an asset and thus wanted to keep it. The costs of storage, however, would soon come to offset the marginal data value gain, so Laney recommended *sampling and limiting data collected.*
- **Velocity:** the increased speed of interactions on e-commerce sites produced a growing constraint on the speed at which data should be ingested and analyzed. The proposed solution to this issue was to develop *architectures* with more bandwidth, caches, and lower latency.
- **Variety:** the most challenging problem that was identified was the large variety of heterogeneous data sources, incompatible data formats, non-integrated data structures and inconsistent data semantics. Various solutions were proposed by Laney, including metadata management solutions and indexing techniques. At that time, data warehousing was deployed more and more widely, so that the solution to variety was viewed in that framework.

Since then, Big Data has become a major news item and a big market for industry: according to Vasanth [34], the market is expected to grow to $53 billion by 2017, with hundreds of billions of dollars potential values in many domains according to McKinsey report [26]. This shows that despite the risks and problems identified in 2001, Big Data has somehow emerged as a big value opportunity.

What allowed this is two-fold: on the hardware side, the "attack of the exponentials" as Driscoll [9] calls it, has seen the costs of storage, CPU and bandwidth over the last decades exponentially dropping, and network access exponentially increasing. What was not cost-effective, or at all possible in 2001, now is, so that all data can actually be collected, stored, and accessed, at low cost [20]. On the software side, tools to handle, store, distribute and mine Big Data have also rapidly developed producing a very dense and complete set of tools [32].

Nowadays, almost 15 years after his original report, Doug Laney, now at Gartner, considers that "adoption of Big Data is simply inevitable" [23]. Today, many organizations are embarking on Big Data projects, but they find many questions on their way. In the next section, we discuss a process for the implementation of projects to derive value from big data.

3   The Process for Big Data Projects

3.1   Machine Learning for Value

Even though Big Data involves many different techniques, Machine Learning is the *major key success factor for delivering value*, and this is because Big Data allows producing better models.

To produce a Machine Learning model, one needs to assemble, from a variety of data sources, a dataset containing a set of observations (for example: customers in marketing applications, patients in health problems …) for which a certain number of variables are available (see Fig. 1). This dataset, in a Big Data setting, will be *deep* with applications involving millions of observations, and *wide* with potentially

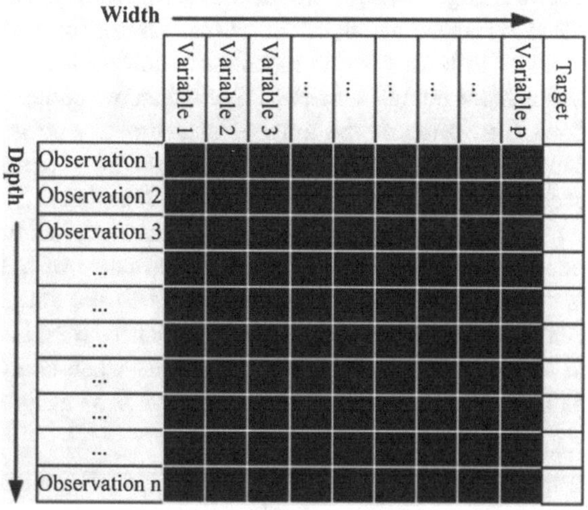

Fig. 1 Dataset for machine learning

Fig. 2 Big Data process

hundreds of thousands of variables (we will indifferently use the words features or variables). The *depth* might be large, for example, if observations are transactions, but not always: for example, if observations are customers of a Bank, the depth will be limited by the total number of customers, which cannot increase *ad libitum*. A large *width* of course comes from the large variety of data sources included in the dataset and we will see in the following section that width can be further increased through various procedures. Finally, the dataset volume can be seen as the product *depth x width*. In many applications, there is a *target* which we want to predict (for example, whether this customer is going to defect next month to a competitor) and the dataset includes the observations' target values.

3.2 The Process

The process for a Big Data project involves a succession of stages (see Fig. 2):

Data collection: first, data needs to be collected from available data sources. The more data there is at this stage, the best it will be for the final model. Increasingly, open data is available, which can be integrated as very useful data sources.[1]

Data cleaning: data needs to be cleaned to improve quality and consistency. At this stage, the various features must be checked for mistakes (for example, misspellings), deduplicated, reconciled and integrated to produce a unique record (a line in the table of Fig. 1) associated to each observation (see [29]).

Feature engineering aims at producing from existing variables additional computed variables, which could be meaningful for the business domain but hard for a model to learn ("hard" meaning: requiring lots of data, large computation time and a more complex model). Such features could be aggregates on sliding windows (for example, the number of claims for an insurance subscriber in the last 6 months, in the last week …), on geographical areas (for example, the number of accidents on a road segment, a town, a region …) or any variable which makes sense for the particular business. Feature engineering is recognized as the most important success factor for the performance of a Machine Learning model [8]. It usually helps producing models, which are simpler, easier and faster to train, while also providing increased performances for a given algorithm as we will show in an example below. In 2007, at a time when analytics Big Data was starting to pick-up, [10] indicated that less than 5 % of analytics projects were using more

---

[1]For example: https://www.data.gouv.fr/en/, http://open-data.europa.eu/en/data/, http://public data.eu/, http://www.data.go.jp/, http://dataportals.org/.

than 1,000 features in their model, while about 50% used less than 40 features. Since then, things have changed a lot. In the various projects we have seen, generating an additional 1,000 features is common, but some projects generate a lot more (a few tens of thousands). Out of these features, about 80% will be standard (time or space aggregates, ratios…), and 20% will be domain-specific. However, since this stage is extremely time-consuming [3], it is interesting to invest in some systematic way to engineer features. Most recent Machine Learning packages are investing in that area (see section Architectures for Big Data below). Note that additional features can also be obtained from outside sources, such as open data sources or private data obtained from partners or data providers. Data from very different sources and semantics will bring more additional value: this is what is reflected in the Variety of data. Of course, increasing the number of features also increases the Variety, and thus the Volume of the dataset.

Modelling: At this stage, we have assembled a dataset with many features (as shown in Fig. 1). To generate a model, we will use one of the many existing Machine Learning algorithms (see for example the book [16]). Choosing from this very large collection of algorithms might seem hard. However, the problem is easier with Big Data: all recent developments in Big Data, [8, 15], have shown that simple models with lots of data are always better than complex models on less data. Hence, one strategy is to choose one relatively simple algorithm, for example logistic regression, and work at increasing data volumes: engineering features is the simple way for that. Simpler algorithms are also easier to explain than more complex ones, so that sometimes one will prefer a simple logistic regression model to a more performant algorithm, such as, for example, random forests because interpretability is much better for the former.

Note that feature engineering produces features which are usually correlated. Hence the algorithm selected should not be sensitive to correlated variables.

Evaluation: Producing a model from data is an iterative process: datasets will be progressively enriched to produce increasingly wider feature sets. Each time, a machine learning algorithm will be trained to produce a model. Usually not all the available dataset will be used, but only a representative sample drawn from it. Then the sample is separated into two parts: the learning sample is used to produce the model and the validation sample (the rest of the sample) is used to validate the model. In the learning phase, one tries to produce the model which best fits the data in the learning sample. In the validation phase, one verifies that the model properly generalizes to new data it has not seen during training: if it does not, it is because it overfits (see Fig. 3).

Producing a model which fits the data is (relatively) easy; producing one which generalizes is much harder. However, generalization is actually what is required if one wants to further use the model [8].

Vapnik's Statistical Learning Theory [33], for example, provides a framework to monitor the learning process so as to achieve generalization. The framework can be summarized as follows:

- Models are restricted to a given family, for example polynomials of degree d, trees of a certain depth, ridge regression, regularized regression etc. The only constraint on the models' family is that it has finite capacity (or Vapnik—Chervonenkis dimension: this is a measure of the complexity of the family).
- The models in the family are explored so as to select the best compromise between learning error and generalization error (errors respectively on the learning and validation samples).
- In the Structural Risk Minimization (SRM) framework, embedded model families are explored, starting with low capacity and progressively increasing capacity, until generalization error starts increasing.

The capacity of the final model is just large enough to produce a good compromise between learning and generalization errors, but not too large, when it would produce overfitting and a larger generalization error (Fig. 3).

In the course of the learning process, the number of features is increased until no further performance increase can be obtained: Fig. 4 shows the typical behavior expected when increasing the number of features. It could take thousands of features before the performance flattening-out appears, depending on the problem. However, not all the features will be retained in the final model: one will try and select the features most significant for the model, sometimes trading a marginal performance increase with a smaller number of features (i.e. a simpler model with lower capacity). In [10], it was reported that while only about 25 % projects used less than 20 features in the original model, about 50 % did in the final version. Explaining a model with more than 20 features is hard and not very intuitive, which is why most final models these days only incorporate a few dozens features.

It is important to note that one set of optimal features at one time might not be optimal anymore later on, when the situation has changed. When retraining a model in later periods, one should thus always start anew from the full set of features.

Fig. 3 Learning and generalization error as capacity increases

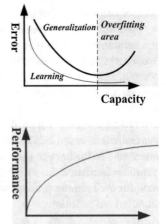

Fig. 4 Performance increases with more features

One should also note that the choice of the optimal features set is not done a priori, or through "expert" knowledge: with big data, it is not reasonable to expect that experts would look at thousands of features and manually evaluate their significance; we need an automatic process for selection. The feature selection process is thus only driven by the data and the algorithm used for producing the model. One should expect different models to produce—usually slightly—different features sets.

A very large number of features can be argued against on the basis of the *curse of dimensionality*: with a fixed-size dataset, the observations occupy an increasingly smaller portion of the input space and the space where the model will apply is increasingly larger when the number of features increases. Generalization thus becomes harder [8]. This issue is lessened in the Big Data situation, since, then, we can increase the size of the sample dataset used for producing the model, while, as always, controlling the capacity of the models' family. However there is generally still a trade-off between the number of additional features we want to use and the increase in sample size necessary for producing the model. At one point, this trade-off will require to stop increasing the number of features, either because the maximum size of the dataset has been reached, or because the performance gain will be too small.

Note that, for many model families, capacity is directly related to the number of features: for example, the family of polynomials of degree 1 in p variables has capacity $p + 1$, while for polynomials of degree 2, capacity is in p^2. So, in that case, increasing capacity through SRM amounts to progressively increasing the number of variables. If one wants to control capacity without restricting the number of features, one would need a family of models for which capacity can be controlled otherwise. For example, the following family F_Θ:

$$F_\Theta = \{f(x, W) = sign\left(\sum_{j=1}^{p}(W_j \times x_j) + 1\right) / ||x|| \leq R, ||W||^2 = \sum_{j=1}^{p} W_j^2 \leq \Theta\}$$

has capacity $C_\Theta : C_\Theta \leq \min(int(R^2\Theta^2); p) + 1$, which can be controlled through a parameter Θ independently of the number of features p.

Thus, in Big Data situations, one can increase the number of features, provided the size of the training sample is large enough and the model capacity is systematically controlled to avoid overfitting.

Deployment: When the final model is ready, with its reduced number of features, it can be deployed to produce predictions on new data. For deployment, one will always prefer simple models, especially when real-time is required. The example of Netflix is famous: after paying 1M $ to the winners of a challenge[2] which required to improve Netflix algorithm's performance by 10%, Netflix finally decided not to put in production the solution delivered to them by the winners: "the additional accuracy gains that we measured did not seem to justify the engineering effort needed to bring

---

[2]http://www.netflixprize.com/.

them into a production environment" [1]. Yet it was 10% additional accuracy! In conclusion, at deployment time, simplicity should always be the goal.

3.3   An Example

Let us present an example from a concrete application to illustrate how feature engineering, especially with different semantics, can increase performance. This was a collaborative project [11] on credit-card fraud detection on Internet, funded by the ANR (the French National Research Agency). Credit card fraud on Internet is a massive, fast-growing phenomenon, in large part at the hands of organized crime. Merchants and banks are faced with the need to implement solutions to detect it as rapidly as possible. In France, GIE Cartes Bancaires is a Group which has been commissioned by French banks to handle the payment process for all transactions made by credit card holders (Visa, MasterCard) from French banks. In 2013 GIE CB managed over 10 billion transactions (up 3.4% compared to 2012), with 584.5 million over the Internet, made by 61.7 million CB cards, totaling 524.3 billion euros [14].

The available transaction data is not very rich: in particular, we do not have information about the card holder (name, address, age, gender ...) or the purchased product (product type, number of products purchased ...), information that only the bank (card data) or merchant (product data) would possess. In contrast, we have all the transactions made by the card-holders on Internet, which would not be the case for a bank (which would only "see" the transactions made by the holders having a card in that bank) or a merchant (who would only "see" the transactions made by the card-holders buying at that merchant). This represented about 50 M transactions per month, made with the card not present (internet or phone). For each transaction we have:

- Information on the card: card number, expiration date, issuing bank ...
- Merchant information: username, SIRET, country, merchant business, merchant bank (acquirer) and country of the bank, terminal used ...
- Transaction information: date of the transaction (local and GMT), amount (in local currency and in euro).

Once a card is blocked, we obtain the blocking date and the reason for the blocking (note that this label can come several months after the fraud occurred). The objective of a model is then to classify a transaction as fraudulent or not fraudulent: this is called *fraud detection*. Transactions classified as fraudulent will be transferred to an investigation team, which will investigate all transactions recently made by the corresponding cards.

From these transactions, we will compute various features, which are aggregates characterizing the history of each card or merchant:

- *Cards Aggregates* at date T: over a sliding window (of various lengths: day, week, month) ending at T, features are computed for each card such as the number and average number of transactions, total and average transaction amount, the difference between the number of transactions at time T, respectively the amount, and the average number of transactions in the window, respectively the average amount etc.
- *Merchants Aggregates* at date T: over a sliding window (of various lengths: day, week, month) ending at T, features are computed for each merchant such as the number and average number of transactions, total and average transaction amount, number of fraudulent transactions and total amount of fraud, difference between the number of fraudulent transactions at time T, respectively the amount, and the average number of fraudulent transactions in the window, respectively the average amount.

We thus obtain 666 aggregates as shown in the first lines of Table 1. We then compute *social features* in the following way:

- We first compute the bipartite network made from cards and merchants nodes, linked when there is a transaction by the card at the merchant (using all the transactions in one month). Through the usual technique to project a bipartite network into two unipartite networks [4, 35], we derive a Cards network and a Merchants network: two cards (resp. merchants) are linked in the Cards (resp. Merchants) network if they have purchased from at least k same merchants (resp. have been visited by the same k' cards), where k and k' are some fixed parameters.
- Then, we compute, for each node (card or merchant), a number of variables in the Cards or Merchants networks, such as the degree of the node, the index and size of its community in the unipartite network (see [4]). For merchants, we also compute the average amount and number of cases of fraud successful in his community/his first circle, the average number of distinct fraudulent cards in his community/his first circle, the average number of transactions accepted/rejected in his community/his first circle etc. This gives 195 *social variables* for cards and 99 for merchants, in addition to the already defined variables (as shown in Table 1).

We now compute three models with an increasing number of features: the baseline model uses the 37 original variables, the second model uses in addition the 666

	Variables	Number
Table 1 Number of features used for the fraud detection model	Original GIE variables	37
	Card aggregates	300
	Merchant aggregates	366
	Card social variables	195
	Merchant social variables	99
	Total	997

Table 2 Performance for fraud detection with an increasing number of variables

Model	Recall (%)	Precision (%)	No variables
Baseline	1.40	8.18	37
Baseline, agg.	9.13	19.00	703
Baseline, agg., social var.	9.09	40.58	997
19 Seg. baseline	5.09	28.21	37
19 Seg. baseline, agg.	7.38	28.82	703
19 Seg. baseline, agg., social var.	16.46	60.89	997

aggregates and the final model uses all 997 variables. We evaluate the performance obtained by these models using Recall and Precision, which have been extensively used in the literature for evaluating classification, information retrieval or recommender systems [17]. *Recall* represents here the portion of fraud captured by the model at a certain threshold on score and *Precision* the proportion of truly fraudulent transactions among those classified as such by the model. The threshold on the score produced by the model is chosen so as to generate a number of cards declared fraudulent compatible with the staff available to further investigate them. This threshold cannot be disclosed for confidentiality reasons. The performances in terms of Recall and Precision at that threshold on score are shown in Table 2 (first three lines). The algorithms used were ridge regressions, regularized through a Vapnik's scheme [12].

As can be seen from the first three line of Table 2, the increase in variety (produced by the increase in the number of features) has a very large impact on Precision, which is multiplied by a factor of about 2 with the addition of aggregates, and again with the social variables. Recall is increased by a factor of about 6 with the addition of aggregates, but slightly decreased with the social variables. So, depending upon the objective of the user, the model with all features will be preferred to the model with aggregates only if Precision is more important than Recall.

One can further increase performance by using a segmented model: when datasets are very large, and the analyzed phenomenon is heterogeneous, one can often improve performance by first identifying homogeneous regions, and then performing a local analysis on each. This will often be the case in Big Data problems. Note that segmentation also has the consequence that each segment will be smaller than the original dataset, which might shorten model computing time, while potentially preventing using too many features (if the depth is too small).

- First we implement a segmentation of cards, supervised by the fraud label (using a supervised k-means): we produced 19 segments.
- For each segment, we compute a model for the three feature sets described in Table 1. When a new transaction needs to be analyzed, we pass it through its card segment model.

As can be shown in Table 2 (last three lines), performances are significantly improved. This time, aggregates and social variables increase Recall, while only social variables increase Precision. This is an illustration of a common finding: it is not possible to tell a priori which feature is going to be significant and bring value. The only way to know is to build a model and look at its performance.

As can be seen from this example, feature engineering can lead to very significant performance increases. This is why, in any Big Data project, the time spent on feature engineering is by far the most important, while the time spent actually producing the model is short: [8] indeed noted that "very little time in a machine learning project is spent actually doing machine learning".

3.4 Big Data Skills

Another discussion branches out from this last remark, related to the importance of statistics skills in Big Data. While it is obvious that Machine Learning relies heavily on statistics, Big Data projects require a set of varied skills, which have come to be known as *data science* [5]. These skills are three-fold: statistics indeed for building, evaluating, analyzing the models; IT to collect data, produce features and deploy models; and obviously business knowledge to correctly frame the problem, identify business-critical features, evaluate the models' performance, and, finally decide on the value of putting the model in production.

As we have seen, the time spent in a project on feature engineering is, by far, the most critical success factor for final performance, which is why IT skills are so important for the success of a Big Data project. But as Fig. 5 shows, all three sorts of skills are required and a total lack of statistics skills may put the IT specialist in a danger zone of producing models which are not valid or which exhibit spurious correlations: the discussion on the dangers of data mining in [21], for example, suggests useful caveats on the uncareful use of data mining.

It should also be clear at this point that Machine Learning used for data science is not the same as pure statistics. Some of the crucial elements necessary for Big Data are just not part of the usual Statistics corpus: these include, for example, learn-

Fig. 5 Big Data skills (from D. Conway's Data Science Venn diagram)

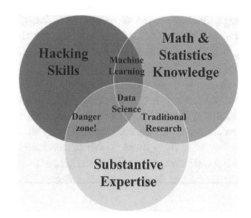

ing and generalization, feature engineering, explainability and capacity for scalable deployment. Another characteristic important in Statistics is absolutely missing in Machine Learning: there is no reasonable assumption which can be made on the data distribution and all the techniques implemented must be agnostic with respect to such assumptions.

Because data science requires a new sort of assorted skills, data scientists are in strong demand. Davenport et al. [6] said that data scientist would be "the sexiest job of the 21st century" and Mc Kinsey [26] predicted that, in the US, demand for data scientists would be 50 to 60 % larger than supply in 2018.

4 The Data Lake

In this section, we will see how companies usually proceed when they want to implement Big Data projects. Obviously, with large companies, that endeavor is global: you do not want to do one project, but many. Companies want to become an analytics competitor [7]: best data and best tools make for the best decisions. Companies thus need to turn their data into an asset.

Of course, all large companies already have multiple data sources in-house to begin with. These data are in silos, application and domain-dependent, and usually contain duplicate, inconsistent versions of the same information. One of the main reasons for this state of affairs is that many large corporations have grown through multiple mergers and acquisitions where each organization merging-in brought its own data system. When starting a Big Data initiative, a company thus has to face a *big data integration problem*, starting with its internal data, continuing with external data (Open Data, partners' data, acquired data …), before finally getting to real Big Data and the ability to exploit it.

Traditional enterprise integration techniques, such as data warehousing or, more generally, *Master Data Management* (MDM) aim at linking all data files into one— *the master file*—providing a *unique version of the truth*. This ensures consistency throughout the various system architectures and applications enterprise-wide and allows sharing data between the various entities in the corporation. However, obtaining a master file requires *Extracting, Transforming and Loading* (ETL) data from the various heterogeneous data sources; developing metadata to describe the data and producing a general data model. Many ETL tools exist on the market (for example Talend Master Data Management, Informatica MDM, IBM InfoSphere, DataStage …). But ETL processes are known to be hard to implement, harder still to maintain and altogether very expensive. Some authors [2] have claimed, for example, that data warehousing projects fail in as much as 50–75 % of the cases. One of the main reasons for such failure is the inability to maintain a manageable scope for the data warehouse, even more so when data sources dynamically change. With Big Data, the problem is going to be even harder than for "simple" warehousing. So we could expect failure rates of the same magnitude at best.

However, MDM might be chasing the wrong rabbit: the master file and data model are efficient for the activities existing or *already defined* in the organization at the time: relational databases and data warehouses are structured in hierarchies and dimensions adapted for the analysis planned at the time of their development. When new needs come up, it is necessary to modify and sometimes completely rebuild the structure to fit the new requirements. With the advent of Big Data, this will be the typical situation: a very wide scope, no known-in-advance goals or analyses, and very dynamic, constantly emerging data sources. A *fixed static data structure cannot* do and data integration on a large scale will still need to be done.

It is to cope with this issue that the concept of *Data Lake* has appeared in the recent years [31]. A Data Lake is a repository of all data collected by an organization, where the data is stored in its *original raw form*. Because no a priori structure or data model is imposed at collection time, all further usage should be possible without having to modify a pre-existing model.

Of course, for any given project, the usual process (described in Sect. 3) will be executed: data will be collected, primarily from the Data Lake and also from other data sources if significant (open data for example). At the cleaning stage, the various data returned by the Data Lake or other sources will need to be deduplicated and reconciled. So this work on data sources reconciliation is not eliminated by the Data Lake: it is postponed to project time when it is needed for a particular objective and restricted to a limited set of data sources. Results of the reconciliation of two data sources can be stored in the Data Lake, as metadata, for further projects which use the same data sources. In this way, data models will emerge progressively from projects developed over time, instead of being imposed at collection time. Hence, data integration is no longer an issue. However, data access is still to be handled: careful tagging and metadata management are required to allow the users to retrieve data interesting for their project, maybe years after the data was collected.

However, more research will certainly be needed to fully exploit Data Lakes: as described in [19], Data Lakes at the present time can only be used by data scientists with significant programming skills. Retrieving data from a potentially enormous (and growing) Data Lake will require adequate skills until adapted tools for navigating the Data Lake are made available. At the present time, it is not possible to simply query a Data Lake, as usual when using SQL for example, since it contains data sources in all possible formats.

As a conclusion, we think that Data Lakes will be increasingly implemented within organizations wishing to implement a succession of Big Data projects, but that additional research will need to take place to help users really, and simply, navigate the Data Lake.

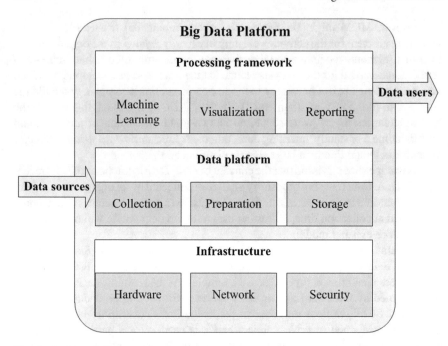

Fig. 6 Big Data platform

5 Architectures for Big Data

5.1 The Big Data Platform

We now turn in this section to a brief description of the possible architectures for Big Data, on both the hardware and software sides. A Big Data platform has the following layers (Fig. 6):

Infrastructure: this includes the hardware, the network, and security components. Obviously, the major requirement for this layer is to allow handling Big Data volumes. Historically, the solution has been to use bigger and bigger servers to handle the increase in data and required processing speed: this is called the *scaling-up* mechanism. However, big servers are not cost-effective when compared to standard PC or low-end commodity machines: a machine with four times the power of a PC costs more than four times the cost of one PC. This is one of the reasons why Google, Yahoo, Amazon, Microsoft and others started to hook together clusters of PC, distributing their data on the machines. With the introduction of Hadoop [27], distributed file system (HDFS) and MapReduce, it became possible to store data in a distributed fashion and run large-scale distributed data processing applications with code distributed where the data is [22]. A cluster of low-end servers with HDFS files is thus very common for storing data, such as, for example a Data Lake. The added benefit of such an architecture is the scaling-out (or horizontal scaling) property: one can elastically grow the cluster by adding new low-end servers, as data volumes grow. This is a lot cheaper than buying up-front a big server, which will, at one point anyway, become too small and will need to be replaced.

In-memory servers, with very large RAMs, have recently appeared on the market: with 512 GB to a few TB RAM (for example, the Bullion machine[3] has up to 24 TB RAM), it becomes possible to collect data, load it in memory and execute all the modelling in-memory, thus very significantly reducing computing times. Apache Spark is a new distributed computing framework which offers in-memory primitives through its Resilient Distributed Datasets (RDDs). In a way, Spark brings a solution with the better of two worlds: Hadoop and in-memory. Performances are much faster than with Hadoop MapReduce.

Data Platform: this layer includes the various tools necessary to access and collect data, to clean prepare and store it. As we have described in the previous section, data is heterogeneous, comes from many sources and in multiple formats. Tools are thus needed to collect data: the ETL family, as well as various data scraping mechanisms, can be used to extract the data. As we have seen before, it is better not to transform data at collection time, but maybe to only retrieve metadata and tags associated to it. Data will be stored in a platform which allows scaling-out: when new data comes in, we need to be able to elastically add to the storing solution to make room for it. This is why, most of the time, data will be stored on a Hadoop cluster.

Very often, data will be stored in NoSQL data bases (Not Only SQL) which offer different structures than relational databases, such as key-value, document, or graph.[4] These databases offer horizontal scaling and may lead to much better performance than relational databases depending on the domain.

Processing framework: this layer includes the tools necessary for visualizing the data, developing machine learning models, and building reports. We will see that there exist a wide range of software packages for running analytics projects on big data. The main challenge, of course, is to be able to scale when handling the data volumes: visualizing small data sets is relatively easy and one can play with many different visualization styles. However, with increasing volume and variety, the requirement for scalability will strongly restrict the visualization effects to the simplest cases.

Note that the Hadoop/MapReduce environment is more adapted to batch processes, and not so much to the iterative development of machine learning algorithms. With the release of Hadoop 2, new tools have been made available, more adapted to machine learning. In particular, Spark allows building specialized Big Data tools such as MLib for machine learning and GraphX for graph processing. Hadoop 1 and 2 are supported commercially by companies such as HortonWorks and Cloudera.

---

[3]http://www.bull.com/download/bullion/B-bullion-2014-enWeb.pdf.

[4]https://datafloq.com/big-data-open-source-tools/os-home/.

5.2 Analytics Software Packages for Big Data

There exist many analytics software packages for big data, a lot being open-source [25]. However, surveys [28, 30] show that most data scientists use but a limited set of tools: R is by far the most wide-spread tool and language, while Python for programming is picking up.

The first generation of tools, SAS and IBM-SPSS, tended to offer a wide range of algorithms, in a proprietary framework. The second generation, such as, for example SAP-KXEN, focused on helping automate the data mining process and opening it to business users, through just one algorithm, a regularized regression [12]. As discussed in Gartner's report [18], SAS and IBM-SPSS still dominate the market with a large installed base. Their position, however, will presumably erode as a new generation of tools gets access to the market.

This last generation focuses on Big Data and attempts at covering the entire Big Data project process described in Sect. 3.2. For example DataRobot[5] focuses on the modeling and deployment stages, automatically generating and comparing thousands of models from various open source libraries (R, Spark MLlib, Python-based scikit learn[6]); Dataiku (see footnote 5) offers through its Data Science Studio tools to load and enrich data, then allows to run models from scikit learn, returning the best performing one; Palantir (see footnote 5) offers tools to *"Integrate, manage, secure, and analyze all of the enterprise data"*, in particular, Palantir has a strong feature engineering tool which helps automate the generation of standard features.

As can be seen, the new generation tools do not try to develop their own machine learning algorithms (as did SAS and SPSS) but, instead, call upon open-source libraries (such as MLlib and scikit learn[6]) which are very actively enriched by active communities.[7] Notice that these two libraries run on different architectures: scikit learn, being based upon Python, runs best on an in-memory server; while MLlib runs with Apache Spark, and thus can be executed on any Hadoop 2 cluster. With the recent development of Spark, MLlib has developed very strongly, taking over the Mahout[8] library which was running on Hadoop MapReduce. As a consequence, there is an effort by the Mahout community to build future implementations on Spark.

It should be expected that more tools will appear in the near future to build upon these healthily competing libraries.

---

[5] http://www.datarobot.com/, http://www.dataiku.com/, https://www.palantir.com/.

[6] https://spark.apache.org/mllib/ is Apache Spark's machine learning library; http://scikit-learn.org/ is a machine learning library in Python.

[7] https://github.com/apache/spark; https://github.com/scikit-learn/scikit-learn.

[8] http://mahout.apache.org/.

6 Conclusion

We have described in this chapter why and how companies implement Big Data projects. The field calls upon a wide variety of techniques, tools and skills and is very dynamically developing. Even though we tried to cover most of the practical issues faced by companies, many topics are still missing here: most notably the privacy and security issues, which would deserve a full chapter of their own.

It is our belief that, in the near future, companies will continue investing in Big Data and the results will bring productivity growths in all sectors of the economy.

References

1. Amatriain, X., Basilico, J.: Netflix Recommendations: Beyond the 5 stars. Netflix Techblog. (6 April 6 2012)
2. Amin, R., Arefin, T.: The empirical study on the factors affecting datawarehousing success. Int. J. Latest Trends Comput. 1(2), 138–142 (Dec 2010)
3. Anderson, M., Antenucci, D., Bittorf, V., Burgess, M., Cafarella, M. J., Kumar, A., Niu, F., Park, Y., Ré, C. & Zhang, C.: Brainwash: A Data System for Feature Engineering. CIDR'13 (2013)
4. Chapus, B., Fogelman Soulié, F., Marcadé, E., Sauvage, J.: Mining on social networks. In: Gettler Summa, M., Bottou, L., Goldfarb, B., F. Murtagh (eds.) Statistical Learning and Data Science, Computer Science and Data Analysis Series. CRC Press, Chapman & Hall (2011)
5. Conway, D.: The Data Science Venn Diagram (2013). Blog. http://drewconway.com/zia/2013/3/26/the-data-science-venn-diagram
6. Davenport, T.H., Patil, D.J.: Data Scientist: The Sexiest Job of the 21st Century. Harvard Bus. Rev. 70–76 (Oct 2012)
7. Davenport, T.H.: Competing on analytics. Harvard Bus. Rev. 84, 98–107 (2006)
8. Domingos, P.: A few useful things to know about machine learning. Commun. ACM 55(10), 78–87 (2012)
9. Driscoll, M.: Building data startups: Fast, big, and focused. Low costs and cloud tools are empowering new data startups. O'Reilly Radar (August 9, 2011)
10. Eckerson, W.W.: Predictive Analytics. Extending the Value of Your Data Warehousing Investment. TDWI Best Practices. Report. Q1, 2007 (2007)
11. Fogelman-Soulié, F., Mekki, A., Sean, S., & Stepniewski, P.: Utilisation des réseaux sociaux dans la lutte contre la fraude à la carte bancaire sur Internet. In: Bennani, Y., Viennet, E. (eds.) Apprentissage Artificiel & Fouille de Données. Revue des Nouvelles Technologies de l'Information, RNTI-A-6. Hermann, pp. 99–119 (2012) (in French)
12. Fogelman Soulié, F., Marcadé, E.: Industrial Mining of Massive Data Sets. Mining massive Data Sets for Security. In: Fogelman-Soulié, F., Perrotta, D., Pikorski, J., Steinberger, R. (eds.) Advances in data mining, search, social networks and text mining and their applications to security, pp. 44-61. IOS Press. NATO ASI Series (2008)
13. Gantz, J.F.: The Expanding Digital Universe. IDC White Paper (March 2007)
14. Groupement des Cartes Bancaires CB: Activity, Report (2013)
15. Halevy, A., Norvig, P., Pereira, F.: The unreasonable effectiveness of data. IEEE Intell. Syst. 24(2), 8–12 (2009)
16. Hastie, T., Tibshirani, R., Friedman, J.: The Elements of Statistical Learning (vol. 2, no. 1). Springer, New York (2009)
17. Herlocker, J.L., Konstan, J.A., Terveen, L.G., Riedl, J.T.: Evaluating collaborative filtering recommender systems. ACM Trans. Inf. Syst. (TOIS) 22(1), 5–53 (2004)

18. Herschel, G., Linden, A., Kart, L.: Magic Quadrant for Advanced Analytics Platforms. Gartner Report G00270612 (2015)
19. Heudecker, N., White, A.: The Data Lake Fallacy: All Water and Little Substance. Gartner Report G00264950 (2014)
20. Hilbert, M., López, P.: The world's technological capacity to store, communicate, and compute information. Science **332**(6025), 60–65 (2011)
21. Leinweber, D.J.: Stupid data miner tricks: overfitting the S & P 500. J. Investing **16**(1), 15–22 (2007)
22. Lam, C.: Hadoop in action. Manning Publications Co (2010)
23. Laney, D.: Big Data's 10 Biggest Vision and Strategy Questions. Gartner Blog (2015)
24. Laney, D.: 3D Data Management: Controlling Data Volume, Velocity, and Variety. Application Delivery Strategies, Meta Group (2001)
25. Machlis, S.: Chart and image gallery: 30+ free tools for data visualization and analysis. Computerworld (2013)
26. Manyika, J., Chui, M., Brown, B., Bughin, J., Dobbs, R., Roxburgh, C., Hung Byers, A.: Big data: The next frontier for innovation, competition, and productivity. Report, McKinsey Global Institute (2011)
27. Olson, M.: Hadoop: Scalable, flexible data storage and analysis. IQT Quart **1**(3), 14–18. (Spring 2010)
28. Piatetsky, G.: KDnuggets 15th Annual Analytics, Data Mining, Data Science Software Poll. KDnuggets (2014)
29. Rahm, E., Do, H.H.: Data cleaning: problems and current approaches. IEEE Data Eng. Bull. **23**(4), 3–13 (2000)
30. Rexer, K.: 2013 Data Miner Survey. Rexer Analytics (2013)
31. Stein, B., Morrison, A.: The enterprise data lake: Better integration and deeper analytics. PwC Technology Forecast: Rethinking integration. Issue 1 (2014)
32. Turck, M.: The state of big data in 2014 (chart). VB News (2014)
33. Vapnik, V.: Estimation of dependences based on empirical data. Springer. Information sciences and Statistics. Reprint of 1982 Edition with afterword (2006)
34. Vasanth, R.: The Rise Of Big Data Industry: A Market Worth 53.4 Billion By 2017 ! Dazeinfo (2014)
35. Zhou, T., Ren, J., Medo, M., Zhang, Y.-C.: Bipartite network projection and personal recommendation. Phys. Rev. E **76**(4), 046115 (2007)

Data Mining in Finance: Current Advances and Future Challenges

Eric Paquet, Herna Viktor and Hongyu Guo

Abstract Data mining has been successfully applied in many businesses, thus aiding managers to make informed decisions that are based on facts, rather than having to rely on guesswork and incorrect extrapolations. Data mining algorithms equip institutions to predict the movements of financial indicators, enable companies to move towards more energy-efficient buildings, as well as allow businesses to conduct targeted marketing campaigns and forecast sales. Specific data mining success stories include customer loyalty prediction, economic forecasting, and fraud detection. The strength of data mining lies in the fact that it allows for not only predicting trends and behaviors, but also for the discovery of previously unknown patterns. However, a number of challenges remain, especially in this era of big data. These challenges are brought forward due to the sheer Volume of today's databases, as well as the Velocity (in terms of speed of arrival) and the Variety, in terms of the various types of data collected. This chapter focuses on techniques that address these issues. Specifically, we turn our attention to the financial sector, which has become paramount to business. Our discussion centers on issues such as considering data distributions with high fluctuations, incorporating late arriving data, and handling the unknown. We review the current state-of-the-art, mainly focusing on model-based approaches. We conclude the chapter by providing our perspective as to what the future holds, in terms of building accurate models against today's business, and specifically financial, data.

Keywords Financial data · Time series · Data streams · Volatility · Stochastic · Marginalisation · Path integral · Bayesian learning · Energy load forecasting

E. Paquet · H. Guo
National Research Council of Canada, Building M-50,
1200 Montreal Road, Ottawa, Canada
e-mail: hongyu.guo@nrc-cnrc.gc.ca

E. Paquet
e-mail: eric.paquet@nrc-cnrc.gc.ca

E. Paquet · H. Viktor (✉)
School of Electrical Engineering and Computer Science, University of Ottawa,
800 King Edward Road, Ottawa, Canada
e-mail: hviktor@uottawa.ca

© Springer International Publishing Switzerland 2016
N. Japkowicz and J. Stefanowski (eds.), *Big Data Analysis: New Algorithms for a New Society*, Studies in Big Data 16, DOI 10.1007/978-3-319-26989-4_7

1   Introduction

Data mining has been successfully applied to many businesses, thus aiding managers to make informed decisions that are based on facts, rather than having to rely on guesswork and incorrect extrapolations. Data mining algorithms allow companies to explore the trends in terms of sales, to predict the movements of financial indicators, and to construct energy-aware buildings, amongst others. Specific data mining (or business analytics) success stories include customer loyalty prediction and sales forecasting, fraud detection, estimating the correlations between stocks and predicting the movements of financial markets. Case studies show that the strength of data mining lies in the fact that it allows for not only predicting trends and behaviors, but also for the discovery of previously unknown patterns in business data.

Making predictions and building trading models are central goals for financial institutions. It is no surprise that this was one of the earliest areas of the application of modern machine learning techniques to real world problems. In this sector, a number of unique challenges need to be addressed. These challenges are brought forward due to the sheer Volume, Velocity (in terms of speed of arrival) and the potential Variety, of the data. In addition, another issue here is that we aim to build an accurate model against uncertain, rapidly changing, and often rather unpredictable, data. That is, the financial sector continuously processes millions, if not trillions, of transactions. For example, the values of stocks are updated at regular intervals, typically every few seconds. These markets require the use of advanced models in order to facilitate trend spotting and to provide some financial trajectory. Ideally, in this scenario, we require just-in-time adaptive models that are accurate even as the data changes, due to concept drifts.

There are many unknowns associated with such financial data, which makes the construction of data mining models a major challenge. Here, analyzing and understanding what attributes and parameters we *do not know* is crucial in order for us to create accurate and meaningful predictions. This fact limits the application of traditional data-driven algorithms, in that we often cannot make assumptions about data distributions or types of relationships. The typical non-parametric way used by most data mining algorithms, to search a large data set to see whether any patterns are exhibited in that set, has limited applicability in a financial setting. Here, the data are susceptible to drift, arrive at a fast rate, may contain late-arriving data, and have parameters that are difficult to estimate. Thus, this type of traditional analysis and model construction may not be ideal when aiming to construct models against big data in finance, where the number of unknowns (and in essence the randomness) is high. Rather, the use of stochastic, model-based approaches comes to mind.

This chapter addresses the above-mentioned issues associated with Volume, Velocity and Variance in big data, while focusing on the financial sector. To this end, we review the state-of-the-art in terms of techniques to mine stocks, bonds, and interest rates. We note that Bayesian approaches have had some success, in which unknown values are integrated out (marginalized) over their prior probability of occurrence. We further describe the special considerations that need to be taken into account when building models against such a vast amount of uncertain and fast-arriving data. Our discussion centers on issues such as handling data distributions with high fluctuations, modeling the unknown, handling potentially conflicting information, and considering boundary conditions (i.e. the prices of the stocks when acquired and sold or the initial and final interest rates) following a path integral approach. We conclude the chapter by providing our perspective as to what the future holds.

We begin this chapter, in Sects. 2 and 3, by setting the stage and by discussing the complexities associated with building predictive models for financial data that are high in Volume and Variety. Section 4 reviews the concepts of bonds and interests rates, while Sect. 5 presents the Black-Scholes model for interest rates. In Sect. 6, we explore the Heath-Jarrow-Molton model for predicting the forward-value of a bond. Next, in Sect. 7, we turn our attention to this issue of Variety, and we discuss the use of social media and non-traditional data sources during model building. Finally, Sect. 8 concludes the chapter and presents our views on the way forward.

2   Business, Finance and Big Data

Our level of indebtedness is unprecedented in history. Whether we like it or not, the finance sector, in general, and the debt sector, in particular, has become paramount to business. In 1965, corporations in the United States of America (US) were earning 12.5 % of their revenues from the financial sector while 50 % of their revenues were coming from manufacturing. In 2007, just before the financial meltdown of 2008, this tendency was completely inverted with 35 % of US corporations' revenues earned from the financial sector, while only 12 % were earned from domestic manufacturing. As a matter of fact, the fraction of corporate earnings from the financial sector has grown more than 400 % over the last 60 years [1].

By all means, finance is big: big by the Volume, Velocity, and Variety of data involved, big by the corresponding amount of money involved (trillions of $), and big by its influence on our lives. Just to present an order of magnitude, on 13 November 2014, a normal trading day, 708,118,734 financial instruments were traded for a total value of $26,847,016,206 at the New York Stock Exchange (NYSE) of which, 641,044 financial instruments were traded with algorithmic programs [2]. (Note that a financial instrument may be defined as a trade-able asset of any kind; either cash, evidence of an ownership interest in an entity, or a contractual right to receive or deliver cash or another financial instrument. For each financial instrument, we keep track of its value as it evolves over time. The market data for a particular instrument would include the identifier of the instrument and where it was traded such as the ticker symbol and exchange code plus the latest bid and ask price and the time of the last trade. It may also include other information such as volume traded, bid and offer sizes, and static data about the financial instrument that may have come from a variety of sources. That is, these massive data streams are in essence time series data.)

It follows that making predictions and building trading models are central goals for financial institutions. For example, a number of researchers have studied the problem of forecasting the volatility of stock markets, through the use of neural networks, decision trees, cluster analysis, and so on [3]. In contrast to econometric approaches, the data-driven modeling approach used in many data mining algorithms makes few assumptions about data distributions or types of relationships. In this framework few (if any) parameters need to be estimated. Neither is there an assumed model form. Instead, the standard non-parametric approach proceeds by searching the data set to see whether any patterns are exhibited in that set. If the patterns found meet certain minimum requirements, then the pattern is recorded for further inspection. The usefulness of the methodology is judged by looking at new data to see whether these patterns also occur there. If so, we say that the data mining model is robust and has found a pattern that holds over time.

However, following a data-driven only approach, as discussed above, may not be ideal when aiming to construct models against big data in finance, in which the number of unknowns, due to the essential randomness, is high. Also, this train-then-test method does not work well for financial data streams that are susceptible to concept drift. To this end, the focus of this chapter is on building models against big data in finance, using a path integral approach. We primarily focus our attention on stocks, bonds, and interest rates from a big data perspective. Stochastic models for the stocks' prices and for the forward rates are introduced. From the knowledge of the probability distribution associated with the noise, it is possible to marginalize our uncertainty about the prices and the rates and to make useful predictions. The lack of knowledge may be leveraged through a framework rooted in the path integral formalism. We show that a thorough understanding of what we *don't know* is instrumental in such a process. In the next sections, we address stock prices, and we then extend our previous analysis to bonds.

3   Finance and Data Mining: Diving into the Unknown

Stock prices and interest rates are time series data that arrive in massive volumes, are fast changing and potentially infinite [3]. In the financial sector, researchers aim to create just-in-time models in order to find similar or regular patterns, to identify trends, to detect sudden concept drifts and to spot outliers, from such big data.

An important task is to find similar series, using either subsequence matching or whole sequence matching [4]. For example, *Selective MUSCLES* as introduced in [5], is an efficient and scalable method for on-line mining for co-evolving time sequences. In their method, they use subset selection and exponential forgetting in order to scale their system up. In addition, trend analysis is often used in order to both gain insights into the underlying forces that generate time series and to predict the future [6]. Here, four main types of analysis are of importance [3]. Firstly, we are interested in modeling long-term movements, e.g. the trend in the behavior of a stock or market over a long period of time. Secondly, there is the study of cyclical movements, which

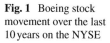

Fig. 1 Boeing stock
movement over the last
10 years on the NYSE

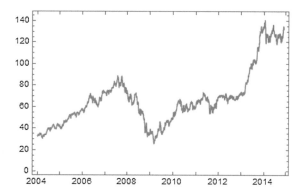

refs to long-term oscillations that may or may not be periodic. Thirdly, seasonal drifts refer to variations that are typically calendar related. For example, there may be an increase in food prices traded out of season. In this case, the seasonal movements are typically very similar from year to year, and we are interested in utilizing this knowledge. The fourth type of movement refers to sporadic motions due to random or chance events, such as a volcanic eruption that disrupts air traffic or some unexpected socio-economic turmoil. These type of movements are also known as sudden concept drift, and the challenge here is to react fast, in order to update the models.

It is often said, in jest, that there are two certainties in life: death and taxes. Finance, on the other hand, is the kingdom of uncertainty, which makes trend analysis a challenge. If it would not be the case, risk-free and high-return investments, would be common place. As we all know, this is far from being the case. In order to obtain knowledge from this type of data stream, we often approach the problem by first making a certain number of hypotheses that could be validated subsequently from historical financial series. These hypotheses, once structured, constitute a model. A question which needs to be thoroughly considered is the following: What do we already know and what information may be utilized?

As an example, Fig. 1 shows the long term movement of the Boeing stocks on the New York Stock Exchange (NYSE) in terms of the value at the time of closure, from 1 January 2004 until 1 December 2014. In Fig. 2, we depict the behavior of the Baskem stocks on the NYSE over the same period of time. The figures show the difference in long term behavior between these two equities, with both experiencing a downturn in the 2008–2009 period.

We further know that stock prices and interest rates are volatile. There may be a function that characterizes such volatility, but its precise form is currently out of reach. We also know that the statistical properties associated with stock prices and interest rates are drifting. Such a concept drift could also be characterized by a function of unknown nature. Furthermore, the fact that stock prices and interest rates are intrinsically uncertain, points toward the existence of random fluctuations (noise). These fluctuations may be characterized by a Gaussian, Lévy, or truncated Lévy probability distribution according to the importance devolved to large fluctua-

Fig. 2 Baskem stock
movement over the last
10 years on the NYSE

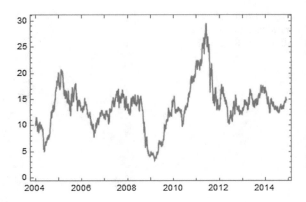

tions [7, 8]. The multidimensional functional Gaussian distribution, for instance, is entirely characterized by its mean and covariance. The model relates all these disparate elements into a common and organic framework. As will be discussed below, in the case of stock prices and interest rates, the model is either a differential or a finite difference equation that relates the target variable (stock price or interest rate), the drift function, the volatility function and the stochastic random fluctuations. For instance, the stock prices may be described by the Black-Scholes model, while the interest rates are depicted by the Heath-Jarrow-Morton model, as explained in the following sections.

Nevertheless, the above-mentioned models are plagued with unknowns. The reasons are threefold. Firstly, as we pointed out, the nature of the functions associated to the drift and the volatility are unknown. Secondly, the boundary conditions are, in all likelihood, unknown. Recall that, by boundary conditions, we mean the price of the stock when acquired and sold or the initial and final interest rate. Finally, stocks and interest rates follow a specific financial trajectory in the sense that the time series associated with these financial instruments take precise values at every time t.

The Bayesian framework has been widely used to study such data. One of the main reasons why Bayesian methods have been so successful is their ability to incorporate information from different sources and also address complex estimation problems. Bayesian methods are based on the principle that probability is subjective, in that the degree of belief may be updated as new information, or data, are acquired [9]. Here, the beliefs based on the current knowledge is referred to as the prior probabilities and the posterior probability represents the updated beliefs. To this end, the Bayesian framework has been used for portfolio allocation [10], asset pricing models [11], and for volatility models [12].

That is, one of the best approaches for marginalizing the unknown functions, boundary conditions and financial trajectories is to be found in the Bayesian framework, as will be illustrated throughout this chapter. Here, unknown values are integrated out (marginalized) over their prior probability of occurrence. Consequently, a model associated with a financial instrument may be constructed as follow. Firstly, unknown functions are associated with the instrument's drift and volatility. The drift, the volatility, and the stochastic fluctuations are combined into a differential equation, which characterized the temporal evolution of the underlying time series. Then, the precise nature of the stochastic fluctuations is determined. For instance, the probability distribution associated with the fluctuation may be a multidimensional Gaussian, which means that it is entirely parameterized by its mean and its covariance. The differential equation acts as a constraint on the probability distribution associated with the noise. The constraint is imposed with a Dirac delta function, a generalized function, or distribution, which is zero everywhere, except at the origin and for which the integral over the entire integration domain is equal to one [13]. The unknown noise and financial trajectory are then integrated out or marginalized. That means that at each instant associated with the time series, the value of the financial instrument and of the corresponding stochastic fluctuation are integrated. If unknown, the boundary conditions must also be marginalized. These calculations allow performing predictions of statistical nature about financial instrument such as their expectation and dispersion.

As will be explained below, if the financial process unfolds as a fair game (or so-called Martingale), it is possible to express the drift as a function of the volatility. In the following, we will analyze the computational aspect of models related to stock prices and interest rates.

4   Finance: A Fair Game ... Most of the Time

A bond is an instrument of debt, while a treasury bond is an instrument of debt with no risk of default. The money is lent in exchange of an interest over the capital, which is the cost for borrowing money. An important concept associated with bonds is the forward interest rate [1]. The forward interest rate, also called the forward rate, $f(t, \tau)$ is the agreed upon future interest rate, at time $t < \tau$, for an instantaneous loan at future time τ. It is typically calculated using a yield curve. For example, the yield on a three-month treasury bill, six months from now, represent a forward rate. The value of a bond is related to the forward interest by

$$B\left(T_i, T_f\right) \equiv e^{-\int_{T_i}^{T_f} d\tau \, f(t,\tau)}, \tag{1}$$

where $B\left(T_i, T_f\right)$ is the value of a bond at time T_i, maturing at time T_f. Bonds have a remarkable property that is shared by other financial instruments and which is called the fundamental theorem of finance. The fundamental theorem of finance states that financial processes follow a martingale [14]. A martingale is a model of a fair game in which the knowledge of past events never helps predict the mean of the future earnings. Mathematically, it may be formulated as follows:

$$E\left[B^{(k+1)} \middle| B^{(1)}, B^{(2)}, \ldots, B^{(k)}\right] = B^{(k)}, \tag{2}$$

At first sight, it seems that the martingale is a rather mild condition. However, as we shall see later, its importance is fundamental in constraining financial models. That is, martingales exclude the possibility of winning strategies based on game history, and thus they are a model of fair games. As will be seen in the next section, the martingale is a strong condition which allows determining the deterministic drift associated with a financial instrument [15]. For instance, in the case of the Black-Scholes (BS) model, the drift is entirely determined by the spot rate, irrespectively of the underlying data as a consequence of the martingale condition. This is something that would be difficult to conclude when following a purely data-driven approach. (Note that the spot rate refers to the price quoted for immediate settlement on a commodity, a security or a currency. The spot rate, also called the spot price, is based on the value of an asset at the moment of the quote.)

To this end, in the next section, we explain how to model evolving equities or stocks. This discussion presents a first step towards the task of modeling interest rates.

5   Black-Scholes Model and Path Integrals or How to Handle the Unknown for Stocks

Before addressing the bond and the forward rate, we consider a rather simpler process, namely the evolution of the price of a financial instrument representing a set of equities or stocks $\{S_i\}_{i=1}^N$. As stated above, this topic has received much attention in the area of time-series data mining [3]. Traditionally, the Black-Scholes model has been used to construct a model of the price evolution of N stocks with a stochastic process [16]:

$$\frac{dS_i(t)}{dt} = \alpha_i S_i(t) + \sigma_i S_i(t) R_i(t),$$ (3)

where $S_i(t)$ is the price of stock i at time t, α_i is the deterministic drift associated with stock i, σ_i is the deterministic volatility associated with stock i, while the Gaussian white noise $R(t)$ has a mean and a variance given by

$$E[R_i(t)] = 0,$$

$$E\left[R_i(t) R_j(t')\right] = \rho_{ij}\,\delta\left(t - t'\right),$$ (4)

$$T_i \leq t, t' \leq T_f.$$

where ρ_{ij} is the estimated correlation in between the various stocks. We shall address the evaluation of the drift and the volatility later in this chapter. For the time being, we concentrate on the stocks per se, and we just mention here that the drift and the volatility may be estimated from historical time series data. The BS model is rather intuitive. That is, we know that the prices of stocks tend to drift; we know that the prices of equities are volatile and we know that the fluctuations associated with financial instruments are of a stochastic nature. The BS model is one of the simplest models that combine all these requirements.

Nevertheless, there are many unknowns associated with financial simulations. For instance, given a financial instrument, the initial and the final value (boundary conditions) of this instrument are unknown. As a matter of fact, this is true for any intermediate state of the instrument. All the intermediate states form a so-called financial trajectory. The only problem, so to say, is that the exact nature of this trajectory is entirely unknown. As it stands, the situation seems rather insoluble. Most of what we know is unknown, but the fact that we know what is unknown, shall prove itself to be crucial. The real question is how should we leverage the unknown? The best answer is that we should consider any possible evolution or path of the stocks. We are not allowed to discard any trajectory, because we do not have any information from which such an action could be justified. What is required is a method to weight the various trajectories in order to extract the expected behaviour of the underlying financial instrument. The weight of a given trajectory may be associated to its probability of occurrence. The white noise, as defined by the previous equation, has a Gaussian distribution. This implies that the probability distribution, as Bayesian prior, associated with a specific noise trajectory is given by:

$$\mathcal{D}R \; \Pr[R] = \frac{\mathcal{D}R \; e^{S[R]}}{Z}, \tag{5}$$

where $S[R]$ is the time integral of the Lagrangian $\mathcal{L}[R]$:

$$S[R] = \int_{T_i}^{T_f} dt \; \mathcal{L}[R]. \tag{6}$$

The latter is a functional that assigns probabilities for the occurrence of the various realisations of the noise and is defined by the quadratic function:

$$\mathcal{L}[R] = -\frac{1}{2} \sum_{i,j=1}^{N} R_i \, \rho_{ij}^{-1} \, R_j. \tag{7}$$

Here, ρ_{ij} is the deterministic factor associated with the correlation in between the various stocks. $\mathcal{D}R$ is the path integral measure, that is, the integration or Bayesian marginalization over all unknown intermediate states along every possible trajectory

$$\int \mathcal{D}R = \prod_{t=T_i}^{T_f} \prod_{i=1}^{N} \int_{-\infty}^{\infty} dR_i(t) \tag{8}$$

and Z is a normalization factor known as the partition function.

The integral over every possible state or trajectory is known as a path integral [13]. The probability distribution associated with Eq. (7) is clearly Gaussian, although it is somewhat different from the distribution we are familiar within the sense that it does not involve a variable but a function: in occurrence the stochastic noise. This is why we don't refer to functions, but to functionals (function of a function). From now on, we shall consider the logarithmic of the stock price

$$z_i \equiv \ln S_i. \tag{9}$$

As stated earlier, the financial trajectories associated with the stocks are governed by Eq. (4). The Bayesian probability associated with a specific trajectory is given by

$$\mathcal{D}z\mathcal{D}R \ \Pr[z, R] = \frac{\mathcal{D}z\mathcal{D}R \ \prod\limits_{t=0}^{T} \prod\limits_{i=1}^{N} \delta\left(\frac{\partial z_i(t)}{\partial t} + \alpha_i - \frac{1}{2}\rho_{ii}\sigma_i^2 + \sigma_i R_i(t)\right) e^{S[R]}}{Z}, \tag{10}$$

where the Dirac delta distribution ensures that the stochastic equation of motion associated with the equities, here the BS model, is always satisfied. The partition function Z, or normalization factor, is obtained by integrating the probability distribution over all possible values of the stocks and of the random variables. The mathematical expectation (mean) of any financial instrument $\mathcal{O}$ is obtained by weighing each occurrence of the financial instrument by its corresponding probability

$$\int \mathcal{D}z\mathcal{D}R \ \mathcal{O}[z, R] \Pr[z, R]. \tag{11}$$

Because the probability distribution associated with the noise is quadratic, we may easily integrate or marginalize the noise out of the equation and obtain a closed-form expression that depends only on the stocks. It follows that closed-form expressions play an important role in the big data era. In finance, these expressions have been successfully used for the pricing of especially exotic derivative products [17]. Indeed, the integration measure typically involves thousands of dimensions. Consequently, for the sake of computational stability and efficiency, numerical evaluation should be strictly restricted to those dimensions that could not be treated analytically. We finally obtained for the expectation value of a given functional of an underlying commodity

$$E[\mathcal{O}[z]] = \frac{1}{Z_{BS}} \int \mathcal{D}z \ e^{S_{BS}[z]} \ \mathcal{O}[z], \tag{12}$$

where the action, the Lagrangian, the partition function and the integration measure are given respectively by [18]

$$S_{BS}[z] = \int_{T_i}^{T_f} dt \, \mathcal{L}_{BS}[z],$$

$$\mathcal{L}_{BS}[z] = -\frac{1}{2} \sum_{i,j=1}^{N} \left[\frac{\frac{\partial z_i(t)}{\partial t} + \alpha_i - \frac{1}{2}\rho_{ii}\sigma_i^2}{\sigma_i} \right] \rho_{ij}^{-1} \left[\frac{\frac{\partial z_j(t)}{\partial t} + \alpha_j - \frac{1}{2}\rho_{jj}\sigma_j^2}{\sigma_j} \right] \qquad (13)$$

and

$$Z_{BS} = \int \mathcal{D}z \, e^{S_{BS}[z]},$$

$$\int \mathcal{D}z = \prod_{t=T_i}^{T_f} \prod_{i=1}^{N} \int_{-\infty}^{\infty} dz_i(t).$$

For instance, the functional $\mathcal{O}[\cdot]$ in question may be an option. An option is a contract that gives the buyer the right, but not the obligation, to buy or sell an underlying asset or instrument at a specified strike price P on or before a specified date. The seller has the corresponding obligation to fulfill the transaction if the buyer (owner) exercises the option. The buyer pays a premium to the seller for this right. For example, options are often used by electricity generators and retailers to protect from price or cost volatility [19]. One type of option that is used in such a setting is the so-called flexibility-of-delivery option, which permits the contract holder to receive any amount of power within a certain range for defined time periods.

Options valuation is a topic of ongoing research in academic and practical finance, due to its importance in financial markets, their complexity and the large Volume of options being exercised. Options contracts have been known for many centuries, however both trading activity and academic interest increased when, as from 1973, options were issued with standardized terms and traded through a guaranteed clearing house at the Chicago Board Options Exchange [16]. Today many options are created in a standardized form and traded through clearing houses on regulated options exchanges, while other over-the-counter options are written as bilateral, customized contracts between a single buyer and seller, one or both of which may be a dealer or market-maker. Options are part of a larger class of financial instruments known as derivatives.

There are a number of ways to model an option. For example, if an investor acquires an Asian option, then the pay-off function depends on the average price of the stock during a given time interval:

$$\mathcal{O}_A[z] = \max \left(P, \frac{1}{\Delta t} \int_{\Delta t} dt \, g[z(t)] \right), \qquad (14)$$

where $g[\cdot]$ is an agreed upon functional.

We can still further improve our model. We know (prior information), from the fundamental theorem of finance [20], that stocks follow a martingale which was defined earlier in Eq. (2–3). That is, it is known that knowledge of past events cannot help us to predict hte mean of future yields. If we compute the mathematical expectation associated with the martingale with Eq. (15) we obtain $\alpha_i = r$. This confirms that, as stated earlier, the drift is entirely determined by the spot rate and is not an independent parameter of the model as might have been expected earlier.

Still, the integration over all possible prices of the stock is not a trivial operation. The value of a stock is a time series, which is updated at regular interval, typically every few seconds. Let us assume that we want to calculate the expectation value of a stock over a period of one week and that the price of the stock is updated every 15 s with a typical trading session lasting from 9:30 until 16:00 local time. Thus the integration measure consists of 7,800 dimensions! This is clearly a big data problem, which is reminiscent of the curse of dimensionality. Nonetheless, such an integral may be calculated efficiently with a Monte Carlo approach, known as the Metropolis-Hasting (MH) algorithm [21]. Instead of systematically integrating over the whole integration domain, the latter is explored with a Markovian process which randomly samples the realizations of the stock. Given a value of the stock $z^{(k)}$, a new value is randomly generated according to

$$z^{(k+1)} = z^{(k)} + R, \tag{15}$$

where R is a Gaussian white noise. The new occurrence of the stock is accepted (or rejected) with probability

$$A\left(z^{(k)} \to z^{(k+1)}\right) = \min\left(1, \exp\left(S\left[z^{(k)}\right] - S\left[z^{(k+1)}\right]\right)\right), \tag{16}$$

where the action S was defined earlier in Eq. (16). This means that the new value is always accepted if its probability of occurrence is higher than the previous one. However, it is nevertheless accepted with a probability that is otherwise equal to $\exp\left(S\left[z^{(k)}\right] - S\left[z^{(k+1)}\right]\right)$. The expectation of a function of the stock is then obtained as the average of this function over the sampled values of the stock

$$E\left[\mathcal{O}[z]\right] \approx \frac{1}{(k_{\max} - k_{\min})} \sum_k \mathcal{O}\left[z^{(k)}\right]. \tag{17}$$

The MH algorithm allows for a more efficient sampling of the integration domain and prevents from integrating over trajectories that have a negligible probability of occurrence. These trajectories tend, generally speaking, to introduce a detrimental numerical noise [21]. In the next section, we extend our previous analysis to the modeling of forward interest rates and bonds based on the well-known Heath-Jarrow-Morton model.

6 Heath-Jarrow-Morton Model and Path Integrals
or How to Leverage Our Ignorance About Interest Rates

The case of a bond, and of the underlying forward rate, is slightly more complicated than the case of a stock. As we saw earlier, the value of a bond is determined by the forward interest rate (cf. Eq. (1)) which is unknown. The forward interest rate depends on both the present time and the future time. Forward interest rates are typically modeled with a stochastic process known as the Heath-Jarrow-Morton (HJM) model [22]. The HJM model is very similar in nature to the BS model (Eq. 4) except that the drift and the volatility are not constant but depend on the current (calendar) time t and on the future time τ:

$$\alpha \Rightarrow \alpha\,(t,\tau) \tag{18}$$

$$\sigma \Rightarrow \sigma\,(t,\tau)\,. $$

It follows that the forward rate is governed by the following stochastic equation:

$$\frac{\partial f\,(t,\tau)}{\partial t} = \alpha\,(t,\tau) + \sigma\,(t,\tau)\ R\,(t)\,, \tag{19}$$

where the white noise $R\,(t)$ was defined earlier [23]. Following the same approach as for the equities (or stocks), the Bayesian probability associated with a specific trajectory of the forward rate is equal to

$$\mathcal{D}f\mathcal{D}R\ \Pr[f,R] = \frac{\mathcal{D}f\mathcal{D}R\ \prod\limits_{(t,\tau)\in\mathcal{T}}\delta\left(\frac{\partial f(t,\tau)}{\partial t} - \alpha\,(t,\tau) - \sigma\,(t,\tau)\ R\,(t)\right)e^{S[R]}}{Z}, \tag{20}$$

where the temporal domain $\mathcal{T}$ is defined as

$$\mathcal{T} \Rightarrow t \in \left[T_i, T_f\right] \quad \cap \quad \tau \in [t, t + T_H], \tag{21}$$

where T_H is the investment horizon: the time, during which an investment may be performed. If the white noise is integrated out, one obtains, for the mathematical expectation, a closed-form expression similar to Eqs. (15) and (16). As in the case of the BS model, one may apply the fundamental theorem of finance and demonstrate that the drift is not an independent quantity but is related to the volatility by [23]

$$\alpha\,(t,\tau) = \sigma\,(t,\tau)\int\limits_{t}^{\tau} d\tau'\ \sigma\left(t,\tau'\right). \tag{22}$$

It then follows that the path integration may be performed with the MH algorithm. This allows for the computation of the expected value of a bond and its standard deviation, together with other quantities of interest. In other words, this calculation enables one to determine whether a specific investment is worthwhile, in addition to evaluating the concomitant risk level and the level of uncertainty.

Note that historic investment data may further be extracted, for instance, from historical yield curves. Consequently, financial institutions may choose to combine model-driven and data-driven approaches. A data-driven approach is particularly suitable when handling late arriving data [24], such as those which result from a manipulation of the interest rates. In follows that the sheer Volume of data requires greater sophistication of statistical techniques in order to obtain accurate results. In particular, recent research has shown that the number of false correlations increases as the data Volume and dimensionality increases [9]. The reader should further notice that the state-of-the art algorithms based on economic theory typically point to long-term investments opportunities as based on trends in historical data. The task to produce efficient results supporting a short-term investment strategy still poses a challenge for current predictive models [8]. Thus, a number of research challenges remain, in this era where financial institutions are increasingly embracing big data analytics.

7 A Word About Variety

In the above-mentioned discussions, we focused our attention on financial data that is high in Volume and Velocity. However, in order to capitalize on the big data opportunity, enterprises should also embrace Variety, that is different types of data from a wide range of fields, including documents, e-mail, web pages, social media forums data, smart devices data, and sensor data, amongst others. This Variety characteristic associated with big data presents rich information for knowledge discovery.

Such Variety may aid the learning processes from different observation angles, and allows exploring correlation across domains and fields. The financial sector is especially susceptible to changes due to socio-economic factors. It then follows that the use of social media data may provide role-players with a competitive advance. For example, recent studies have shown that the evaluation of large-scale Twitter feeds may be used to accurately predict stock market indicators for markets such as Dow Jones, NASDAQ, and S&P 500 [25, 26]. Specifically, the results in [25] indicate that the accuracy of Dow Jones Industrial Average (DJIA) closing predictions can be significantly improved by the inclusion of specific public mood dimensions.

As another example, we turn our attention to the case of Smart Cities, which has increasingly become of importance in the financial sector [27]. Energy usage costs accounts for approximately 19 % of total expenditures for a typical building in the US [28]. In the European Union (EU), buildings account for approximately 40 % of final energy consumption in 2008 [29]. To this end, both governments as well as the owners of commercial buildings have moved to time-of-use pricing and are exploring ways to balance demand and response signals [27]. Smart energy consumption models, however, heavily relies on accurate short-term load predictions.

In order to generate accurate energy load forecasts, a number of factors from a variety fields need to be taken into account. For instance, the building's routine schedules such as the office hours and daily occupancy information present useful knowledge on how the building is occupied. This knowledge thus provides basic energy usage patterns. Also, the weather condition throughout a day (e.g. the hourly temperatures) is strongly correlated to a building's energy consumption curve [30]. Another factor is the pricing fluctuations, which are further complicated by uncertain energy price policies and uncertainty about fossil fuel prices. Real-time pricing quotes from power grid utilities can force a building to dramatically change its energy consumption behaviors. In addition, related social events (e.g. local sport activities and political news) can significantly shift the energy usage and pricing patterns. Also, recall from above, that electricity supplies and consumers increasingly make use of options in order to optimize their financial gains [19].

Another important data source for accurate energy load prediction is the building's daily operations. For example, actions being taken to reshape energy usage curves have a significant impact on the building's short term energy load. Consider a building with an energy storage unit. After having initial short-term predictions, the building managers often aim to reduce buildings' energy usage during peak energy demand periods, which often impost high-energy usage rates for consumers and large load demand for utility grids. In such scenarios, energy storages such as an ice bank, chiller, boiler, and battery, etc., are often used. An ice bank, for instance, is typically used to build ice in summer when the electricity is cheap, and the ice is then used to cool the building, rather than using electricity, when the price of the energy load is high. In order to have accurate short-term load forecasts, features or sensors related to such reschedulable energy-intensive units have to be taken into account. In short, integrating difference sources of data into the learning will allow the mining methods to figure out key components which impact the energy consumption. In particular, it enables the learning algorithm to explore the multiple interconnected data so that important data or attributes (factors) are not excluded [30].

8 Conclusions

This chapter focused on recent advances in data mining in the financial sector, within the context of big data. The Volume and Verocity of such massive datasets, as well as the large number of unknowns and volatility, led to the use of model-driven approaches. To this end, this review mainly centred around model-based approaches currently used when analyzing stocks, bonds, and interest rates. We also turned our attention to the issue of Variety, and briefly reviewed current advances in terms of using social media data to augment and strengthen current predictions.

When the amount of data is relatively small or when the framework in which they evolve is either well understood or deterministic, it is legitimate to primarily use our prior knowledge about the data and not to pay too much attention about what we *don't know*. In such a setting, the use of data-driven modeling approaches, following the standard training, testing, and validation model construction process holds value. The situation is quite different in a big data framework in which our prior knowledge about the data is often rather marginal and has to be supplemented with an assumed model form. Another issue that needs mention is that we often also need to handle late arriving data, i.e. there is a need to incorporate retroactive data as they arrive. Many of these models must be stochastic in order to make allowance for the random nature of the underlying data. As demonstrated, what should be determined carefully is what we *don't know* and how such drawbacks may be marginalized. The path integral approach provides an efficient and coherent framework to marginalize the unknown. This is possible since it considers every possibility and weighs them according to their probability of occurrence, which may be determined from the concomitant model. Despite the fact that the amount of data is big, it does not mean that closed-form expressions are outdated. As a matter of fact, they are more essential than ever, especially in order to reduce the massive dimensionality associated with the problem.

We further believe that the current surge in the area of data stream mining [31] may hold the key to build accurate, just-in-time models to be used by the financial sector. That is, adaptive learning algorithms that build incremental models from asynchronous streams have much application in the financial sector [24]. Specifically, techniques for building dynamic probabilistic models for streaming data [32] have shown to produce high quality results against data that both contain temporal trends and are susceptible to noise and unknowns. Indeed, these types of models may yet prove to be ideal for exploring financial data.

References

1. Baaquie, B.E.: Interest Rates and Coupon Bonds in Quantum Finance. Cambridge University Press, Cambridge (2010)
2. NYSE Market Data, Data Products, Product Summaries. http://www.nyxdata.com/Data-Products/Product-Summaries. Accessed 14 Nov 2014
3. Han, J., Kamber, M.: Data Mining Concepts and Techniques, 2nd edn. Morgan Kauffman (2008)
4. Shasha, D., Zhu, Y.: High Performance Discovery In Time Series: Techniques and Case Studies. Springer (2004)
5. Yi, B.K., Sidiropoulos, N., Johnson, T., Jagadish, H.V., Faloutsos, C., Biliris, A.: Online data mining for co-evolving time sequences. In: Proceedings of 2000 International Conference on Data Engineering (ICDE 2000), pp. 13–22. San Diego, CA (2000)
6. Shumway, R.H., Stoffer, D.S.: Time Series Analysis and Its Applications. Springer (2005)
7. Oshaug, C.A.J.: Lvy Processes and Path Integral Methods with Applications in the Energy Markets. Norwegian University of Science and Technology, Thesis (2011)
8. Paquet, E., Viktor, H.L., Guo, H.: Learning in the presence of large fluctuations: a study of aggregation and correlation, new frontiers in mining complex patterns. LNCS **7765**, 49–63 (2013)
9. Rachev, S.T., et al.: Bayesian Methods in Finance. Wiley, Hoboken (2008)
10. McNeil., A.J., Wendin, J.P.: Bayesian inference for generalized linear mixed models of portfolio credit risk. J. Empirical Finan. **14**(2), 131–147 (2007)

11. Garlappi, L., Uppal, R., Wang, T.: Portfolio selection with parameter and model uncertainty: a multi-prior approach. Rev. Finan. Stud. Oxford J. **20**(1), 41–81 (2007)
12. Jacquier, E., Polson, N.G., Rossi, P.E.: Bayesian analysis of stochastic volatility models. J. Busin. Econ. Statis. **20**(1), 69–87 (2002)
13. Masujima, M.: Path Integral Quantization and Stochastic Quantization. Springer, Berlin (2009)
14. Campbell, J.Y., Low, A.S., Mackinlay, A.C.: The Econometric of Financial Markets. Princeton University Press, Princeton (1997)
15. Lloyds fined 218m over Libor rate rigging scandal. http://www.bbc.com/news/business-28528349?print=true. Accessed 21 Dec 2014
16. Merton, R.C.: Continuous Time Finance. Blackwell Publishing, Oxford (1990)
17. Lemmens, D., Wouters, M., Tempere, J.: Path integral approach to closed-form option pricing formulas with applications to stochastic volatility and interest rate models. Phys. Rev. E **78**, 016101-1–016101-8 (2008)
18. Baaquie, B.E.: Quantum Finance: Path Integrals and Hamiltonians for Options and Interest Rates. Cambridge University Press (2004)
19. Dalakouras, G.V., Kwon, R.H., Pardalos, P.M.: Semidefinite programming approaches for bounding Asian option prices. In: Computational Methods in Financial, Engineering, pp. 103–116 (2008)
20. Devreese, J.P.A., Lemmens, M., Tempere, J.: Path integral approach to asian options in the BlackâĂŞcholes model. Physica A: Statis. Mech. Appl. **384**, 780–788 (2010)
21. Binder, K., Heermann, D.W.: Monte Carlo Simulation in Statistical Physics: An Introduction. Springer, Berlin (2010)
22. Heath, D., Jarrow, R., Morton, A.: Bond pricing and the term structure of interest rates: a new methodology for contingent claim valuation. Econometrica **60**, 77–105 (1992)
23. Baaquie, B.E.: Financial modeling and quantum mathematics. Comput. Math. Appl. **65**, 1665–1673 (2013)
24. Krempl, G., Z̃liobaite, I., Brzezinski, D., Hullermeier, E., Last, M., Lemaire, V., Stefanowski, J.: Open challenges for data stream mining research. ACM SIGKDD Explor. Newslett. **16**(1), 1–10 (2014)
25. Bollen, J., Mao, H.: Twitter mood as a stock market predictor. Computer **44**(10), 91–94 (2011)
26. Zhang, X., Fuehres, H., Gloor, P.: Predicting stock market indicators through twitter ("I hope it is not as bad as I fear"). Procedia: Social Behav. Sci. **26**, 55–62 (2011)
27. Ferrer, J.N., Olivero, S., Medarova-Bergstorm, K., Rizos, V.: Financing models for smart cities. November 2013. http://eu-smartcities.eu/sites/all/files/GuidelineFinancingModelsforsmartcities-january.pdf. Accessed 15 Dec 2014
28. NationalGridUS, Managing Energy Costs in Office Buildings. https://www.nationalgridus.com/non_html/shared_energyeff_office.pdf. Accessed 15 Dec 2014
29. European PPP Expertise Centre (EPEC): Guidance on energy efficiency in public buildings. http://www.eib.org/epec/resources/epec_guidance_ee_public_buildings_en.pdf. Accessed 15 Dec 2014
30. Guo, H.: Modeling short-term energy load with continuous conditional random fields. In: ECML/PKDD 2013 machine learning and knowledge discovery in databases. In: Lecture Notes in Computer Science, vol. 8188, pp. 433–448 (2013)
31. Bosnić, Z., Demsar, J., Kešpret, G., Rodrigues, P.P., Gama, J., Kononenko, I: Enhancing data stream predictions with reliability estimators and explanation. Eng. Appl. Artif. Intell. **34**, 174-188 (2014)
32. Kanagal, B., Deshpande, A.: Online filtering, smoothing and probabilistic modeling of streaming data. In: IEEE 24th International Conference on Data Engineering, 2008, ICDE 2008, pp. 1160–1169 (2008)

Industrial-Scale Ad Hoc Risk Analytics Using MapReduce

Andrew Rau-Chaplin, Zhimin Yao and Norbert Zeh

Abstract Modern reinsurance companies hold portfolios consisting of thousands of reinsurance contracts covering millions of individually insured locations. To ensure capital adequacy and for fine-grained financial planning, these companies carry out large-scale Monte Carlo simulations to estimate the probabilities that the losses incurred due to catastrophic events such as hurricanes, earthquakes, etc. exceed certain critical values. This is a computationally intensive process that requires the use of parallelism to answer risk queries over a portfolio in a timely manner. We present a system that uses the MapReduce framework to evaluate risk analysis queries on industrial-scale portfolios efficiently. In contrast to existing production systems, this system is designed to support arbitrary ad hoc queries an analyst may pose while achieving a performance that is very close to that of highly optimized production systems, which often only support evaluating a limited set of risk metrics. For example, a full portfolio risk analysis run consisting of a 1,000,000-trial simulation, with 1,000 events per trial, and 3,200 risk transfer contracts can be completed on a 16-node Hadoop cluster in just over 20 min. MapReduce is an easy-to-use parallel programming framework that offers the flexibility required to develop the type of system we describe. The key to nearly matching the performance of highly optimized production systems was to judiciously choose which parts of our system should depart from the classical MapReduce model and use a combination of advanced features offered by Apache Hadoop with carefully engineered data structure implementations to eliminate performance bottlenecks while not sacrificing the flexibility of our system.

A. Rau-Chaplin (✉) · Z. Yao · N. Zeh
Risk Analytics Lab, Dalhousie University, Halifax, Nova Scotia, Canada
e-mail: arc@cs.dal.ca

Z. Yao
e-mail: yao@cs.dal.ca

N. Zeh
e-mail: nzeh@cs.dal.ca

© Springer International Publishing Switzerland 2016
N. Japkowicz and J. Stefanowski (eds.), *Big Data Analysis: New Algorithms for a New Society*, Studies in Big Data 16, DOI 10.1007/978-3-319-26989-4_8

1   Introduction

The financial management of the risk associated with catastrophic events such as earthquakes, hurricanes, and large-scale floods falls largely to insurance and reinsurance companies [9, 21, 23]. Their risk portfolios often consist of thousands of reinsurance contracts covering millions of individually insured locations. To quantify risk and to help ensure capital adequacy, each portfolio must be evaluated with respect to a range of risk metrics that take the uncertainty associated with both event order and magnitude into account [1, 37]. These risk metrics represent a probability distribution over the expected losses in a given year. Since this probability distribution is a mix of probability distributions for a large number of individual events, obtaining closed-form expressions for these risk metrics is computationally infeasible. Therefore, reinsurance companies employ large-scale Monte Carlo simulations to closely approximate the portfolio-level loss distribution from the losses incurred over a large number of trials. This is referred to as *aggregate analysis* and is both computationally and data-intensive.

Production risk analysis systems exploit parallelism and aggressively aggregate intermediate results in order to boost performance. The results they produce summarize risk in terms of a small set of standard metrics on the entire portfolio that are important to regulatory bodies, ratings agencies, and an organization's risk management team. These include Probable Maximum Loss (PML) [36, 37] and Tail Value-at-Risk (TVaR) [18, 19]. While production systems can efficiently aggregate potentially terabytes of simulation results into a small set of key risk metrics, they typically do not support ad hoc queries that can help actuaries or underwriters to better understand the multiple dimensions of risk that can impact a portfolio, such as spatial correlation, seasonality, peril features or financial terms.

In this paper, we present a system that allows users with extensive mathematical and statistical skills but perhaps limited programming background, such as risk analysts, to pose a rich variety of complex ad hoc risk queries using an SQL-like syntax and answer these queries efficiently. We implemented our system using Apache's Hadoop implementation of the MapReduce framework, along with a number of associated technologies such as Apache Hive. Our experimental results show that the level of flexibility offered by our system and the ease of exploiting parallelism using MapReduce come at a much smaller performance penalty than expected. Our experiments demonstrate that an industrial-scale risk analysis with 1,000,000 simulation trials, 1,000 events per trial, on a portfolio consisting of 3,200 layers with an average of 5 event loss tables per layer can be carried out on a 16-node Hadoop cluster in just over 20 min, which is close to the time a production system would take to answer such a query.

The key to achieving this level of performance was the elimination of performance bottlenecks we encountered by judiciously departing from a straight MapReduce implementation, exploiting extensions to the MapReduce framework offered by Hadoop, and carefully engineering the core data structures used in our system. The amount of effort involved in this type of performance engineering was substantial but still much less than the years of development that might go into a highly optimized production system.

The design of our system is flexible enough that it should be possible to adapt it to other application areas that employ large-scale Monte Carlo simulations.

The remainder of this paper is organized as follows. Section 2 gives an overview of reinsurance risk analysis. Section 3 describes our risk analysis system. Section 4 considers the implementation of various example queries using our system. Section 5 describes implementation details and algorithmic optimizations we employed to significantly improve the performance of the key data structure in our system. Section 6 presents a performance evaluation of our system. Section 7 presents conclusions and discusses future work.

2 An Overview of Reinsurance Risk Analysis

The analysis of catastrophic risk and reinsurance is a growing area of research with computational models for region peril specific catastrophic risk evaluation [7, 9, 15, 21], spatial exposure analysis [11], financial treaty optimization [13, 34], and portfolio risk analysis [5, 30–32] playing a growing role. In this section we focus on portfolio risk analysis.

A reinsurance company typically holds a *portfolio* of programs that insure primary insurance companies against large-scale losses, like those associated with catastrophic events. Each *program* contains data that describes (1) the buildings to be insured (the *exposure*), (2) the modelled risk to the exposure (the *event loss tables*), and (3) a set of risk transfer contracts (the *layers*).

The *exposure* is represented by a table containing one row per building covered, which lists the building's location, construction details, primary insurance coverage, and replacement value. The modelled risk is represented by an *event loss table* (ELT) that is the output of a region peril catastrophe model [15, 21]. This table lists for each of a large set of possible catastrophic events the expected loss that would occur to the exposure should the event occur. Finally, each *program* is described by the set of layers it covers and by a set of financial terms that include aggregate deductibles and limits (i.e., deductibles and maximal payouts to be applied to the sum of losses over the year) and per-occurrence deductibles and limits (i.e., deductibles and maximum payouts to be applied to each loss in a year), plus other financial terms. For a full discussion of reinsurance contractual terms and their application, see [23]. Each layer may also specify separate financial terms for the individual ELTs it covers.

Consider, for example, a Japanese earthquake program. The exposure might list 2 million buildings (e.g., single-family homes, small commercial buildings, and apartments) and, for each, its location (e.g., latitude and longitude), construction details (e.g., height, material, roof shape, etc.), primary insurance terms (e.g., deductibles and limits), and replacement value. An event loss table might, for each of 100,000 possible earthquake events in Japan, give the sum of the losses expected to the exposure should the event occur. Such ELTs are the output of stochastic region

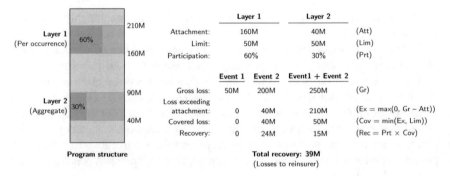

Fig. 1 An earthquake program with two layers and without inuring between the layers. Layer 1 covers 60 % of the losses between 160 and 210 M per earthquake event. Layer 2 is an aggregate layer covering 30 % of the total earthquake losses beteen 40 and 90 M throughout the year. If there are two earthquakes incurring losses of 50 and 200 M, respectively, Layer 1 is applied to each event individually, and Layer 2 is applied to the total losses incurred by both events. Layer 1 does not result in any recovery to the primary insurer for Event 1 because the losses it incurs are below the layer's attachment (deductible). For Event 2, it generates a recovery of 24 M. Through Layer 2, the primary insurer recovers 15 M of their aggregate losses during the year. Note that the recovery is the payout by the reinsurer to the primary insurer, that is, it constitutes the loss incurred by the reinsurer

peril models [21] and typically also include additional primary insurance financial terms. Finally, the program might consist of two layers. The first layer may be a per-occurrence layer that pays out a 60 % share of losses between 160 and 210 million associated with a single catastrophic event. The second layer may be an aggregate layer covering 30 % of losses between 40 and 90 million that accumulate due to earthquake activity over the course of a year. The structure of this program and the payout to the primary insurer in the case of two earthquakes with 50 and 200 million in losses, respectively, is illustrated in Fig. 1.

The most fundamental type of query on a reinsurance company's portfolio computes an exceedance probability (EP) curve, which represents, for each of a set of user-specified loss values, the probability that the total claims a reinsurer will have to pay out in a given year exceeds this value. Not surprisingly, there is no computationally feasible closed-form expression for computing such an EP curve over hundreds of thousands of events and millions of individual exposures. Consequently, a simulation approach must be taken. This is done using a *year event table* (YET) storing a large number of *trials*, each representing one possible sequence of catastrophic events that might occur in a given year. This YET is generated by an event simulator based on the expected occurrence rate of each event plus other hazard information like seasonality. The process to generate the YET is beyond the scope of this chapter, and we focus on the computationally intensive task of computing the expected loss distribution (i.e., EP curve) for a given portfolio from a given YET. The loss value for a particular trial can be computed from the sequence of events in this trial, and the loss distribution is obtained from the losses computed for all trials in the YET.

While computing the EP curve for the company's entire portfolio is critical in assessing a company's solvency, analysts are often interested in posing a wide variety of more fine-grained queries with the goal of understanding such things as cash flow throughout the year, diversity of the portfolio, financial impact of adding new contracts to the portfolio, and many others. The following is a representative, but far from complete, set of example queries.

EP curves with secondary uncertainty: As described above, each entry in an ELT lists the expected loss to the exposure if a certain event occurs, that is, a single loss value. In practice, the loss incurred if a certain event occurs varies based on many exposure and hazard parameters whose interactions are hard to predict. This is called *secondary uncertainty*, in contrast to the *primary uncertainty* whether the event will occur. In order to take secondary uncertainty into account in aggregate analysis, each entry in an ELT is in fact represented as a probability distribution over possible loss values incurred by an event rather than just its expectation.

Performing aggregate risk analysis with secondary uncertainty is computationally intensive due to the statistical tools employed—for example, the beta probability distribution is employed in estimating the loss using the inverse beta cumulative density function [31]—but is essential in many applications.

Return period losses (RPL) by line of business (LOB), class of business (COB) or type of participation (TOP): A layer defines coverage on different types of exposures and the type of participation. Exposures can be classified by class of business (COB) (e.g., property, business interruption or liability coverage) or line of business (LOB) (e.g., marine, property or engineering coverage). The way in which the contractual coverage participates when a catastrophic event occurs is defined by the type of participation (TOP) (e.g., excess of loss or quota share). Decision makers may want to know the loss distribution of a specific layer type in their portfolios, which requires the analysis to be restricted to layers covering a particular LOB, COB or TOP.

Region/peril losses: This type of query calculates a loss distribution for a set of geographic regions (e.g., Florida or Japan), a set of perils (e.g., hurricane or earthquake) or a combination of region and peril. This allows the reinsurer to understand what types of catastrophes add the most risk to their portfolio, and in which regions of the globe they are most heavily exposed to these risks. This type of analysis helps the reinsurer to understand the spatial and peril diversity present in their portfolio and ensure that the portfolio meets high-level underwriting rules established by the managing board.

Multi-marginal analysis: Given a small set of potential new contracts, a reinsurer has to decide which contracts to add to the current portfolio. Adding a new contract means additional cash flow but also increases the exposure to risk. To help with the decision which contracts to add, multi-marginal analysis calculates the difference between the loss distributions for the current portfolio and for the portfolio with any subset of these new contracts added. This allows the insurer to choose contracts or to price the contracts so as to obtain the right trade-off between added cash flow and added risk.

Periodic Loss Distribution: Many natural catastrophes have a seasonal component, that is, do not occur uniformly throughout the year. For example, hurricanes on the Atlantic coast occur between July and November. As a result, the reinsurer may be interested in how their potential losses fluctuate throughout the year, for example to reduce their exposure through reduced contracts or increased transfer of risk to other parties during riskier periods. To aid in these decisions, a periodic loss distribution represents the loss distribution for different periods of the year, such as quarters or months.

Stochastic exceedance probability (STEP) analysis: The final example we consider here differs from the previous examples in that it does not focus on estimating the losses to the portfolio expected due to a series of events but rather on estimating the losses to be expected from an event that has not been modelled before. After the occurrence of a natural disaster not in their event catalogue, catastrophe modelling [21] vendors attempt to estimate the distribution of possible loss outcomes for that event, in order to include the event in an updated event catalogue. One way of obtaining such a loss distribution is to find similar events in existing stochastic event catalogues and propose a weighted combination of the distributions of several events that best represents the actual occurrence. A simulation-based approach allows for the simplest method of producing the resulting combined distribution. To perform this type of analysis, a customized YET must be produced from the selected events and their weights. In this YET, each trial contains only one event, chosen with a probability proportional to its weight. By aggregating the losses in the different trials in a manner similar to a standard portfolio-level risk analysis, a loss distribution for the new event, including various statistics such as mean, variance, and quantiles, is obtained.

3   QuPARA: Large-Scale Query-Driven Portfolio Aggregate Risk Analysis

In this section, we describe the design of our system for large-scale query-driven portfolio aggregate risk analysis, QuPARA. Before doing so, we describe the steps involved in answering an aggregate query sequentially. This will be helpful in understanding the parallel evaluation of aggregate queries using QuPARA.

The loss distribution is computed from the portfolio and the YET in two phases. The first phase computes a *year-event loss table* (YELT). For each trial in the YET and each event in this trial, the YELT contains a tuple ⟨trial, event, time, loss⟩ recording the time the event occurs during the year and the loss incurred by this event, given the financial terms of the layer and the sequence of events in the trial up to the current event. The second phase then aggregates the entries in the YELT to compute the final loss distribution. Algorithm 1 shows the sequential computation of the YELT.

Algorithm 1: Sequential Aggregate Risk Analysis

Input: Portfolio and YET
Output: YELT

---

1 **for** *each trial T in the YET* **do**
2   **for** *each event X in T* **do**
3          $l_X \leftarrow 0$
4   **for** *each program P in the portfolio* **do**
5   **for** *each layer L in P* **do**
6                  $l_L \leftarrow 0$
7   **for** *each ELT E covered by L* **do**
8   Lookup X in E to determine the loss l_E associated with X.
9   Apply the financial terms associated with ELT E under layer L to l_E.
10                      $l_L \leftarrow l_L + l_E$
11   Apply layer L's financial terms to l_L.
12                  $l_X \leftarrow l_X + l_L$
13   Append a tuple $\langle T, X, t_X, l_X \rangle$ to the YELT, where t_X is the time of event X in the trial T.

For each trial T, each event X, each program P, and each layer L in this program, the algorithm first looks up the losses incurred by a given event X in the ELTs covered by L, applies the financial terms L associates with these ELTs to the individual losses incurred under each ELT, sums up the resulting losses to obtain a *layer loss* l_L, and finally applies L's financial terms to l_L. The resulting layer loss is then added to the total loss l_X incurred by event X in the current trial T under all programs in the portfolio. Once this computation has been applied to all programs in the portfolio, the final loss value l_X is recorded by adding a tuple $\langle T, X, t_X, l_X \rangle$ to the YELT. Note that applying aggregate financial terms to individual events requires a lookup table (e.g., a hash table) to keep track of the total losses incurred in the current trial under each layer and possibly under each ELT covered by a given layer.

In order to answer ad hoc aggregate queries on industry-size portfolios efficiently, QuPARA provides a parallel implementation of the above algorithm using MapReduce. The YELT is computed by the mappers, while the final loss distribution(s) are computed by the reducer(s).

Figure 2 visualizes the design of QuPARA. The system is split into a front-end offering a *query interface* to the user and a back-end consisting of a *distributed file system*, a set of *filters*, and a *query engine*. The distributed file system is implemented using Hadoop's HDFS and stores the portfolio and YET. The filters are responsible for selecting the records from the portfolio to which the query is to be applied, based on the user's query. These filters are implemented using Hive. The query engine finally evaluates the query over the selected set of records using Apache Hadoop. Next we discuss these components in more detail.

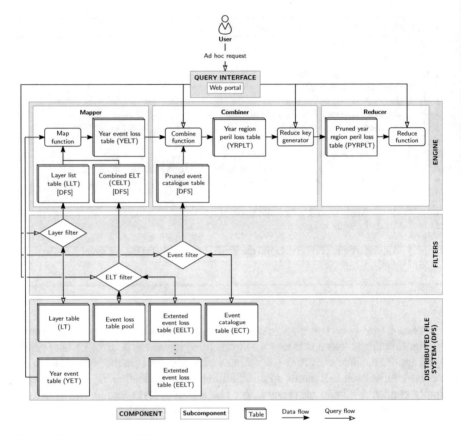

Fig. 2 The design of QuPARA

3.1 Query Interface

The query interface offers a web-based portal through which the user can issue ad hoc queries in an SQL-like syntax. The user enters the query in the form of multiple sub-queries; each sub-query controls the operation of one of the functions in the query engine or of one of the filters. Although this division of the query into sub-queries directly corresponds to the different components of QuPARA, it also provides a natural way to logically structure a query.

3.2 Distributed File System

The distributed file system is implemented using Hadoop's HDFS and stores the portfolio and YET in the form of the following tables (see Table 1 for a summary).

Table 1 The different tables used by QuPARA divided into tables that are part of the portfolio (top) and intermediate tables produced while answering a query and discarded afterwards (bottom)

Table	Description
YET	Year-event table produced by an event simulator. Stores a sequence of trials, that is, possible event sequences for a given year
LT	Layer table storing the one record per layer in the portfolio, including the layer's financial terms and list of event loss tables
ELTP	Event loss table pool associating a region and peril with every ELT in the portfolio
(E)ELT	(Extended) event loss table stores a loss distribution for every event, capturing the probable loss should the event occur
ECT	Event catalogue table associating a region and peril with every event
LLT	Layer list table: reduced version of the LT containing only the layer records relevant to the query, with fields used only for identifying relevant layers removed
CELT	Combined event loss table, stores all the records in the ELTs used by the query, used to improve query performance
YELT	Year event loss table storing one loss record for every trial (year) and every event in the selected set of ELTs
YRPLT	Year region peril loss table storing one loss record for every combination of time period, region, and peril

For each table, we give a description of the record fields relevant to the discussion in this paper and then briefly mention additional fields that are needed by a production system but do not change the overall flow of the computation:

- The YET contains tuples ⟨trial_ID, event_ID, time_Index⟩, trial_ID is a unique identifier associated with each of the 1 million trials in the simulation. event_ID is a unique identifier associated with each event in the event catalogue. time_Index determines the position of the occurrence of the event in the sequence of events in the trial.
  In addition, each tuple stores a random number z_PE specific to the event occurrence that controls the level of correlation of the losses to the different properties affected by the event. The values of z_PE associated with different occurrences of the same event in the same or in different trials are independent of each other.
- The layer table (LT) contains tuples ⟨layer_ID, cob, lob, top, elt_IDs, occ_Ret, occ_Lim, agg_Ret, agg_Lim⟩. layer_ID is a unique identifier of each layer in the portfolio. cob is an industry classification according to perils insured and the related exposure. It groups homogeneous risks. lob defines a set of one or more related products or services where a business generates revenue. top describes how reinsurance coverage and premium payments are calculated. elt_IDs is a list of event loss table IDs that are covered by the layer. occ_Ret is the occurrence retention or deductible of the insured for an individual occurrence loss. occ_Lim is the occurrence limit or coverage the insurer will pay for occurrence losses in excess of the occurrence retention.

`agg_Ret` is the aggregate retention or deductible of the insured for an annual cumulative loss. `agg_Lim` is the aggregate limit or coverage the insurer will pay for annual cumulative losses in excess of the aggregate retention.

- The event loss table pool (ELTP) contains tuples ⟨`elt_ID`, `region`, `peril`⟩ associating a particular type of peril and region with the ELT with ID `elt_ID`.
- An extended event loss table (EELT) contains tuples ⟨`event_ID`, `mean_Loss`, `sigma_I`, `sigma_C`, `max_Loss`⟩. `event_ID` is the unique identifier of each event in the event catalogue. `mean_Loss` and `max_Loss` denote the mean loss and maximum expected loss incurred if the event occurs, respectively. `sigma_I` represents the variance of the loss distribution for this event assuming all losses to properties affected by the event are completely indepentent. `sigma_C` represents the variance of the loss distribution for this event assuming all losses to properties affected by the event are completely correlated.

  In addition, each record stores a random number `z_E` that controls the level of correlation of the losses to the different properties affected by the event, but different occurrences of the same event across trials share the same `z_E` value.
- The event catalogue table (ECT) contains tuples ⟨`event_ID`, `region`, `peril`⟩ associating a region and a type of peril with each event.

3.3   Filters

QuPARA uses three filters that allow the user to focus their queries on specific geographic regions, types of peril, etc. These filters select the appropriate entries from the data tables stored in the distributed file system for further processing by the query engine. The predicate used by each of these filters to select entries from the table it operates on is provided by the user through the query interface.

Layer filter. The *layer filter* allows the selection of layers from the portfolio based on their attributes in the layer table (LT). Once the filter has been applied, attributes such as `cob`, `lob` or `top` are no longer needed for the analysis. Thus, the layer filter prunes these fields from the extracted records to produce a *layer list table* (LLT) containing one ⟨`layer_ID`, `elt_IDs`, `occ_Ret`, `occ_Lim`, `agg_Ret`, `agg_Lim`⟩ tuple for each extracted record. The LLT is distributed to the mappers via HDFS. The union of the lists of ELT IDs in the records of the LLT is passed to the ELT filter for retrieval of the ELTs needed by the query.

ELT filter. The *ELT filter* is used to select, from the pool of EELTs, the set of ELTs required by the query. To do this it uses the list of ELT IDs received from the layer filter, as well as a user-specified predicate, to filter ELTs based on the region and type of peril they cover, as stored in the event loss table pool (ELTP). The resulting set of ELTs is collected and sent to the mapper via HDFS.

Event filter. The *event filter* selects the features of events, such as region or type of peril, from the event catalogue table (ECT), to be used for grouping of event losses based on these features. After pruning all but the event attributes required by the query from the records in the ECT to produce a pruned event catalogue table, it sends this table to the combiners via HDFS.

Note that both the ELT filter and the event filter allow the query to be focused on specific types of peril and regions. The difference between the two filters is that the ELT filter uses this information to decide which ELTs in the portfolio to restrict the query to while the event filter does not really perform any filtering but only selects the set of attributes of events, not ELTs, used for grouping of losses in the final output.

3.4 Query Engine

The query engine is implemented using Hadoop and evaluates the query using a single MapReduce round consisting of a map/combine step and a reduce step. During the map step, the engine uses one mapper per trial in the YET, in order to construct a YELT from the YET. This is a parallelization of Algorithm 1 using one mapper per iteration of the outer loop. The combiner and reducer collaborate to aggregate the loss information in the YELT into the final loss distribution for the query. There is one combiner per mapper. The combiner pre-aggregates the loss information produced by this mapper, in order to reduce the amount of data to be sent across the network to the reducer(s) during the shuffle step. The final aggregation can then be carried out either by the reducers or in a post-processing step. In our implementation we chose to use trivial reducers that simply write the data records they receive to output files, one per reducer. A post-processing step of these files then produces the final loss distribution. The reason for this design choice was that it may be desirable to produce multiple outputs from the same analysis. By producing these outputs in a post-processing step, it is not necessary to re-run the analysis. There is no performance penalty compared to carrying out the aggregation in the reducer because the amount of data to be processed in the post-processing step is negligible compared to the work carried out by the mapper and combiner. In most queries, which require only a single loss distribution as output, there is a single reducer. Multi-marginal analysis is an example where multiple loss distributions are to be computed, one per subset of the potential contracts to be added to the portfolio. In this case, we have one reducer for each such subset, and each reducer writes the output data necessary for producing the loss distribution for its corresponding subset of contracts.

Mapper. Each mapper operates on the set of layers and ELTs provided by the layer filter and the ELT filter via HDFS. The mapper constructs a *combined ELT* (CELT) from the given set of ELTs, which associates a loss with each ⟨event, ELT⟩ pair. It then iterates over the sequence of events in its trial, looks up the ELTs recording non-zero losses for each event, and generates the corresponding ⟨trial, event, time, loss⟩ tuple in the YELT, taking each layer's financial terms into account. Algorithm 2 shows the details of the mapper, excluding the construction of the CELT, which will be discussed in more detail in Sect. 5.

Algorithm 2: Mapper in parallel aggregate risk analysis

Input: $\langle T, E := \{(E_1, t_{E_1}), (E_2, t_{E_2}), \cdots, (E_m, t_{E_m})\}\rangle$, where m is the number of events in a trial, E_i is the ith event, and t_{E_i} is the time at which event E_i occurs.

Output: The list of YELT entries $\langle T, E_1, t_{E_1}, l_{E_1}\rangle, \langle T, E_2, t_{E_2}, l_{E_2}\rangle, ..., \langle T, E_m, t_{E_m}, l_{E_m}\rangle$ for trial T

1 Construct a CELT from the list of ELTs received from the ELT filter.
2 **for** *each event E_i in E* **do**
3      $l_{E_i} \leftarrow 0$
4   Retrieve the losses $L_{E_i} = \{l^1_{E_i}, l^2_{E_i}, \cdots, l^n_{E_i}\}$ associated with event E_i in the CELT, where $ELT_1, ELT_2, ..., ELT_n$ are the ELTs in the CELT and $l^j_{E_i}$ is the loss recorded for event E_i in ELT_j.
5   **for** *each layer L in the LLT* **do**
6          $l_L \leftarrow 0$
7   **for** *each ELT ELT_j covered by L* **do**
8   Lookup $l^j_{E_i}$ in L_{E_i}
9   Apply the financial terms layer L associates with E_i to $l^j_{E_i}$.
10             $l_L \leftarrow l_L + l^j_{E_i}$
11   Apply L's financial terms to l_L.
12         $l_{E_i} \leftarrow l_{E_i} + l_L$
13   Emit($\langle T, E_i, t_{E_i}, l_{E_i}\rangle$)

Combiner. The aggregation to be done by the combiner depends on the query. In the simplest case, a single loss distribution for the selected set of layers and ELTs is computed. In this case the combiner sums the loss values in the YELT it receives from its mapper and sends the aggregate value to the reducer. A more complicated example is the computation of a weekly loss distribution. In this case, the combiner would aggregate the losses corresponding to the events in each week and send each aggregate to a different reducer. Each reducer is then responsible for computing the loss distribution for one particular week. The combiner may also aggregate losses based on peril or region. The result of this aggregation is a *year region peril loss table* (YRPLT) that stores one loss value for each combination of time period, type of peril, and region the query distinguishes.

To carry out this grouping of loss values, the combiner first joins the portion of the YELT it receives from its mapper with the pruned event catalogue table received from the event filter, in order to annotate each event with the set of features relevant to the query. Then the events are grouped according to these features as specified by the user and one aggregate loss value for each group is added to the YRPLT. This computation is carried out by the combine function in Fig. 2. In order to be able to produce the final output, each entry in the YRPLT has to be sent to the right reducers. In QuPARA, there is one reducer for each combination of time period, type of peril, and region of interest to the query. Thus, each YRPLT entry is to be sent to a different reducer and the reduce key generator in Fig. 2 annotates each YRPLT entry with the ID of the reducer that corresponds to its combination of event attributes.

Reducer/postprocessing. As already mentioned, each reducer simply produces a file that stores the input records it receives from the different combiners. The final output of the analysis is produced from each file in a postprocessing step. For example, to generate an exceedance probability curve, the postprocessing step sorts the received loss values in increasing order and, for each loss value v in a user-specified set of loss values, reports the percentage of trials with a loss value greater than v as the probability of incurring a loss higher than v.

4 Posing Queries to QuPARA

In Sect. 4.1 we demonstrate the flexibility of QuPARA by discussing the expression of a complex risk analysis query using the sub-queries in QuPARA's query interface. In Sect. 4.2 we briefly discuss how each of the sample queries from Sect. 2 can be expressed using QuPARA's query interface.

4.1 An Example Query in Detail

As an example, consider generating a report on seasonal loss value-at-risk (VaR) with a confidence level of 99 % due to hurricanes and floods that affect all commercial properties in different locations in Florida. In other words, the goal is to bound the loss due to hurricanes and floods in each of the four seasons of the year; the desired bound is the lowest loss value that is exceeded in only 1 in 100 trials in the YET. The query is broken down into five SQL-like sub-queries passed to the different filters and functions:

Layer filter: We are interested in all layers covering commercial properties. This translates into the following SQL query passed to the layer filter:

```
SELECT * FROM LT
WHERE lob IN commercial
```

ELT filter: We are interested in all ELTs covered by layers selected by the layer filter (LAYER_ELTS) and which cover Florida (FL) as a region and hurricanes (HU) and floods (FLD) as perils:

```
SELECT elt_ID FROM ELTP
WHERE elt_ID IN LAYER_ELTS
AND region IN FL
AND peril IN HU, FLD
```

Event filter: The grouping of losses in our query does not require any information from the ECT, so we provide a trivial event filter:

```
SELECT event_ID FROM ECT
```

Grouping query (combine function): Since we are interested in loss values by season, the combiner produces one loss value per season by grouping the losses in the YELT by season and aggregating the losses in each group:

```
SELECT trial_ID,
SUM(estimated_Loss)
FROM YELT
GROUP BY SEASON(time_Index)
```

Reduce function/postprocessing: The seasonal loss Value-at-Risk (VaR) with 99 % confidence level can be computed from the trial losses by the reducer or post-processing step using the following query:

```
SELECT *
FROM YRPLT
WHERE ROWNUM = 0.01 * (
   SELECT COUNT(*) FROM YRPLT
)
ORDER BY loss
```

4.2   A Sample of Typical Risk Analysis Queries

The following section briefly describes how to implement each of the example queries from Sect. 2 in QuPARA:

EP curves with secondary uncertainty: We ignore the treatment of secondary uncertainty here, as the difference is simply in the treatment of loss "values", which are probability distributions when taking secondary uncertainty into account. This query analyzes the entire portfolio, so the layer filter selects all layers in the LT, the ELT filter selects all ELTs in the portfolio, and the event filter does not select any attributes from the ECT because no grouping on even attributes is performed. The combiner sums the loss values of all YELT entries it receives from its mapper, and all combiners assign the same reduce key to the loss values they produce. The result written to disk by the reducer is a list of loss values, one per trial. By sorting these loss values and counting the fraction of loss values above user-specified critical values, we obtain the exceedance probabilities for these critical values.

Return period losses by LOB, COB or TOP: Similar to the previous query, a single query distribution is to be produced, but this time it is restricted to layers covering a particular LOB, COB, TOP or combination thereof. The layer filter is used to select the layers relevant to the query. The remainder of the query is identical to a full-portfolio analysis as described above.

Region/peril losses: In this case the layer filter extracts the complete set of layers in the LT as the query is not restricted to a particular LOB, COB, or TOP. The ELT filter, however, extracts only the ELTs that cover events of the peril type(s) and/or affecting the region(s) of interest to the query. If only a single loss distribution is to be produced for the selected set of regions and perils, then the remainder of the query is once again identical to a full-portfolio analysis. If the losses are to be grouped by region and/or peril, the event filter extracts region and/or peril information from the ECT, the combiner groups events in the YELT based on their associated region/peril information extracted from the ECT, produces one loss value per group, and sends each loss value to a different reducer by tagging it with a different reduce key. Each reducer writes its loss values to disk and the postprocessing step produces an EP curve from the set of values written by each reducer.

Multi-marginal analysis: For this query, the layer filter can be seen as extracting two layer lists, one containing all layers of the base portfolio and one containing the candidate layers to be added to the portfolio. In reality, the layers are all in one single list but are tagged as *base layers* or *additional layers*. The ELT filter once again extracts all ELTs covered by the extracted layers. The event filter does not extract any information from the ECT because no grouping based on peril or region is to be performed. The mapper constructs a standard YELT from the list of base layers and outputs one layer loss per event for each additional layer. The combiner aggregates the base portfolio losses of all events into a single portfolio loss. The reduce key generator generates a distinct reduce key for each possible combination of additional layers, tags every base portfolio loss with each of these keys, that is, sends each base portfolio loss to every reducer. Every layer loss is tagged with the reduce keys corresponding to all combinations of additional layers that include this layer. Each reducer then produces the loss distribution for one possible combination of additional layers by aggregating the base portfolio and trial losses it receives into a single loss distribution.

Periodic loss distribution: This query is applied to the entire portfolio, so once again the layer filter selects all layers in the LT and the ELT filter selects all ELTs in the portfolio. The grouping is done based on the times the events occur, so no information from the ECT is needed and the event filter does not select any attributes from the ECT. The combiner then translates the time of occurrence of each event in the YELT into a time period appropriate for the query, such as season or month, groups the YELT entries by their periods, and sums the loss values in each group to obtain one loss value per period. Each loss value is tagged with a reduce key corresponding to the period it belongs to. Each reducer once again writes all records it receives to disk, and the postprocessing step produces an EP curve from the records written by each reducer.

STEP analysis: This analysis is implemented in the same way as a standard EP curve calculation for the entire portfolio. The difference is that the analysis is applied to a custom YET, generated as discussed in Sect. 2, and that the postprocessing step does not produce an EP curve from the set of loss values written by the reducer but instead produces a description of the loss distribution in terms of expected loss and variance.

5 Implementation

Given that most of the work in QuPARA is carried out by the mappers, the key to performance was the efficient implementation of the mapper and in particular the lookup of loss values in the ELTs. Even in our optimized final implementation, these lookups account for a significant portion of the running time of a QuPARA query, so improvements in lookup performance have an immediate impact on the overall query performance.

The first step to improving lookup performance was the construction of a CELT by each mapper, which it uses for subsequent lookups of loss values. We explain in Sect. 5.1 why the use of a CELT improves lookup performance substantially. Alas, a CELT cannot be used for industrial-scale portfolios within the QuPARA system as described in Sect. 3 because the CELT is too big to fit in the memory of a single mapper. Therefore the second step was to refine the design of QuPARA to split the evaluation of queries into multiple MapReduce rounds operating on only a portion of the portfolio each, so the CELT used in each round does fit in the memory of a mapper. We discuss this in detail in Sect. 5.2. The third step focuses on improving the performance of the CELT itself. In particular, we carefully engineered a CELT implementation that is space-efficient and at the same time optimizes lookup performance. Since lookups account for a significant portion of the query time, improved lookup performance has a direct impact on the running time of a QuPARA query. A more space-efficient CELT representation has the effect of increasing the size of the portion of the portfolio whose CELT fits in memory and thus leads to a reduction in the number of required MapReduce rounds, which also improves performance significantly. The details of our CELT implementation are discussed in Sect. 5.3.

5.1   The Case for a Combined ELT

Given that determining the losses associated with each event in a trial needs to take the order of these events into account, each mapper iterates over the event sequence in its trial and, for each event, looks up associated losses in the ELTs covered by the current query. If these lookups are performed directly on the ELTs, they incur one access to DFS per lookup because each lookup accesses a different ELT. Even if the ELTs were first copied into the memory of the mapper, the cost per lookup would be substantial because ELTs are large.

The pool of ELTs can be viewed as a two-dimensional table that allows lookups primarily by ELT ID (by locating the ELT of interest) and secondarily by event ID (by performing an event lookup in the chosen ELT). The CELT transposes this table and restricts it to only the event IDs that occur in the current trial. Specifically, the CELT stores a loss value for every ⟨event, ELT⟩ pair, where the event occurs in the current trial and the ELT is one of the ELTs covered by the current query. The CELT is indexed primarily by event IDs and secondarily by ELT IDs. Thus, for a given event, only a single lookup is required to locate the row of the CELT corresponding to the event. For each layer covered by the query, this row is then searched using the ELT IDs covered by this layer, in order to calculate the layer loss associated with the current event. This substantially improves performance because the individual lookups by ELT ID operate on a much smaller table than lookups by event ID in an entire ELT, and these lookups are localized in memory, which improves cache efficiency.

The CELT, viewed as an event-ELT matrix as just described, is very sparse, that is, the loss value associated with most event-ELT pairs is 0. This is true because most events are covered by only a small number of ELTs. Thus, an obvious optimization to reduce the size of the CELT is to store only its non-zero entries. This can be achieved by organizing the CELT as a two-level structure consisting of a table indexed by event IDs whose entries are (pointers to) tables indexed by ELT IDs. Each such "secondary table" associated with a given event ID stores loss values only for ELTs that report a non-zero loss for this event. Even using this optimization, the CELT does not fit in the memory of a single mapper. In the next two subsections, we discuss how we (i) refined the design of QuPARA to employ multiple MapReduce rounds so that the CELT used in each round does fit in memory and (ii) carefully engineered the implementation of this two-level CELT representation to improve its size and lookup cost.

5.2 Multiple Phases

The computation of the YELT produces one loss value for each ⟨trial, event, ELT⟩ tuple, where the event occurs in the trial and the ELT is one of the ELTs covered by the query. Dependencies exist only between loss values associated with events in the same trial and covered by the same ELT or layer (due to the ordering of events and the application of aggregate financial terms). This allows us to divide the layers covered by a query into *batches* such that the YELT entries corresponding to one batch can be produced completely independently of the YELT entries corresponding to other batches. The portion of the YELT corresponding to each batch can then be produced in a separate MapReduce round. If the batch size is chosen small enough, the CELT needed for producing the YELT entries for this batch fits in the memory of a single mapper. Indeed, a single trial typically contains around 1000 events and a single layer covers 5 ELTs on average. A 1000×5 CELT can be held in memory even as a full matrix. Using our space-efficient CELT implementation described in Sect. 5.3, we can choose a batch size of between 100 and 200 layers and thus have between 500 and 1000 ELTs on the hardware we use to run our experiments.

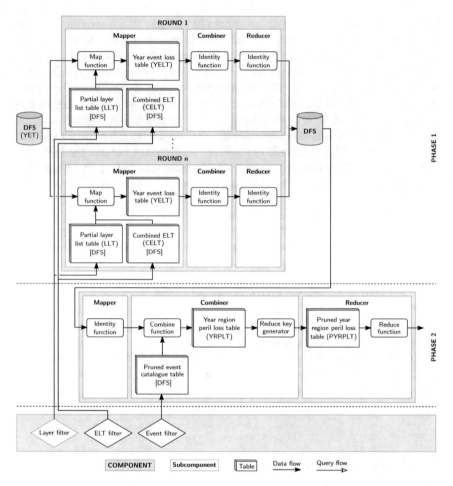

Fig. 3 Splitting the implementation of QuPARA over multiple MapReduce rounds reduces the memory requirements of the CELT. Each *grey box* in Phases 1 and 2 is a separate MapReduce round. The query interface and tables stored on the DFS are not shown as they are the same as in Fig. 2

While the production of the YELT can be split into multiple batches, the production of the final query output requires all YELT entries. Thus, the final query answer can only be produced once the MapReduce rounds for all batches have ended. We therefore split the evaluation of a query into two phases (see Fig. 3). The first phase consists of as many MapReduce rounds as there are batches. In each round of this phase, the mappers produce the YELT portions corresponding to the current batch as described above. The combiners and reducers in each round of this phase are trivial, that is, the combiners do nothing, and the reducers simply write the tuples they receive to disk. The second phase consists of a single MapReduce round with trivial mappers and where the combiners and reducers produce the final query answer as described in Sect. 3. Specifically, every mapper reads the YELT entries corresponding to one trial and simply passes the read entries to its associated combiner. The combiner then performs grouping and preaggregation of loss values based on, for example, event features. The reducers then either produce the final loss distribution from the tuples they receive from the combiners or, as in our implementation, write the tuples they receive to disk for consumption by a postprocessing step that produces the final query answer.

5.3   An Efficient CELT Implementation

The remainder of this section describes the efficient CELT implementation we engineered in order to improve the performance of QuPARA. The choices we made in this implementation strongly relied on insights gained from exploratory performance experiments conducted as part of the development process. For this reason, we start with a discussion of the experimental setup we used.

Experimental setup. All experiments in this section were performed on a 2.66 GHz Quad Core Intel Xeon X3350 processor with 4 GB DDR2 RAM and three 1 TB 7,200 rpm SATA disk drives. The operating system was CentOS 6.3 with Java version 1.7.0_03. The Java heap size was limited to 2 GB. The system evaluation in Sect. 6 was performed on a cluster of 19 of these machines configured as a Rocks cluster [29] connected using Gigabit Ethernet. For the performance evaluation of QuPARA, the Java heap size per node in the cluster was limited to 1 GB, in order to set aside memory needed for the Hadoop runtime system [33, 35]. Hadoop and HDFS was provided as part of the Cloudera BigData platform version 4.7.3 [10], which provides Hadoop version 2.0.0-cdh4.5.0, HIVE version 0.10.0, and Pentaho version 4.8-CE. One of the 19 nodes in the cluster was configured as a master node running the job tracker and job queue and serving as the major name node for HDFS, while the remaining nodes were worker nodes (and data nodes for HDFS) with a total of 72 cores available to run MapReduce jobs. The maximum capacity of HDFS on our system was 20 TB.

The data sets used in our experiments were sampled uniformly at random from a real-world portfolio consisting of 1,600 layers with 5 ELTs per layer and 10,000 loss entries per ELT. Each loss entry consisted of one integer and four doubles representing the event ID and various loss perspectives (e.g., total damage incurred by the event or total losses to the primary insurers incurred by the event), respectively. QuPARA currently uses only the final perspective in its analysis, which reflects the losses the reinsurer has to cover after primary insurance terms have been applied and support, for example, from governmental catastrophe relief programmes has been applied. The YET used for the experiments consisted of 1,000,000 trials, each containing 1,000 events.

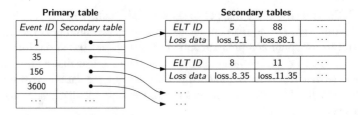

Fig. 4 Two-level structure of the CELT

High-level data structure. The CELT is conceptually a matrix that associates loss information with ⟨event, ELT⟩ pairs. In theory, the fastest access time is achieved by storing the CELT as a 2d array, which allows trivial constant-time lookups using simple index arithmetic. However, since the valid range of event IDs and ELT IDs is the range of 32-bit integers, the size of such a representation would be astronomical. Even if we performed index mapping on the row and column IDs of the array so we store only columns corresponding to ELTs included in the CELT and only rows corresponding to events included in at least one of these ELTs, the table would be extremely sparse, as most events are covered by only very few ELTs.

A simple space-efficient representation of a sparse matrix with expected constant lookup time consists of a primary hash table indexed by row indices; the value associated with each row index is a reference to a secondary hash table storing the non-zero entries in this row indexed by their column indices. As already discussed, we chose event IDs as row indices, and ELT IDs as column indices (see Fig. 4). Our *baseline implementation* realizes this two-level structure and implements the hash tables using a Java STL HashMap [27].

Updates and lookups on this data structure are straightforward. An insertion of a value v with key (e, t) first looks up the event ID e in the primary table. If this event ID is found, its associated value is a secondary table, into which v is inserted with key (ELT ID) t. If e is not found in the primary table, we first create a new secondary table and insert it into the primary table with key e. Then we insert v with key t into the secondary table just created.

Similarly, a lookup operation with key (e, t) first looks up the event ID e in the primary table. If the event ID is not found, the operation returns immediately and reports that there is no value associated with key (e, t) in the CELT. If the event ID is found, its associated value is a secondary table and we return the result of the lookup with key t in this table (which may be that no value is associated with key t in this table).

In QuPARA, all CELT lookups with the same event ID e (but different ELT IDs) are consecutive. Thus, we can optimize lookups further by performing only a single lookup with key e in the primary table for all these lookup operations. If this lookup fails, we can report failure for the entire batch of lookup operations. If it succeeds, we use the returned secondary table for lookups using the ELT IDs in this query batch. This optimization reduces the total lookup cost in the primary table during a QuPARA run to a minimum, so the lookup cost in the secondary tables dominates the total time spent on CELT lookups. Given that the space usage of the CELT representation is also dominated by the space occupied by the secondary tables, our focus is mostly on optimizing the insertion and lookup times and the space usage of the secondary tables.

Choice of HashMap implementation and JVM parameters. There exist a number of open-source alternatives to the Java standard library (STL) whose data structures have been optimized for performance. The HashMaps in the two-level structure we have just described can be implemented using the HashMap implementations provided by any of these libraries. In our experiments, we evaluated the Java STL HashMap [27], the TIntObjectHashMap class provided by the GNU Trove library [17], and the IntObjectOpenHashMap class provided by the high-performance primitive collections (HPPC) library [28].

In order to maximize the performance of each implementation in terms of construction and lookup time, we tuned various JVM parameters that impact the garbage collection overhead involved in these operations. These parameters include the initial heap size and the ratio between the amounts of space allocated to the young and old generations in Java's generational garbage collector.

A higher initial heap size makes the program use more memory initially but reduces the number of times the heap is resized as the heap space currently allocated becomes insufficient to hold newly created objects. Since QuPARA processes layers in batches chosen to fill the available memory, the heap always fills the entire memory eventually, so there is no space penalty to allocating the maximum available heap space also as the initial heap size, but the performance benefit of avoiding heap resizing altogether is significant. We verified experimentally that setting the initial heap size to the maximum heap size (2 GB on our system) results in the best performance for all three HashMap implementations.

A higher ratio between the young and old generation's space allocations reduces the frequency of minor garbage collection runs (which sweep only the young generation) but may lead to old objects remaining in the young generation's space if there is no room left in the old generation's space. When this happens, old objects are swept repeatedly during minor GC runs, which hurts performance. In QuPARA, a very large number of permanent objects (which all eventually become old) are created as part of the CELT. Thus, allocating more space to the old generation should be beneficial. We verified experimentally that the STL HashMap achieves the best overall performance (lookup and insertion performance) with a young-old ratio of 1:3. For the Trove and HPPC HashMaps, the optimal ratios were 1:2 and 1:3, respectively.

In our performance comparisons, we ran each implementation using its optimal JVM parameters determined above. We compared the space usage and the times taken to process batches of insertions and lookups. Since a QuPARA query first constructs the CELT by inserting loss values into it one by one and then performs lookups on the constructed CELT without changing it further, these experiments are representative of the performance characteristics of these HashMap implementations as part of a QuPARA run.

Figures 5, 6, and 7 compare the total insertion times, the total times taken to process 10 m lookups, and the memory footprints of the HashMap implementa-

Fig. 5 Insertion times of the different HashMap implementations

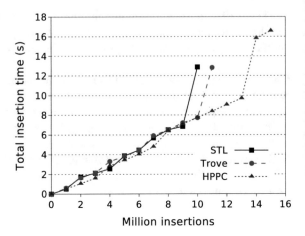

tions as functions of the number of inserted elements, respectively. The insertion costs of all three implementations differed little up to 9 m elements. However, the higher memory footprints of the STL and Trove HashMaps compared to the HPPC HashMap made the garbage collector thrash as these two implementations ran out of memory after 10 and 11 m insertions, respectively. The HPPC HashMap also runs out of space eventually, but this happens only after 15 million insertions. The Trove HashMap implementation achieved the lowest lookup cost, which was 4 % lower than the lookup cost of the HPPC HashMap, and 24 % lower than the lookup cost of the STL HashMap. Given that a smaller memory footprint enables us to process fewer, larger batches of layers in QuPARA, the substantially higher number of elements that can be handled by the HPPC HashMap implementation in the 2 GB of memory we had available compared to the Trove HashMap implementation outweighs the 4 % increase in lookup performance, and we concluded that the HPPC implementation is

Fig. 6 Total time for 10 m lookups using the different HashMap implementations

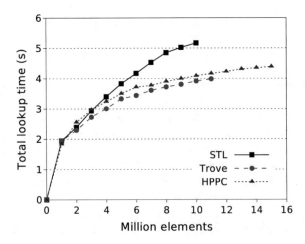

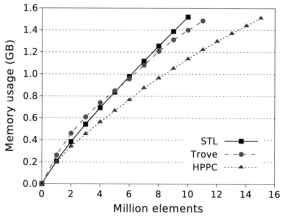

Fig. 7 Memory usage of the different HashMap implementations

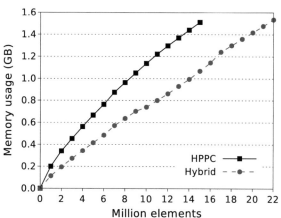

Fig. 8 Memory usage of the HPPC HashMap CELT and the hybrid CELT

the best choice of HashMap implementation to be used in our CELT implementation.

Hybrid CELT implementation. An obvious approach to further reduce the space used by the secondary tables (and, hence, of the entire CELT), in order to increase the size of the batches QuPARA can process, is to implement the secondary tables as ArrayLists rather than HashMaps. Since we continue to use an HPPC HashMap to implement the primary table, we refer to this as a *hybrid* CELT implementation. As can be seen in Fig. 8, the hybrid CELT implementation uses substantially less space than the CELT implementation using only HPPC HashMaps, which we refer to as the HPPC CELT implementation. The construction of the hybrid CELT proceeds in two phases: The first phase inserts elements one by one as in the HPPC CELT but simply appends each key-value pair inserted into a secondary table to the end of this table. Once all loss values have been inserted into the CELT, we sort the entries in each secondary table by their keys to enable lookups using binary search in these

Fig. 9 Insertion times of the HPPC HashMap CELT and the hybrid CELT

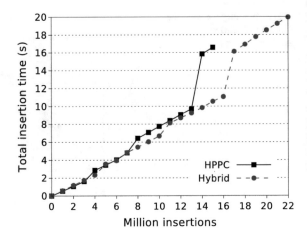

Fig. 10 Total time for 10 m lookups on the HashMap CELT and the hybrid CELT

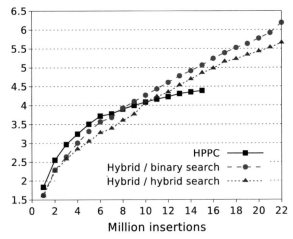

tables. Lookup operations on the hybrid CELT differ from the lookups on the HPPC CELT only in that they use binary search on the secondary tables.

As can be seen in Fig. 9, the insertion times of the hybrid CELT and the HPPC CELT differed only insignificantly until the HPPC CELT started thrashing at its limit of 15 m elements. As can be seen in Fig. 10, the lookup cost of the HPPC CELT is lower than that of the hybrid CELT. However, this penalty is outweighed by the increase in the batch size that can be processed using the hybrid CELT implementation (22 vs. 15 m elements, a 46 % increase).

The performance comparisons in Figs. 8, 9, and 10 were again performed using the optimal JVM parameters for each implementation. For the hybrid CELT implementation, we verified experimentally that an initial heap size of 2 GB once again yielded the best performance, as did a young-old ratio of 1:2.

Fig. 11 Distribution of
secondary table sizes

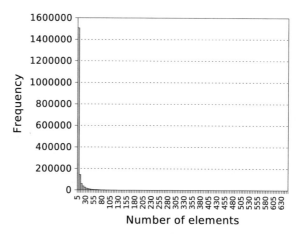

Fig. 12 Lookup times using
binary and linear search

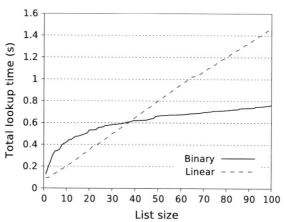

   Figure 11 shows the distribution of secondary table sizes. The heavy bias towards
small secondary tables in the CELT suggested another performance improvement
we can apply. As shown in Fig. 12, for tables up to around 35 elements, linear search
is faster than binary search, that is, linear search is faster for most of the secondary
tables in our CELT. Thus, we implemented a hybrid search strategy that employs
linear search for secondary tables of up to 35 elements and binary search for larger
secondary tables. Figure 10 includes a comparison of the lookup times achieved by
the two search strategies (binary or hybrid) for the hybrid CELT implementation.
As expected, the hybrid search strategy improved the competiveness of the hybrid
CELT implementation in terms of its lookup cost. Combined with the substantially
larger batch size enabled by the hybrid CELT and its competitive construction time,
we concluded that the hybrid CELT implementation with a hybrid search strategy is
the best choice of CELT implementation to be used in QuPARA.

6 Complete Performance Evaluation

In order to evaluate QuPARA's ability to fully utilize its available resources, we measured its speed-up, size-up, and scale-up on the Hadoop cluster and data sets described in Sect. 5.3.

Speed-up. The *speed-up* of a parallel program is the ratio between the running time it achieves on a single core and the running time it achieves on P cores. *Linear speed-up* means that the speed-up for P cores is P, that is, the work is perfectly balanced across the cores. For the speed-up test, we fixed the input size at 1,600 layers and increased the number of worker nodes from 1 to 18, that is, the number of cores from 4 to 72. Figures 13 and 14 show the running times and speed-up values achieved, respectively. Up to 24 cores (6 nodes), the speed-up was almost linear. Beyond 24 cores (6 nodes), the speed-up started to decrease. Due to the substantially decreased overall computation time, the fixed overhead involved in starting Hadoop jobs started to account for a greater fraction of the total running time at this point. Even so, our implementation achieved a speed-up of 64 with 72 nodes, an efficiency of 88 %. This can be considered a very good speed-up result, particularly given that a MapReduce implementation is less fine-tuned than a carefully handcrafted parallel risk modeling system.

Size-up. The *size-up* shows how the running time increases with the input size for a fixed number of cores. Ideally this increase should be linear. For the size-up test, we fixed the number of cores at 72 and increased the input size from 100 to 1,600 layers. Figure 15 shows the running time as a function of the number of layers. The running time increased linearly with the input size.

Scale-up. The *scale-up* measures the running time of the system while keeping the ratio between input size and cores fixed. If the running time remains constant, this demonstrates that the system is able to scale to larger input sizes, given a proportional

Fig. 13 Running time of QuPARA on 1,600 layers using between 4 and 72 cores

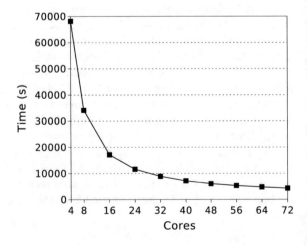

Fig. 14 Speed-up of
QuPARA on 1,600 layers
using between 4 and 72 cores

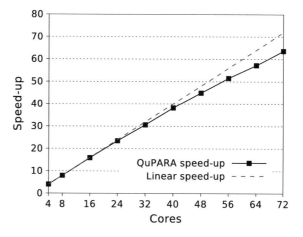

Fig. 15 Running time of
QuPARA on 100–1,600
layers using 72 cores

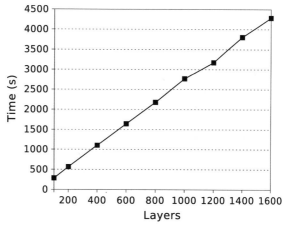

increase in available resources. For the scale-up test, we fixed the input size to 100
layers per node (4 cores) and increased the number of nodes from 1 to 18. Thus, up to
9,000 ELTs were processed. Figure 16 shows a very slow increase in the total running
time, despite the constant amount of computation to be performed by each node. This
is due to the increase in the setup time required by the Hadoop job scheduler and
the increase in network traffic. Nevertheless, the increase in running time was very
slight, which suggests that, given sufficient hardware resources, QuPARA can scale
to handle very large inputs.

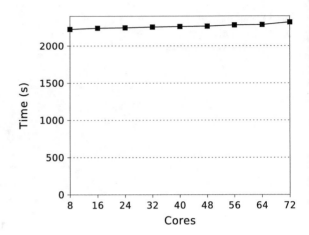

Fig. 16 Running time of QuPARA on 25 layers per core using between 4 and 72 cores

7 Summary and Future Directions

Typical production systems performing aggregate risk analysis in the industry are efficient at generating a small set of key portfolio metrics required by rating agencies and regulatory bodies and essential for decision making. However, these systems do not support ad hoc queries that provide a broader view of the many dimensions of risk that can impact a reinsurance portfolio.

In this paper, we presented QuPARA, a flexible system, implemented using the MapReduce framework, that allows arbitrary ad hoc queries to be posed in an SQL-like language and nevertheless achieves performance close to that of highly tuned production systems on industry-size data sets. As an example, a portfolio analysis on 3,200 layers and using a YET with 1,000,000 trials and 1,000 events per trial took less than 20 min.

The key to combining flexibility with high performance was the exploitation of advanced features beyond the standard MapReduce model offered by Hadoop combined with carefully engineering an efficient implementation of the key data structure used by the mappers in QuPARA. Based on our experience with QuPARA we believe that many simulation-based portfolio analysis systems might be engineered to take advantage of Hadoop's flexible and extensive software ecosystem without sacrificing production-level performance.

While QuPARA is a step in the right direction, there is much research to be done in the area of risk portfolio analysis. Data is the life blood of the insurance industry, but until recently much of the data available to reinsurers has been at a highly aggregated level. This however is changing. The volume, veracity, variety, and velocity of data available within the insurance industry is increasing rapidly.

Location-level exposure data is becoming the norm in many jurisdictions, and its level of detail in terms of location, construction, occupancy, and coverage is growing. Catastrophe models are becoming steadily more complex as they embrace location-level modeling techniques to better capture secondary uncertainty and spatial correlations, and begin to take multi-tiered and multi-event approaches to risk management. As a consequence of these changes, they are generating ever more data that can further enhance risk portfolio analysis.

Not only is the volume of data increasing, but the very nature of that data is changing. Real-time weather feeds, public and private geospatial data repositories, and streams of data being generated from sensors and mobile devices are becoming available to enrich portfolio-level analysis and allow reinsureres to access and manage their risk portfolios in a more dynamic and responsive way. For post-event analysis, when reinsurers are trying to make important reserving decisions, unstructured data in the form of text on social media platforms, news feeds, videos, offer the opportunity to provide decision makers a much better picture of their exposure in near real-time.

The future of risk portfolio analysis lies in harnessing increasing data volumes, incorporating a growing variety of data types and sources, and leveraging real-time data for post-event analysis. The risk landscape faced by reinsurers is constantly changing as the inventory of global insured properties changes. Big data approaches to risk portfolio analysis are going to be crucial to reinsurers looking to make better business decisions and manage their catastrophe risk more effectively.

References

1. de Alba, E., Zúñiga, J., Corzo, M.A.R.: Measurement and transfer of catastrophic risk. ASTIN Bull. **40**(2), 547–568 (2010)
2. Amazon Elastic MapReduce (Amazon EMR). http://aws.amazon.com/elasticmapreduce. Accessed 25 May 2013
3. Anderson, R.R., Dong, W.: Pricing catastrophe reinsurance with reinstatement provisions using a catastrophe model. In: Casualty Actuarial Society Forum, pp. 303–322 (Summer 1998)
4. Apache Hadoop. http://hadoop.apache.org. Accessed 25 May 2013
5. Bahl, A.K., Baltzer, O., Rau-Chaplin, A., Varghese, B.: Parallel simulations for analysing portfolios of catastrophic event risk. In: Proceedings of the International Supercomputing Conference (SC12). Workshop on High Performance Computational Finance, pp. 1176–1184. Salt Lake City, Utah, USA (Oct 2012)
6. Berens, R.M.: Reinsurance contracts with a multi-year aggregate limit. In: Casualty Actuarial Society Forum, pp. 289–308 (Spring 1997)
7. Byrne, M., Dehne, F., Hickey, G., Rau-Chaplin, A.: Parallel catastrophe modelling on a Cell B.E. J. Parallel Emergent Distrib. Syst. **25**(5), 401–410 (2010)
8. Capriolo, E., Wampler, D., Rutherglen, J.: Programming Hive, 1st edn. O'Reilly Media (2012)
9. Castella, H., de Montmollin, G., Rüttener, E.: Catastrophe Portfolio Modeling: A Complete View. PartnerRe (2009)
10. Cloudera. http://www.cloudera.com. Accessed 14 June 2015
11. Coelho, M., Rau-Chaplin, A.: eXsight: An Analytical Framework for Quantifying Financial Loss in the Aftermath of Catastrophic Events. In: Proceedings of the Workshop ISSASiM (DEXA 2014), Munich, Germany (2014)
12. Condie, T., Conway, N., Alvaro, P., Hellerstein, J.M., Elmeleegy, K., Sears, R.: MapReduce online. EECS Department, University of California, Berkeley, Technical Report No. UCB/EECS-2009-136, October 2009
13. Cortes, O.A.C., Rau-Chaplin, A., Wilson, D., Cook, I., Gaiser-Porter, J.: Efficient optimization of reinsurance contracts using discretized PBIL. In: International Conference on Data Analytics (Data Analytics 2013), Porto, Portugal, pp. 18–24 (2013)

14. Dean, J., Ghemawat, S.: MapReduce: simplified data processing on large clusters. Commun. ACM **51**(1), 107–113 (2008)
15. Dong, W., Shah, H., Wong, F.: A rational approach to pricing of catastrophe insurance. J. Risk Uncertainty **12**, 201–218 (1996)
16. Eaton, C., Deroos, D., Deutsch, T., Lapis, G., Zikopoulos, P.: Understanding Big Data: Analytics for Enterprise Class Hadoop and Streaming Data. McGraw Hill (2012)
17. Eden, R.: GNU Trove: High Performance Collections for Java. http://trove.starlight-systems.com. Accessed 19 Jan 2013
18. Gaivoronski, A.A., Pflug, G.: Value-at-risk in portfolio optimization: properties and computational approach. J. Risk **9**(2), 1–31 (Winter 2004–2005)
19. Glasserman, P., Heidelberger, P., Shahabuddin, P.: Portfolio value-at-risk with heavy-tailed risk factors. Math. Finance **12**(3), 239–269 (2002)
20. Google MapReduce. https://developers.google.com/appengine/docs/python/dataprocessing/overview. Accessed 25 May 2013
21. Grossi, P., Kunreuter, H.: Catastrophe Modelling: A New Approach to Managing Risk. Springer (2005)
22. Hadoop Distributed File System. http://hadoop.apache.org/docs/r1.0.4/hdfs_design.htmlS. Accessed 25 May 2013
23. Harrison, C.: Reinsurance Principles and Practices. American Institute for Charter Property Casualty Underwriters (2008)
24. HiveQL. http://hive.apache.org. Accessed 25 May 2013
25. Lee, K.-H., Lee, Y.-J., Choi, H., Chung, Y.D., Moon, B.: Parallel data processing with MapReduce: a survey. SIGMOD Rec. **40**(4), 11–20 (2011)
26. Meyers, G.G., Klinker, F.L., Lalonde, D.A.: The aggregation and correlation of reinsurance exposure. In: Casualty Actuarial Society Forum, pp. 69–152 (Spring 2003)
27. Oracle. Java Platform SE 7 HashMap. http://docs.oracle.com/javase/7/docs/api/java/util/HashMap.html. Accessed 28 Jan 2014
28. Osiaski, S., Weiss, D.: HPPC: High Performance Primitive Collections for Java. http://labs.carrotsearch.com/hppc.html. Accessed 19 Jan 2013
29. Rocks Cluster Distribution. http://www.rocksclusters.org/. Accessed 14 June 2015
30. Rau-Chaplin, A., Varghese, B.: Accounting for secondary uncertainty: efficient computation of portfolio risk measures on multi and many core architectures. In: Proceedings of the 6th Workshop on High Performance Computational Finance (WHPCF), Denver, USA, No. 3, pp. 1–10 (2013)
31. Rau-Chaplin, A., Varghese, B., Yao, Z.: A MapReduce framework for analysing portfolios of catastrophic risk with secondary uncertainty. In: Proceedings of the Workshop of the International Conference on Computational Science (2013)
32. Rau-Chaplin, A., Varghese, B., Wilson, D., Yao, Z., Zeh, N.: QuPARA: query-driven large-scale portfolio aggregate risk analysis on MapReduce. In: Proceedings of the IEEE International Conference on Big Data (IEEE BigData 2013), IEEE Comp. Soc. Dig. Library (2013)
33. Shvachko, K., Hairong, K., Radia, S., Chansler, R.: The Hadoop distributed file system. In: Proceedings of the 26th IEEE Symposium on Mass Storage Systems and Technologies, pp. 1–10 (2010)
34. Salcedo-Sanz, S., Carro-Calvo, L., Claramunt, M., Castañer, A., Mármol, Maite: Effectively tackling reinsurance problems by using evolutionary and swarm intelligence algorithms. Risks **2**(2), 132–145 (2014)
35. White, T.: Hadoop: The Definitive Guide, 1st edn. O'Reilly Media (2009)
36. Wilkinson, M.E.: Estimating probable maximum loss with order statistics, pp. 195–209. Casualty Actuarial Society, Forum (1982)
37. Woo, G.: Natural catastrophe probable maximum loss. Br. Actuarial J. **8**(5), 943–959 (2002)

Big Data and the Internet of Things

Mohak Shah

Abstract Advances in sensing and computing capabilities are making it possible to embed increasing computing power in small devices. This has enabled the sensing devices not just to passively capture data at very high resolution but also to take sophisticated actions in response. Combined with advances in communication, this results in an ecosystem of highly interconnected devices referred to as the Internet of Things—IoT. In conjunction, the advances in machine learning have allowed building models on this ever increasing amount of data. Consequently, devices all the way from heavy assets such as aircraft engines to wearables such as health monitors can all now not only generate massive amounts of data but can draw back on aggregate analytics to "improve" their performance over time. Big data analytics has been identified as a key enabler for the IoT. In this chapter, we discuss various avenues of the IoT where big data analytics either is already making a significant impact or is on the cusp of doing so. We also discuss social implications and areas of concern.

Keywords Internet of things · IoT · IoTS · Big data · Industrial analytics · Industrial internet

1 Introduction

In recent years, technological advances have opened up entirely new opportunities for both collecting and processing large-scale data. The capability to build algorithms that can generalize and do inductive inference has also increased significantly. This has resulted in advancing the state-of-the-art in traditional research fields that relied on huge quantities of data but were challenged by limited data acquisition capability or computing power. Research fields such as astronomy, physics, neurosciences, as well as medical genomics are some immediate examples (see, for example, [19, 27]). Further, largely driven by problems such as *search* and then those pertaining to social

M. Shah (✉)
Research and Technology Center - North America, Robert Bosch LLC,
Palo Alto, USA
e-mail: mohak@mohakshah.com

© Springer International Publishing Switzerland 2016
N. Japkowicz and J. Stefanowski (eds.), *Big Data Analysis: New Algorithms for a New Society*, Studies in Big Data 16, DOI 10.1007/978-3-319-26989-4_9

media, novel data- and compute-architectures as well as learning algorithms have also appeared in recent years. This has further propelled the prospects of building value added offerings.

In conjunction, there have been immense developments in sensing technologies resulting in "smart" devices that are constituted of sensors, actuators as well as data processors. We are at the cusp of a revolution in terms of how humankind interact with the technology in that an ever-increasing number of devices that we use, operate or interact with (even passively) are capable of collecting these actions, and more, in the form of data. As [74] note, "...concentration of computational resources enables sensing, capturing, collection and processing of real time data from billions of connected devices serving many different applications including environmental monitoring, industrial applications, business and human-centric pervasive applications." Such sensing technology is becoming pervasive and ubiquitous, and will be able to collect data through intermittent sensing, regular data collection as well as Sense-Compute-Actuate (SCA) loops. Hence, data can be collected at desired resolution all the way from continuous monitoring, to event or action captures. Moreover, such devices, be they appliances at home, heavy assets such as aircraft engines in the field, or wearables and mobile devices, do not function in isolation. More and more such devices or "things" are being "interconnected" resulting in an ecosystem referred to as the *Internet of Things-IoT*. This interconnectivity offers opportunities for enhanced services and efficiency optimization that can supplement each other by means of derived and abstracted insights—higher level of observations and inferences made from data arriving from multiple interconnected devices. Note that this interconnectivity need not be a device-to-device or machine-to-machine interconnectivity but can also be achieved via common platforms. Moreover, this can both be (near-) real-time as well as passive (data collected and analyzed over time).

Gartner estimates that, by 2020, this network of interconnected devices will grow to about 26 billion units with an incremental revenue generation in excess of \$300 billion, primarily in services. Furthermore, global economic value-add through sales into diverse end markets would reach \$1.9 trillion [47]. Consequently, the data resulting from these devices will grow exponentially too resulting in new business opportunities as well as posing novel challenges to managing and processing it for value gain. The data of the digital universe is slated to grow 10 folds by 2020. Various research and analysis firms have confirmed the scale of these projections in addition to the Gartner report. IDC further notes that data just from embedded systems, i.e. sensors and physical systems capturing data from physical universe, will constitute 10 % of the digital universe by 2020 (this currently stands at 2 %) and represent a higher percentage of target-rich data [64]. These technologies are also resulting in novel business models as well as new revenue sources, diversification of revenue streams in addition to increasing visibility and operational efficiency. Businesses will increasingly focus on services' aspects enabled by an increased understanding of utilization and operation of assets, consumer interests and behaviors, along with usage patterns and contextual awareness. Consequently, the IoT in specific contexts has also been referred to as the Internet of Things and Services (IoTS). It is also referred to as the Industrial Internet to highlight the applications in the world of heavy industrial assets. We will, however, stick to the general term IoT to look at the opportunities that cut across domains as well as services.

1.1 Chapter Focus

In this chapter, we will review some important aspects of the intersection of big data analytics and the internet of things. Even though we will briefly discuss the connectivity, communication and data acquisition issues, this is not the main focus of the chapter. We would rather like to focus on the novel opportunities and challenges that the new world of interconnected devices offer, along with some advancements that are being made on various fronts to realize them. Importantly, we will also discuss social implications as well as some of the, possibly underappreciated, areas that need responsible consideration as we move forward with a technology with a profound impact on society.

As a consequence of potentially billions of connected devices, the landscape of both handling and learning from data will undergo massive change. Further, the speed and scale at which the edge devices[1] will produce data will dwarf those of the current big data enablers such as social media, let alone manual data generation. This, previously unseen speed and scale of data, of course introduces challenges not only to the data and computing infrastructure but also pose a challenge to conventional learning methodologies and algorithms. As [2] rightly note, *scalability*, *distributed computing* and *real time analytics* will be critical for enabling the data-driven approaches to generate value.

We would also like to reiterate the point made by [2] that the concept of the internet of things goes beyond those of RFID technology and social sensing. While the former can be considered as a key enabler of the IoT, this technology is not the sole source of data acquistion as we noted above. Similarly, social sensing referring to peoples' interactions via embedded sensor devices, is a subset of IoT whereby this concept is not limited to people but also extends to machines and devices. Furthermore, we would also like to bring into discussion the resulting services based offerings that would be generated of this network. This is not just servitization, that refers to "the strategic innovation of an organization's capabilities and processes to shift from selling products, to selling an integrated product and service offering that delivers value in use" [38, 65]. In our view, IoT goes beyond integrated product and service offerings to enable novel business and revenue models. A variety of views on the IoT have been proposed based on different contexts. Aggarwal et al. [2] categorize these views in three broad categories: things-oriented vision (focusing on devices), internet

---

[1]Note that we use the term "edge devices" loosely to encompass not just the devices such as RFID tags but also other sensors esp. MEMS, including embedded sensors, monitoring and diagnostic sensors aboard industrial assets and so on.

oriented vision (focusing on communication and interconnectivity), and semantic oriented vision (focusing on data management and integration).[2]

We would like to discuss a *functional vision of the IoT*, a vision where the resulting data and insights, and not the enablement mechanisms, plays a central role. From a functional perspective, we discuss the basic components of an enablement stack and also current and some future areas where we envision witnessing the immediate impact. It should be noted that it is impossible to cover a topic such as the IoT, even in the context of big data, in its entirety in a book chapter. The aim of this chapter is to familiarize the reader with how big data analytics is a major part of the IoT vision and will be *a*, if not *the*, key player in deriving business and societal value. Finally, big data does not refer only to volume aspect of data but also to the variety and velocity—the three important V's used to describe big data all of which pose novel challenges.

The rest of the chapter is organized as follows: We discuss major components of a big data analytics stack in the context of IoT in Sect. 2. Section 3 then details various domains that stand to benefit from big data and the IoT, followed by recommendations on what steps organizations need to take in order to harness this value in Sect. 4. We then focus on the social implication issues as well as areas of concerns in Sect. 5 and present some concluding remarks in Sect. 6.

2   Big Data Analytics Stack for the IoT

We highlight in this section the major areas relevant to enabling analytics to leverage the value from the IoT as well as allow a general model to scale. These are also crucial to the broader ecosystem that would allow for devices to house analytical capabilities themselves. Any IoT application, whether it manifests at the user level or cloud level would need existence of an end-to-end analytics stack to support its functionalities. The offerings from IoT applications will be contingent on how each of these building blocks are realized. Of course, the levels at which each of these components will play a role in any specific application is subjective but they are a necessary condition nonetheless.

2.1   Data Acquisition and Protocols

Most of the data acquisition in the context of the IoT happens through edge devices. Edge devices are referred to as such since these typically reside at the edges of the network. That is, they are present at the point at which either a human or an asset interacts with the rest of the network and it is through these devices that the initial data

---

[2]We do briefly cover some of these categories since they are indeed critical and the effectiveness of IoT applications and capabilities are highly contingent on effective solutions in these areas.

will be acquired and possibly re-transmitted back to the network. Examples include health sensors on patients, activity monitors such as Jawbone, control and advanced monitoring sensors on industrial assets, weather sensors, movement sensors in a home, visual, sonar and laser cameras aboard an autonomous vehicle, diagnostic sensors on appliances, sensors embedded on mobile devices, etc. Radio Frequency Identification (RFID) tags were one of the first mechanisms of acquiring such data but there have been other devices including sensors such as microelectro mechanical sensors (MEMS), mobiles and wearables that have vastly expanded the possibilities of large-scale, high-resolution data acquisition.

Efforts have been underway to establish proper channels to acquire and persist data collected from the edge devices. While currently there is no agreed upon protocol for such acquisition, domain-specific mechanisms are appearing. There is certainly a need for accepted protocols for communication for these devices both to each other and to a central capability such as cloud to enable aggregate analytics. Technologies involving wired or wireless communication of homogenous devices as well as capture and transmission of sensor data for storage and processing by applications are referred to, broadly, as Machine-to-Machine (M2M) technologies. Some companies are going the proprietary route while there have also been announcements of open-source efforts (e.g., Bosch, ABB, LG and Cisco's joint venture announced recently to cooperate on open standards for smart homes; see Appendix). Similarly, there have been other joint efforts trying to bring more standardization to the IoT including Open Interconnect Consortium (OIC), AllSeen Alliance, Thread group, Industrial Internet Consortium (IIC) and IEEE P2413 [35]. Developing an open-source ecosystem has its advantages since broader community can contribute to the efforts. Moreover, given that the user community is involved in the development, adoption becomes relatively easier and wide-ranging. Since these devices collect high dimensional and high frequency captures of device states, this will in turn also require high-bandwidth connectivity. For instance, an aircraft engine can send data through 10's to 100's of sensors at millisecond-resolution and can generate multiple GB's of data per flight.

2.2 Data Integration and Management

One of the most distinguishing aspects of the IoT is the fact that the data is acquired from a variety of sources. In order to provide useful services the data from edge devices typically need to be combined with external data sources including business data, utilization data of assets, geographical data, weather data, etc. Consequently, the data quality and management issues also grow exponentially. Combining and analyzing heterogeneous data is a major challenge. Efforts have been made to standardize data characterization so that a communication protocol can be developed for data exchange. The Open Geospatial Consortium, for instance, has developed various such protocols under the Sensor Web Enablement initiative allowing for interoperability for sensor resource usage. Some of the standard interfaces proposed as a part of the initiative include O&M (Observations and Measurements, to encode the real-time measurements from sensors), SML (Sensor Model Language, to describe sensor systems and processes), Transducer Model Language (TML, to describe transducers and supporting real-time streaming of data), Sensor Observation Service (SOS, standard web service interface for requesting, filtering and retrieving sensor system observations), Sensor Alert Service (SAS, for publishing and subscribing alerts from sensors), Sensor Planning Service (SPS, for requesting user-driven acquisitions and observations), and Web Notification Service (WNS, for delivery of messages or alerts from SAS to SPS). See [2] for more details. Further, in order to integrate and annotate the sensor data, the World Wide Web Consortium (W3C) has initiated the Semantic Sensor Networks Incubator Group (SSN-XL) with a mandate to develop semantic sensor network ontologies. These efforts have constituted a big part of the semantic web effort, further defining ontological frameworks such as the Resource Description Framework (RDF) and the Web Ontology Language (OWL) that enable defining ontologies such as SSN (Semantic Sensor Network) and SWEET (Semantic Web of Earth and Environmental Terminology) to express identifiers and relationships in various contexts.

Further, given that the data is acquired in real time and field settings, there are myriad of issues around missing values, skewness and noise. The high resolution temporal nature of such data further makes it difficult to align multiple sources as well as devise strategies to learn from them in conjunction with static data sources. In the context of assets, the data is also accompanied by derived attributes—ones whose values are calculated from the raw data using a conversion mechanism. However, the protocols for obtaining the derived quantities are not uniform or standardized even within a given domain let alone across domains. Data integration becomes more difficult since it requires the reconciliation of such derived quantities. In addition to formulaic data transformations, there can also be hurdles in data management arising from issues such as privacy and security resulting in deidentified and/or encrypted data.

From a storage perspective, classical relational databases are no longer enough since the data is not only from disparate sources but it also appears, or needs to be organized, in native forms such as documents, graphs, time-series, etc. The whole paradigm around data organization that addresses the set of requirements around big data is broadly referred to as NOSQL (standing for Not Only SQL) databases. This includes columnar data stores such as BigTable, Cassandra, Hypertable, HBase (inspired by the BigTable); key-value and document databases such as MongoDB, Couchbase server, Dynamo and Cassandra (also supports documents); stream data stores such as Eventstore; graph based data-stores such as Neo4j and so on. Each of these have associated technologies for efficiently querying and processing data from respective stores and have unique advantages and capabilities. For instance, services such as Flume and Sqoop allow for ingestion and transfer of big data while languages such as Hive, Pig, JAQL and SPARQL enable efficient querying of big data in various forms including ontologies such as the RDF. From an integration perspective, the classical approaches of business-to-business (B2B) data integration do not apply either since such data cannot generally be organized using a master database schema. A combination of these storage strategies are typically employed depending on the types of data and customized views can be created depending on the application requirements.

A data persistence strategy is also needed since many times storing such high resolution data in massive quantities is neither viable nor needed. Strategies involving data summarization and sampling along with storing accompanying metadata can be quite effective especially when the data has very high level of redundancies. Recall how sparse format allowed to store and process data files with few non-zero values much more efficiently in the case of very high dimensional data. These data structures, for instance, are a regular offering in various analytics toolsets and libraries such as Pandas [See appendix].

2.3 Big Data Infrastructure

The massive amounts of acquired data necessitates powerful infrastructure to support not just storing and querying, but also extracting insights from such data. Various categories of learning that need to be performed on such data exert unique set of requirements. For instance, one of the most common requirement is that of being able to perform batch analytics over historical data to build aggregate models. However, this can be a complicated endeavor given that the data does not necessarily reside on the same network let alone the same machine. Hence, parallel learning algorithms as well as distributed learning capabilities are needed depending upon the size, location and other data characteristics in addition to the communication constraints. Frameworks such as Hadoop have shown significant promise when it comes to distributed data analysis including efficient search, indexing as well as learning [2]. Hadoop is a distributed storage and processing framework for large scale data relying on a Hadoop distributed file system (HDFS) with an aim to "take compute to the data". This is in contrast to the classical parallel high performance computing (HPC) architectures that relied on parallel file system where the computation would require high-speed communication mechanism to the data. Over past few years, Hadoop has developed as an ecosystem (see Appendix) with various applications and services supporting functions on the core architecture allowing for efficient data storage and organization, search and retrieval (including querying), processing, as well as services such as resource scheduling and maintenance. Various data-stores as well as querying languages mentioned above form a part of this ecosystem.

One of the major limitations in the distributed settings such as Hadoop has been that of performing analytics with low latency requirements including model deployment, real-time, iterative, or interactive analytics. In such cases, especially when multiple passes on the data are required (e.g., for many machine learning algorithms), Hadoop framework can be quite costly in terms of communication to the underlying HDFS. Frameworks such as Spark were developed to address these issues on Hadoop and since then have grown into its own ecosystem. These frameworks, especially Spark, have shown significant promise and are being investigated for their suitability in the IoT scenarios. Spark enables in-memory primitives for cluster computing as opposed to Hadoop's MapReduce which is a two-stage disk-based paradigm and hence allows for faster performance on applications with low-latency requirements mentioned above. While both Hadoop and Spark offers streaming API's, Storm is a computational framework designed with streaming analytics as its objective. Since each of these paradigms have their strengths and limitations, choosing the right storage as well as computional paradigm involves an in-depth analysis of requirements for the use-case in which these would be employed. However, due to their open-source nature, there has been significant effort in promoting interoperability of these frameworks. For instance, both the Spark and Storm frameworks can operate on Hadoop clusters and hence provide for easy integration. Hadoop commercial providers such as Cloudera and Hortonworks have also announced support for Spark and Storm respectively. Companies such as Databricks are already providing commercial version of the Spark framework. Finally, to effectively deploy and scale the analytics models, standardization and benchmarking mechanishms for analytics are available that can allow for efficient communication of these models. Predictive Model Markup Language (PMML) provides one such mechanism. There have been successful commercialization of such standards from vendors such as Zementis that provides not just an encoding mechanism but a full deployment capability. This includes an execution and scoring engine, namely Adaptive Decision and Predictive Analytics (ADAPA) that can run PMML specified models allowing for modular and efficient model deployment. Capabilities such as Velox also target machine learning model management and serving at scale [13].

2.4   Machine Learning and Data Mining

The natural subsequence to the handling, management and integration of IoT data, is the actual insight discovery step which is the ultimate goal of the network. Even though each step starting from the data acquisition onwards poses a variety of challenges for the IoT, the ultimate value from these steps is realized only when useful and generalizable insights can be derived from this data. The current use of edge devices (at least in the consumer domain) seem to be predominantly point-use, that is, operationalization at the single user level. However, as more and more devices get interconnected this will inevitably change. In fact, there are various use cases where this is already visible as we will discuss in the next section. Learning from IoT data is particularly interesting and challanging at the same time. Classical machine learning methods need to be extended and adapted to cope with the challenge of scale, diversity and the distributed nature of the data. The volume, acquisition speed and temporal nature of the sensor and other related data is already highlighting the limitations of traditional approaches to learning. Some of the major challenges include learning in distributed settings, learning from very high dimensional, high resolution temporal data and learning from heterogeneous and complex data. Novel frameworks such as the alternating direction method of multipliers (ADMM) [7] have appeared to enable optimization, a core functionality of many learning algorithms, in such distributed settings. Furthermore, advances have also enabled versions of successful machine learning algorithms such as topic modeling via Latent Dirichlet Allocation (LDA) [68, 75], convolutional neural nets, Restricted Boltzmann Machines (RBM's) [14, 59], Support Vector Machines, Regression and so on (see, e.g., [24, 41, 56]) for large scale settings. Online versions of various classical learning algorithms have also appeared allowing for faster execution times on large datasets.

Consequently, this has also necessitated extensions of the evaluation approaches to the learning algorithms [29] to be extended to large scale settings. Some promising approaches for resampling in large scale settings such as the bag-of-little-bootstraps (BLB) [32] have appeared that also provides a theoretical framework characterizing them. In addition, there have been advancements in methods aimed at analyzing streaming data at scale for event prediction, change point detection, time-series forecasting and so on owing to the use cases that require online learning or where the models need to be adapted to evolving realities (see, for instance, [40]). Feature discovery is also one of the issues that has resurfaced since it is no longer feasible for learning-features to be designed or discovered in conventional manner. Novel approaches are enabling automated feature discovery and learning in cases where generalized models can be built from extremely large distributed datasets. One of the most prominent developments has been in learning sophisticated networks and autoencoders via Deep Learning methods [14]. Deep learning has shown significant promise in domains such as image classification, speech recognition and text mining [4, 33, 36, 61].

In addition to these, there have also been efforts to scale up the deployment of large scale classifiers in hardware and embedded systems. Specific chip designs inspired by both the machine learning and cognitive computing fields have appeared to this end. Some prominent examples include IBM's SyNAPSE, NVidia's Tegra X1 and Qualcomm's Zeroth (see Appendix for links).

2.5 Bringing the Building Blocks Together

From an organization or application level, it is clear that an end-to-end IoT stack is needed. The components of this stack will include data acquisition right from the M2M layer, data-processing, data-sharing through interconnected network, insights' discovery and capability to relay results both to devices (for potential actions) as well as to (automatic or manual) decision makers. Various teams and companies are identifying the nature and structure of such an IoT stack that can provide infrastructure, platform and services for both front-end application and solution development, as well as back-end computing and support (e.g., via cloud). Commerical vendors such as EMC, Microsoft, Amazon and IBM offer building blocks of this stack that can be instantiated by organizations or service providers based on their specific requirements. Increased modularity and interoperability will further speed-up the adoption and scaling of these capabilities. For example, being able to choose the desired infrastructure, platform and software selectively from a combination of vendors can address specific needs of IoT applications. Consequently, Infrastructure-, Platform-, and Software-as-a-service (IaaS, PaaS and SaaS respectively) are becoming increasingly desirable (see, for example, offerings from Cloud Foundry, Microsoft Azure; link in Appendix). Lambda architecture [43] has shown promise as a basis that allows for batch and real-time analytics together. This also allows to account for the volume, velocity and variety of big data. Lambda architecture already underlies many Hadoop and Spark instantiations. Architectures for specific cases, such as embedded systems and sensor networks, are also being proposed (see, for instance, [25, 62, 63, 70]).

3   Domains Impacted by Big Data Analytics and the IoT

The applications within the IoT domains depend highly on the respective business drivers leading to multiple manifestations of business cases through such network. Various works have attempted to paint a picture of the application landscape for the IoT. For instance, [11], categorize the applications in two broad categories: (i) Information and Analysis, consisting of tracking behavior, enhanced situational awareness, and sensor-driven decision analytics; and (ii) Automation and Control, consisting of process optimization, optimized resource consumption, and complex autonomous systems. Another categorization comes from [42] who categorize these in five categories viz. predictive maintenance, product and service development, usage behavior tracking, operational analysis and contextual awareness. [12] also discusses some of the opportunities in the IoT.

IoT potentially goes beyond the possibilities mentioned in above reports in that it will also enable sophisticated services capabilities as mentioned earlier. In this respect, another categorization is quite illustrative that divides these opportunities in consumer-facing and business-facing opportunities [39].

To provide a flavor of the type of some specific applications, let us look at some illustrative use cases from different IoT-related domains. Note that we are reviewing these domains from a big data and analytics perspectives. There are many more applications as a consequence of advancements in sensing technologies and hyperconnectivity achieved in the IoT. As the readers will notice, our categorization has overlaps with various above-mentioned efforts. However, looking at the applications and opportunities from a domain perspective can provide a more coherent picture.

3.1   Manufacturing

Lee et al. [37] describes manufacturing as a 5M system consisting of Materials (properties and functions), Machines (precision and capabilities), Methods (efficiency and productivity), Measurements (sensing and improvement) and Modeling (prediction, optimization and prevention). In this context, additive manufacturing can be considered as a process for creating products using an integrated 5M approach. Recent advances have significantly improved sensing capabilities and data gathering around various aspects of this 5M system. However, in traditional-, as well as in many cases advanced-, manufacturing setup such information gathering had a preventive or control purpose and hence didn't necessarily serve an analytics-oriented insight discovery objective. Even in traditional sense, it can be argued that big data has been utilized for quite some time especially in the context of modeling. However, this usage typically corresponds to modeling based on data under simulated or nominal conditions in which the product or manufactured industrial asset is run under a controlled environment. For a manufacturing process, big data can enable functions such as correlating controller and inspection data. When this is combined with the traditional overall equipment efficiency (OEE) providing the production efficiency status, insights into the relationship between performance and the cost involved in a sustained OEE level can be obtained. This is particularly timely as there is a significant initiative, referred to as *Industry 4.0*, to increase digitization of manufacturing with a goal to build an *intelligent factory*, with cyber-physical systems[3] and the IoT as the basis.

The opportunity landscape in the manufacturing domain is vast, in addition to the OEE and performance optimization. Big data analytics can help (and has started to do so) in areas such as cycle time reduction, scrap reduction, product defect detection (e.g., to improve quality ratios, or detecting products that may lead to quality issues later), identifying and resolving issues with machine failure and optimizing material and design choices (see [6, 34] for some examples in various manufacturing domains). As smart factories move beyond sole control-centric optimization and intelligence, big data can enable further optimizations by taking into account interactions of surrounding systems as well as other impact factors. For instance, the production cycle quality assurance can benefit not only from the quality data of current cycle, but can also analyze quality data from the previous steps (e.g., quality and monitoring data from parts-suppliers) or feedback (e.g. quality reports and issue notifications from consumers). It should be noted that while such benefits result in immediate value, they also have significant indirect advantages. For instance, while slight increase in the quality as a result of improved defect detection may seem to be a marginal improvement for advanced manufacturing facilities, these can translate into new business opportunities for companies and sometimes can be a major deciding factor for the clients. Similarly, identifying defective or potentially defective parts right at the manufacturing or quality testing stages can mean reduction in quality claims at later stages. Such benefits have big multiplication factors in terms of business values associated with them, of course not to mention intangible benefits such as credibility and brand building for manufacturers. In addition to the above opportunities directly related to the manufacturing process, big data and predictive analytics can have a significant impact on making cyper-physical systems much

---

[3]Cyber-physical systems consist of computational and physical components that are able to perceive real-time changes as a result of seamless integration [53].

more effective and efficient. Predictive analytics are also poised to address important issues in areas such as capacity planning due to uncertainty in downstream capacities, inventory and supply-chain management by reducing uncertainities around material and part availabilities, and by reacting to (or anticipating) market and customer demand changes. Importantly, this can also help understand and address product design and performance issues, and can help to complete the loop with respect to the manufacturing process at the material and design stages. This final aspect in fact has immense significance and leads us to the next area of operation and maintenance of (heavy industrial) assets.

3.2   Asset and Fleet Management

Over the past few years, most organizations have undergone a major change in their business models or are in the process of doing so referred to as servitization. This basically emphasizes a customer focus in product and service delivery, and is already a major factor in consumer-oriented companies (e.g., home appliances and electronics). However, this has taken on even higher importance in asset-heavy organization such as manufacturers and operators of heavy assets like aircraft engines, locomotives, turbines, mining and construction equipments. The renevue models of these organizations have undergone drastic changes over past few decades. While manufacturers drew majority of their revenues from sale of these assets earlier, they do so now by selling service agreement and performance guarantees over the lifetime or usage of these assets. Moreover, such agreements rely heavily on asset utilization and hence it is imperative for the organizations that they have a very high visibility into asset operation so that they can not only quickly address but also effectively anticipate any major impending failure (at least at the asset level but ideally at the component level) that can jeopardize operational efficiency of the operators. Achieving an optimal efficiency and availability of asset are critical to both these manufacturers and their clients. Moreover, this also enables effective planning on the maintenance actions, performing fast and effective root-cause analysis as well as detecting and anticipating warranty issues as early as possible (this is analogous to the requirement at the manufacturing plants discussed above).

For instance, in the context of aviation, capabilities for predicting anomalies as well as prognostics can be potentially integrated in the flight controls. The IoT can further help in correlating these anomalies with additional information such as weather, particulate matter, altitudes as well as put this in context with fleet level statistics [8]. Big data acquisition capabilities are further enabling high resolution monitoring of aircraft engines. While traditionally field engineers were able to monitor snapshot data from engines, now full flight data can be reliably analyzed. Further, large scale analysis on multiple sensors can be performed that facilitate tasks such as event detection, signature discovery and root cause analysis. Other industrial domains stand to benefit from these approaches as well. Data is captured at different stages during the life and operation of assets. This data is organized in disparate forms and is typically disconnected over the stages and actions taken in maintaining and operating an asset. Some of these major data sources include condition monitoring data (present and historical), controller parameters, digitized machine performance data, machine and component configuration, model information, utilization and operational data, as well as maintenance activities. Leveraging such data would not only allow building of aggregate models for the fleet but also for more customized models for unique (set of) assets. Further, such information can be combined at the fleet level in order to understand aspects such as asset-deterioration patterns and behavior under varying operating conditions. Also, the underlying physical models that otherwise explain the behavior of assets under nominal conditions and effect of operations and utilization patterns on engine life, can be enhanced. In turn, this can significantly impact the maintenance of the engines taking us closer to condition based maintenance.

Furthermore, data-driven insights would allow for enhanced capabilities to manage and contain unanticipated field events (e.g., asset failures or malfunctions) by enabling their localization, subsequent root cause analyses and identification of the most efficient resolution mechanisms as well as future design changes. Remote maintenance is an opportunity that is already being realized in some cases. Lee et al. [38], for instance, discuss a case study on the remote maintenance of Komatsu smart bulldozers used in mining and construction. Such capability would have significant impact on asset reliability, maintenance scheduling as well as reduction of unplanned downtimes [38, 72] with the ultimate goal of having self-aware and self-maintenance machines. This would in turn benefit fleet operations, scheduling as well as optimizations.

Fleet-level analytics would also enable more efficient fleet management as well as better user experience, and hence is not limited to the heavy asset industry; connected cars and networked electromobility are examples. For instance, the BMW group, Bosch, Daimler, EnBW, RWE and Siemens have come together to take on the "Hubject" initiative (see Appendix) with the aim of optimizing electromobility through convenient access to a charging infrastructure. Connecting electromobility service providers, charging station operators, energy suppliers, fleet managers, and manufacturers, as well as utilizing analytics to provide value-add services (e.g., identify closest charging station, and suggest charging routines), is a demonstration of end-to-end IoT enabled capability for consumers. [6] outlines another example use case in the context of mobility and automation using telematics data. Similarly, as cars evolve (they typically have the computing power of 20 PC's processing about 25 GB's of data per hour [45]) to a connected world, they are moving beyond optimizing internal functions. The connected car initiative aims at developing the car's ability to connect to external network and not just enhance in-car experience but also to self-optimize its operation and maintenance.

Other advantages of connected vehicles will obviously be for fleet management and companies that rely on such vehicle fleet for their operations or even entire business models. Examples include postal and courier services, delivery industries, servicing companies (e.g. consumer appliance services) and so on that would look for connected vehicles to optimized different value drivers like gas consumption, route optimization, service time reduction, resource allocation and fleet efficiency.

3.3 Operations Management

In the above subsection the focus of our discussion was mainly mobile assets. However, big data analytics capabilities are also impacting operations and maintenance of stationary assets such as energy turbines, and plants (production, generation and so on). Various monitoring, positional and control sensors in plants are enabling more effective plant maintenance, identifying sub-optimalities as well as safety and security. Predictive maintenance of machinery and plants can allow for higher availabilities, as well as better guarantees on quality and for reducing process variability [28, 34]. By intelligently instrumenting the plants with advanced sensors as well as integrating information from existing control and monitoring sensors, we can develop a better understanding of plant's operational status, anticipate and predict failures, and identify sensor correlations to understand (and in some cases ascertain) sensor interdependence as well as adjust working set-points for the equipments. This can enable improved stability in plant operations as well as a reduction in the high alarm rates that can hamper the operations. Garcia et al. [21] propose such a monitoring model for the case of an oil processing plant. In addition to improvement in plant utilization (efficiency) and availability (less downtime), benefits will appear in terms of a reduction in operating costs as well as the ability to make real-time decisions.

Similarly commercial facilities maintenance (e.g., large buildings, campuses or company facilities) in broader context can also be seen as a part of this topic. However, most efforts in those areas are currently focused on energy optimization and hence we will cover those a little later.

3.4 Resource Exploration

Resource exploration industry such as oil and gas, mining, water and timber constantly face challenges in terms of finding renewable reserves of natural resources, and balance them with volatility in demand and price. The goals for these industries are often achieving a delicate balance of increasing production and optimizing costs while at the same time reduce the impact of environmental risks (e.g., reduction in carbon footprint). Among these various industries, the oil and gas industry, and particularly upstream sector of it is a complex business that rely heavily on data.[4] Even before the advent of "big data", these industries have made use of data from various sources of information whether it is in the context of studying soil composition during mining, or monitoring deep sea or sub-surface assets' health through traditional prognostics and health management methods. However, these industries face new challenges as the data grows exponentially in volume, resolution (or speed of

---

[4]Oil and gas industry can be viewed in three different segments: Upstream (concerned with exploration, drilling/development and production), Midstream (concerend with trading, transportation and refining) and Downstream (concerned with bulk distribution and retail). Refining step has components in both midstream and downstream sectors.

capture) and in variety. For instance, seismic data such as wide azimuth offshore data results in very high volumes. In addition to the seismic data, structured data comes from sources such as well-heads, drilling equipments and multiple types of sensors, such as flow, vibrations and pressure sensors, to monitor assets. This needs to be combined not only with drilling and production data but also those from unstructured data sources such as maps, acoustic data, image and video data and well logs. Further, these data are used, studied, analyzed and processed by various business segments that in turn generate a variety of derived data such as reports, interpretations and projections. The industry stands to gain deep and meaningful insights if these data can be efficiently and effectively *managed*, *integrated* and *reconciled*. Like many other areas, the foremost challenges are those of data management, preprocessing and more importantly ascertaining data quality. Working successfully with such data sources can mean a significant increase in production, possibly at lower risks to the environment and safety, reduction in costs, as well as speed to *first resources*.

While many companies in the oil and gas sector such as Chevron and Shell, have started looking into leveraging big data analytics, much more effort is needed to benefit from these opportunities in various areas. Baaziz and Quoniam [3] identify areas that stand to benefit specifically in the upstream oil and gas industry (see [18, 26, 52, 60] for further details):

1. *Exploration*: Enhancing exploration efforts (e.g., by helping experts verify field analysis assumptions where new surveys are restricted by regulations); Improved operational efficiency by combining enterprise data with real-time production data; Efficient and cost-effective assessment of new prospects by more efficiently utilizing geospatial data; Early identification of potentially productive seismic trace signatures; and building new scientific models via insights discovery from multiple data sources (e.g., mud logging, seismic, testing and gamma ray).
2. *Drilling and Completion*: Building more robust drilling models from current and historial well data and subsequent integration into the drilling process; Adaptive models to incorporate new data; Improved drill accuracy and safety by analyzing continuous incoming data for anomalies and event prediction; Reduction in Non Productive Time (NPT), one of the major concerns in the industry, by early identification of negative impacting factors of operations while increasing foot-per-day penetration; Models for optimal cost estimation; Predictive maintenance for increased asset availability, reduction in downtime as well as managed maintenance planning.
3. *Production*: Mapping reservoir changes over time for adaptation of lifting methods for enhanced oil recovery (e.g. to guide fracking in shale gas plays); More accurate production forecasts across the wells for quicker remediation of ageing wells; real-time production optimization by allowing the producer to optimize resource allocation and prices; Increased safety through earlier anticipation and prediction of problems such as slugging and WAG gas breakthroughs;
4. *Equipment Maintenance*: We covered this more broadly in the previous subsection. In the current context, this refers to preventing downtime, optimizing field scheduling as well as maintenance planning on shop floor.

5. *Reservoir Engineering*: More accurate engineering studies and a better understanding of subsurfaces by more efficiently analyzing data and subsurface models.

IoT related opportunities also exist in the midstream and downstream industries.
For instance, [60] further identifies opportunities in environmental monitoring (e.g.,
by analyzing real time sensor data for regulatory as well as company control com-
pliance; maintenance prediction based on pollution levels), reducing set up times
at refineries by quicker crude assay analysis for oil quality prediction, and predictive and condition-based maintenance on assets in the transportation and refinement
facilities (both in mid- and downstream). Similarly, opportunities also exist in retail
optimization (e.g., gas station automation).

While the above opportunity areas are detailed in the context of oil and gas industries, these areas, opportunities and challenges are quite similar in other resource
exploration industries too (see, for instance, [48] for a broader discussion).

3.5 Energy

Energy is another sector that can be transformed as a result of big data analytics.
While utilities have been identified as one of the biggest stakeholders, various other
industries whether they are direct energy producers, services companies performing campus and facilities' energy management or sectors relying on and impacted
by energy consumption can foresee significant potential in increasing production,
improving energy demand prediction, reducing uncertainities in energy supply, better resource management through efficiency gains and energy waste reduction. These
can yield benefits not just from a financial and efficiency perspective but can also
be instrumental from an environmental perspective. For instance, a study on energy
efficiency from McKinsey and Co. concluded that a holistic program could result in
energy savings in access of $1.2 trillion and a reduction in end-use consumption by
9.1 quadrillion BTUs while eliminating up to 1.1 gigaton of greenhouse gas every
year by 2020 [49].

The opportunity areas in the energy domain that stand to benefit from big data
analytics include but are certainly not limited to:

1. *Asset and Workforce Management at Utilities*: Increasing availability (e.g., in
weather conditions like storms) by reduction in downtime and maintenance optimization as well as identifying potential hazardous situations, outage management, wind-farm management by turbine optimization both at an asset- and an
aggregate-level, reducing energy thefts, etc.
2. *Grid Operations*: For example load forecasting and load balancing primarily for
peak-shaving which is an immediate priority, outage management and voltage
optimization, optimizing network and energy trading and incorporation of distributed smart grid components into the storage system, proactive management
of distribution network, combining energy facilities into virtual power plants,[5]

---

[5]https://www.bosch-si.com/solutions/energy/virtual-power-plant/virtual-power-plant.html.

incorporating renewable energy sources into the grid, increasing grid flexibility and scalability, optimization of energy production and supply as well as efficiency gains for utilities.

3. *Transportation*: For example reducing energy consumption by dynamic pricing for road use and parking, frequent updated traffic information and route optimization.

4. *Infrastructure*: Smart cities (smart parking, traffic monitoring and control, structural health systems); Infrastructure management, smart grids, etc. Please refer to [5, 10, 73] for a broader discussion on the implications and uses of big data and IoT in cities.

5. *Residential and Commercial Facilities*: reducing energy consumption by optimizing utilization (e.g., of HVAC and heating operations), smart meters to track consumption and subsequent load balancing, smart appliances and lighting for need-based operation, reduction in energy waste by identifying energy holes and sinks; smart systems for water, lighting, fire, power, cooling, security and notifications resulting in cost savings, preventative maintenance of critical systems and in environmental benefits [58].

6. *Operations*: efficiency gains and cost reductions in company operations, e.g., identification of energy sinks such as running unutilized resources, optimization of energy usage in data centers, etc.

Please refer to [54] for a discussion on many of these perspectives. Further, [49] covers a variety of opportunities as well as challenges as cities both grow and get "smarter" through more sensors, and advanced network and communication capabilities.

3.6 Healthcare

Reddy [58] points out that one of the major changes in the healthcare domain as a result of IoT is the ability to monitor staff and patients, and the ability to locate and identify the status of healthcare equipment/asset resulting in improved employee productivity, resource usage and efficiency gains, and cost savings. Further, as [50] note, "big data analytics can improve operational efficiencies, help predict and plan responses to disease epidemics, improve the quality of monitoring of clinical trials, and optimize healthcare spending at all levels from patients to hospital systems to governments". Many edge devices have been introduced to patients in particular, and a wider population in general. For patients, these range from temperature monitors, blood glucose-levels monitors, fetal monitors, electrocardiograms (ECG), and even electroencephalography (EEG)[6] devices. Not only this, efforts are underway to move beyond monitoring towards comprehensive health management. In addition to the patient, an increasing section of healthy population are routinely using health

---

[6]See http://www.mybraintech.com/.

and activity monitoring devices such as the jawbone and fitbit, as well as various applications through sensors on the mobile devices. Consequently, this also allows for various value-added application that can guide healthy living practices. Moreover, the fact that these devices are connected to the cloud also enables anonymized aggregate analyses at population segment levels as well as across other dimensions such as geographies and demographics. While this means a more healthy lifestyle for healthy populations, the ability to monitor patients' conditions on a continuous basis through these devices has very significant advantages. For instance, combining remote monitoring capability and distant communication technologies (e.g., low-cost video-conferencing), can allow for efficient remote healthcare in areas where direct access to medical personnel is difficult or time-consuming. Many other benefits such as reducing the number of at-risk patients, a reduction in readmission risk, epidemic monitoring, mobile healthcare for home and small clinics [22], ambient assisted living [16] and chronic patient monitoring [55] can be realized. There are also advantages in pharmaceutical drug trials, treatment effectiveness, as well as chronic disease management. Big data and the IoT can further allow for valuable insight discovery and knowledge extraction from personal health record, or PHR (current and historical) data [57].

In addition to being able to provide higher quality of healthcare, providers will also benefit from the other previously discussed benefits of IoT including predictive maintenance and real-time asset monitoring capabilities resulting in high availability levels, reduction in operating costs as well as increasing supply chain efficiencies. See [9, 17, 74] for further related discussions.

One of the major concerns in such large scale data analyses is with respect to the data privacy and security, mainly the protected health information (PHI). Mechanisms to ensure proper data de-identification and anonymization as well as ascertaining data security in communication channels is an extremely high priority. While this also applies to areas discussed above, these issues are critical in the healthcare domain and if ignored can have serious implications for both individuals and population at large.

3.7   Retail and Logistics

We are moving towards a shopping experience in connected supermarket. One of the widely discussed IoT use cases involves the pre-specification of shopping lists that can be communicated to the superstore so that the checkout wait-times (one of the major problems in retail stores today) can be reduced. However, there are more interesting and compelling use cases that will be built on top of the resulting data that is generated. For instance, shoppers' purchase patterns can be mined for recommendation of relevant items, and in addition sales and discounts can be highlighted. Combined with social data and other preferences that a consumer may make available, a consumer-centric experience can be created tailored to consumers' unique preferences. These capabilities can also enable effective monitoring of shopper traffic across stores, targeted marketing as well as product placement.

Further, being able to track product movement, e.g., through technologies such as RFID tags, will allow retailers to have a more accurate and efficient inventory management, increase inventory accuracy and reduce thefts as well as administrative costs. As [58] points out, big data capabilities along with IoT will result in stock-out prevention as a result of connected and intelligent supply chains, as well as real-time tracking of parts and raw materials allowing to preempt problems and address demand fluctuations. Naturally, this information can be fed back into the manufacturing and distribution channels for further optimizations leading to a reduction in required working capital, efficiency gains as well as avoiding disruptions. [6] illustrates another use case in retail where inventory can be tracked as it moves from shelf to basket allowing the retailer to enable analytics for optimizing available supply to predicted demand, reducing uncertainties and fluctuations through warehouse operations and the supply chain.

Waller and Fawcett [67] discuss opportunities for big data analytics in general in the logistics industry highlighting potential areas of opportunity as real-time capacity availability, time of delivery forecasting, optimal routing and reduction in driver turnover, all when it comes to carrier optimization. More specifically, in the context of fleet management, logistics companies will rely increasingly on the capabilities offered by big data analytics and the IoT to harness benefits in various areas. Jeske et al. [30] discuss areas relevant to the intersection of big data and IoT:

1. Optimization of service properties like delivery time, resource utilization, and geographical coverage—an inherent challenge of logistics.
2. Advanced predictive techniques and real-time processing to provide a new quality in capacity forecast and resource control.
3. Seamless integration into production and distribution processes for early identification of supply chain risks leading to resilience against disruptions.
4. Turning the transport and delivery network, as a result of efficient sensor instrumentation, into a high-resolution data source. In addition to fleet management by network optimization, this data may provide valuable insight on the global flow of goods allowing the level of observations to a microeconomic viewpoint.
5. Real- or near real-time insights into (changes in) demographic, environmental, and traffic statistics by analyzing the huge stream of data originating from a large delivery fleet.

As IoT enables self driving vehicles, the logistics industry anticipates a large impact on end-to-end logistics operations as highlighted by [31]. Please also refer to *Delivering Tomorrow*, for some studies on how such new trends are anticipated to impact the logistics industry in the future.[7]

---

[7]http://www.delivering-tomorrow.com/.

3.8 Other Opportunities

In addition to the major sectors that we discussed above, IoT has many opportunities directly targeted to consumers. While the above discussion also identifies ultimate opportunities for consumers, such as higher quality and possibly more accessible healthcare for patients, better retail experience, smart homes and energy efficiency gains, the IoT will become a part of daily lives via direct interaction with a lot of devices; *Wearables and Assistant devices* is one such important area. For instance, activity trackers such as jawbone and fitbit are already becoming a routine part of our lifestyle. *beddit*, *withings Aura* as well as advanced versions of *jawbone* can now also perform sleep monitoring, *Being* does better tracking than regular accelerometers while devices such as *Vessyl* can monitor what we drink. Connected wearables such as *Ego LS*, a wearable camera, can stream live video while *Tzoa* can do real-time environment tracking including pollution and UV exposure. Monitoring children is also made easier by devices such as *Pacif-i* and *Sproutling*. In addition, there are many assistant devices that intend to make our lives easier including home assistant robots (e.g., *Jibo*), automatic lawn mowers (e.g., Bosch's *Indego*), home interaction devices (e.g., Amazon's *Echo*) as well as self-driving cars. These are of course just a few illustrative examples. There are a myriad of devices in the market today.

3.8.1 Integrated Systems and Services

Metz [46] describes the wearables revolution of sorts as a result of IoT-enabled devices like the ones mentioned above. However, when it comes to consumers, we believe that most of the benefits from these developments will come not from individual offerings but from integrated systems. This requires not just connected, but interconnected devices. That is, the interaction doesn't just happen via the cloud but also between devices allowing them to adapt their bevahior as per requirements. For instance, a complete home automation systems that can not only control temperature, lighting and energy consumption of home appliances, but can also connect, communicate and coordinate with assistant devices such as *echo*, vacuum or lawn mower as well as other aspects such as cars for a seamless and integrated experience to the user and optimization of resource use. One can similarly think of an integrated health management capability. Even though self driving or autonomous cars can be considered an exception here since they can be self-contained, they would also draw benefits from these capabilities. These would be beneficial in a range of areas includng transportation, logistics and mobility. Similarly, in industrial facing applications, this would mean more responsive, self-monitoring and potentially self-maintaining assets. For instance, wind-turbines can adapt their performance not just in relation to the wind and local weather but also in relation to the a global optimization at the aggregate level of a wind farm. Similarly, assets such as aircraft engines can be responsive in relation to their peers (e.g., assets operating under similar operational and utilization conditions).

On protocols for data transfer and communication too, a variety of standards currently exist either due to various disparate efforts (to avoid dependencies) or due to companies developing proprietary offerings. Services companies will probably fill the gap created by non-existence of a common communication standard for various devices. The market will see a growth not only in such interconnection and integration services but also value-added services resulting from such integration. For instance, *Tado* is providing an interfacing through a variety of heating systems from multiple manufacturers for a smart thermostat system. Beyond increasing interoperability of standards or devices, such services will also generate new business and revenue models and value-add capabilities allowing for better operations (e.g., improving availability, ensuring higher quality of service of systems [15]), and financial risk modeling (e.g., better pricing and term-structure based on field operations).

Finally, many other sectors such as insurance (more informed risk modeling by utilizing real-time information), sustainability, social good, and security stand to gain with advancements in the big data and IoT technologies. The future holds even more promises such as opportunities with nano robots that can cure diseases, or in near-term, drones for various applications including deliveries and integrated surveillance functions.

4 Harnessing Value: What Do Organizations Need?

From organizations' perspective, harnessing value not just from IoT related big data analytics, but data science in general, requires foundational capabilities to be set in place before useful insights discovery can begin. The *analytics readiness* requirements include some of the capabilities discussed in Sect. 2 such as efficient storage and compute infrastructure, data acquisition and management mechanism, machine learning and data modeling capabilities as well as efficient deployment and scaling mechanisms. In addition, organizations also need to facilitate interfacing between engineering or domain experts and data scientists for efficient and productive knowledge transfer, agreed-upon validation as well as adoption and integration mechanism for analytics.

There are three most important objectives that an organization needs to achieve to realize these gains as they transform to be more data-driven.

1. *Data and Analytics Strategies* that align with the business vision. While a lot of data science activities and modeling exercises can be done in a bottom-up fashion, a coherent strategy can guide how the individual scattered efforts come together. Such a strategy should take a comprehensive view of how analytics can be a part of the decision making and insights generation process in the light of existing as well as future business directions, the enablement channels, and required skillsets. In the absence of a sound strategy and an execution plan, the isolated analytics efforts can quickly go adrift since it would be almost impossible to ask the "right" questions. While this topic is not the focus here, it is still important to recognize the need for such strategy.

2. *Culture Change* to accept insights from *validated, verified* and *principled* data-driven analytics into decision making at *all* levels. Even though a lot of capabilities in analytics are being commoditized, it is extremely important that users, both the ones performing analytics and the ones ingesting the resulting insights, are aware of the assumptions and constraints of the methods applied, as well as the ranges in which these should be interpreted. Moreover, such a culture-change is not unidirectional. Data divisions also have the onus of understanding the domains and their operational constraints better to be able to deliver value and to complement the domain and engineering experts.

3. *Innovation* to address open problems especially in the context of respective business applications. Organizations need to invest in innovation since differentiation will result from novel capabilities and well engineered integration.

5 Societal Impact and Areas of Concerns

While the technological feasibility of big data analytics for the IoT has been demonstrated in limited contexts, much more needs to be done to realize the broader vision. Not only the existing technology needs to be perfected, further innovation is needed to solve current bottlenecks as well as address longer term requirements. On the IoT end, this can mean increasing efficiency and affordability of data acquistion devices while reducing energy consumption as well as standardization of M2M service layer. Efforts are also needed in building common communications standards (while efforts are underway we do not have any consensus yet) and improving interoperability across data, semantics and organizations. In addition to the sources listed earlier, see [66] for a discussion on some additional aspects of IoT as well as architectural approaches in different contexts.

On the big data processing and analytics, we have just scratched the surface. Improved solutions are needed for problems such as analyzing massive temporal data, automated feature discovery, robust learning, analyzing heterogeneous data, efficiently managing complex, as well as meta-data, performing real-time analytics and handling streaming data (see, for instance, [76, 77]).

However, from a social point of view, there are also some major areas of concern that need to be addressed. We broadly divide these concern areas into two categories. The ones in the first category are *technological challenges*: research community has been sensitized to these and work is currently underway to better understanding and addressing them. However, it must be mentioned that these areas warrant more attention and effort than they currently receive. Main areas in this first category include:

1. *Privacy Issues*: Machine learning and data mining communities as well other fields including policy, security and governance have been working on these issues for some time. From an analytics perspective, privacy preserving data mining has developed into a subfield and considerable effort has gone into studying privacy

challenges in data mining [44, 51], data publishing [20] and, to some extent, integration and interactions of sensors [1]. However, these efforts have focused mainly on the data and analytics layers. Better protocols are also needed for other layers in the IoT stack. For instance, privacy and de-identification at the data acquisition layer needs to be efficiently addressed. For every application, there are also specific requirements, both regulatory and technological, that should govern privacy concerns. For instance, in the US, HIPAA[8] governs the majority of the requirements in dealing with medical data in many scenarios. Clear data governance and handling policies are needed to guide the efforts in the desired direction.

2. *Security Issues*: Security is always a concern in the case of large distributed systems. The more access points a network has, the more vulnerable it becomes. In the absence of clear and agreed upon standards and protocols, the security challenges are increased exponentially. In fact, security issues in the IoT are already a reality. For instance, [69] discusses top security mishaps in various contexts of IoT. Some work in this direction is already being done (e.g., [23, 71]) and needs to continue and expand.

3. *Interpretability Issues*: When employing analytics models in practice, we need to confirm how much we can rely on abstract models generalized based on non-linearities in the data and what aspects require interpretability of these models. Some requirements can be imposed due to the nature of application field (e.g., due to regulations) while in others interpretability can be needed to make use of the findings (e.g., gene identification). Sophisticated models can undoubtedly leverage more information from data compared to their simpler, interpretable counterparts. However, better evaluation and validation mechanism should be put in place to guarantee generalizability.

4. *Data Quality Issues*: Often, it is seen that the acquired data does not support desired analysis. For example, in a lot of cases, the acquired data from sensors is not intended to performed inductive inference at scale but rather is aimed to target a specific aspect such as safety, or reliability. Such cases would need an enhanced understanding of what use can be made of available data in the analytics context and how data quality can be ascertained.

The second category of issues is even more important in our opinion. We call these *adoption challenges*, referring to the issues resulting from inevitable, pervasive and ubiquitous adoption of analytics in various domains. This should not be viewed as an argument against more integration of analytics. Just as any other technology, analytics is a neutral force and the implications of its integration and use would rely on responsible choices made while trying to leverage it. Our aim is to sensitize the community so as not to overlook these as we move towards a new paradigm. Even though it is not possible to have immediate answers, we would like to highlight the issues to raise awareness of them during decision making processes as well as evolving strategies:

---

[8]Health Insurance Portability and Accountability Act.

1. *Model Reliability, Validation and Adaptation*: This is possibly the most widely discussed issue in the current list. Just by statistical chance, given that the models operate on vast amounts of data, correlations *will* be found. How should these correlations be validated? Standardization and agreement is required to evaluate these models and understand the associated risks. Principled forward testing mechanisms will be required, especially in cases of rare events such as asset failures. Backward testing and validation set-based evaluations are limited. Further, robustness of the models needs to be ascertained in changing environments either via model adaptation or via regular evaluation and requirements caliberation. Moreover, as these models interact with the environment and do not operate in isolation, their validation and verification becomes all the more crucial. This is especially important since the cost of doing "wrong" analytics may be significant for certain areas such as physical and mission-critical systems.

   (a) *The Risk of Over-Sophistication*: Extreme fine-tuning may result in models that can be very effective, but only for a very short period. If analytics has to be integreted into the process, it should be long lasting and adaptable. This requires more than just models that take into account evolution of the data or labels (e.g. concept drift) but also refers to how these models are utilized, how the expectations change over time and how the process responds to the results.

2. *Integration and Reconciliation with Our Physical Understanding of the World*: As IoT grows, analytics will increasingly be integrated in the environment, whether embedded in devices or assisting in decision making based on aggregate analytics. It is extremely important that we can reconcile these capabilities with the basis that we use to build and operate the physical devices (e.g., physics-based models). An argument can be made to restrict the models to "interpretable" ones when it comes to analytics. However, this trades off the knowledge that can be had from non-linear models in deriving non-obvious relationships hidden in the data. We need better mechanisms to integrate these models and to validate their findings.

3. *Human-Analytics Interaction*: As technology becomes pervasive, it tends to have an *assumed truth effect*, meaning that over time the users take the results with ever increasing trust. Consequently, in scenarios where decision-making will move closer to automated approaches, we should be mindful of their advantages as well as limitations. For instance, automated approaches have the potential to reduce the variations resulting from manual approaches. However, in some cases such variations are desired, even required, so that we can advance our understanding through a multitude of perspectives. It is timely to start seriously discussing about how humans will interact with analytics moving forward; how would this impact the decision making; would this lead to undesired uniformity? will we be able to notice inconsistencies and errors in the suggested decisions as our reliance on these models increase? How would these models respond to evolving realities of the world? How would the automated decision making impact policy?

4. *Potential for Systemic Errors and Failures*: Another aspect to consider is how much of a threat do automated decision making models pose to systematic as well as systemic failures as they become pervasive; Can the errors of individual pieces multiply resulting in system-wide risks? Will they have potential to bring down the whole system? Can massive interconnectivity result in a system-wide spread of failures, threats or even attacks? Note that the individual risks can be small and gradual but taken together they may have serious implications. Consequently, a risk containment mechanism will be needed in interconnected systems.

   (a) *Localization of "Failures"*: If system-wide events were to happen, would we have the ability to locate the sources? will we be able to quarantine a part of the system? Moreover, what effect would this have on the users since these systems will be an integrated part of peoples' lives? How would the necessary and important services be affected?

5. *Personalization Versus Limitation of Choice*: There can be intended and unin-tended, but nonetheless undesirable, consequences of "personalization" of ser-vices to individual lifestyles. On the unintended side, can over-personalization limit choice? For instance, as an effort to recommend the most relevant options, a subset of possible options is presented to the user. However, over time, and with increasing reliance on these recommendations, the users' exposure to possibilities outside of these recommendation-ranges can potentially be adversely impacted. Such systems can then potentially be used for malicious purposes such as social engineering around issues. Just as policy should take into account these aspects as technology grows, technologists also share the responsibility to contribute to addressing these issues.

6 Concluding Remarks

In this chapter, we discussed how big data technologies and the internet of things are playing a transformative role in the society. The pervasive and ubiquitous nature of such technologies will profoundly change the world as we know it, just as the indus-trial revolution and the Internet did in the past. We discussed opportunities in various domains both from an industrial and from the consumers' perspective. Given the data acquisition capabilities that are in place in the context of monitoring physical assets, the immediate opportunities are bigger from an industrial perspective. On the con-sumer end, we are currently undergoing a transformation as physical devices capable of advanced sensing become part of our routine life. Consumer applications will start witnessing a rapid growth in integrated services and systems, which we believe will generate much more value in contrast to one-off offerings as noted in Sect. 3.8, once a critical mass of such interconnected devices is reached in various domains. The capabilities in leveraging big data in both of these contexts are already transitioning from performing *descriptive analytics* to *predictive analytics*. For instance, based on real-time sensor data, we can predict certain classes of field events (e.g., failures or malfunctions) for heavy assets such as aircraft engines and turbines more reliably; this complements physics-based models employed in such cases. As these technologies mature, they will enable another transition from predictive to *prescriptive analytics* whereby recommendations on resolutions of such events could be made. This may develop to the extent of devices themselves taking corrective actions, and thus making them self-aware and self-maintaining. Even though we are already witnessing a paradigm shift, more needs to be done on various fronts, such as advancements in big data technologies, analytics, privacy and security, and policy making. In addition, the requirements at an organizational level in terms of readiness to harness the value resulting from analytics are discussed.

We then discussed broad social implications and highlighted areas of concerns as these technologies become pervasive. We organized these concerns into two categories: *technological challenges* that are relatively better understood, even if not entirely resolved, and *adoption challenges* that we believe are more unclear. As the adoption and integration of such technologies grow, so will our understanding of the implications evolve. However, the pace of change is fast indeed, and we will need to be quick in understanding this evolving landscape, analyzing the resulting changes and defining proper policies and protocols at various levels. Factors such as *human-analytics interaction* will also play an important role in how responsibly and effectively analytics complement our decision-making ability as well as how much autonomy these systems eventually attain.

Finally, we should reiterate that technologies are neutral. Any technology will have implications on society. The onus is on us to define how the technology is adopted in a responsible manner.

Appendix

Links to entities referred to in the article (in alphabetical order):

- Amazon AWS for IoT: http://aws.amazon.com/iot/
- Amazon Echo: http://www.amazon.com/oc/echo/ref_=ods_dp_ae
- Beddit: http://www.beddit.com/
- Being: http://www.zensorium.com/being
- Bosch, ABB, LG and Cisco's joint venture announced recently to cooperate on open standards for smart homes: http://www04.abb.com/global/seitp/seitp202.nsf/0/9421f99d7575ceccc1257c1d0033fa4a/file/8364IR_en_Red_Elephant_20 131024_final.pdf
- Bosch Indego: https://www.bosch-indego.com/gb/en/
- Cloud Foundry: http://www.cloudfoundry.org/about/index.html
- Cloudera and Hortonwork's real-time offering: http://www.infoq.com/news/2014/01/Spark-Storm-Real-Time-Analytics
- Ego LS: http://www.liquidimageco.com/
- Fitbit: http://www.fitbit.com/

- Hadoop Ecosystem: See, for instance, http://hadoopecosystemtable.github.io/
- Hubject, a joint networking mobility initiative of the BMW group, Bosch, Daimler, EnBW, RWE and Siemens: https://www.bosch-si.com/solutions/mobility/our-solutions/hubject.html
- IBM SyNAPSE: http://www.research.ibm.com/cognitive-computing/neurosynaptic-chips.shtml\#fbid=CAQQuy4xAkK
- Jawbone: https://jawbone.com/
- Jibo: http://www.myjibo.com/
- Microsoft's IoT offerings: http://www.microsoft.com/en-us/server-cloud/internet-of-things.aspx
- NVidia's Tegra X1: http://www.nvidia.com/object/tegra-x1-processor.html
- Pacif-i: http://bluemaestro.com/
- Pandas: http://pandas.pydata.org/
- Predictive Model Markup Language (PMML): http://www.dmg.org/v4-1/General Structure.html
- Qualcomm Zeroth: https://www.qualcomm.com/news/onq/2013/10/10/introducing-qualcomm-zeroth-processors-brain-inspired-computing
- Spark: https://spark.apache.org/
- Sproutling: http://www.sproutling.com/
- Storm:https://storm.apache.org/
- Tado: https://www.tado.com/
- Tzoa: http://www.mytzoa.com/#homepage
- Vessyl: https://www.myvessyl.com/
- Withings Aura: http://www.withings.com/us/withings-aura.html
- Zementis: http://zementis.com/

References

1. Aggarwal, C.C., Abdelzaher, T.: Integrating sensors and social networks. In: Aggarwal, C.C. (ed.) Social Network Data Analytics, pp. 379–412. Springer, US (2011). doi:10.1007/978-1-4419-8462-3_14; ISBN:978-1-4419-8461-6; http://dx.doi.org/10.1007/978-1-4419-8462-3_14
2. Aggarwal, C.C., Ashish, N., Sheth, A.: The internet of things: a survey from the data-centric perspective. In: Managing and Mining Sensor Data, pp. 383–428. Springer (2013)
3. Baaziz, A., Quoniam, L.: How to use big data technologies to optimize operations in upstream petroleum industry. Int. J. Innov. (IJI) 1(1), 30–42 (2013)
4. Bengio, Y., Ducharme, R., Vincent, P., Jauvin, C.: A neural probabilistic language model. J. Mach. Learning Res. 3, 1137–1155 (2003)
5. Bettencourt, L.M.A.: The uses of big data in cities. Santa Fe Institute working paper 2013-09-029, September 2013. http://www.santafe.edu/media/workingpapers/13-09-029.pdf
6. Bosch MongoDB white-paper: IoT and big data. Technical report, October 2014. http://info.mongodb.com/rs/mongodb/images/MongoDB_BoschSI_IoT_BigData.pdf
7. Boyd, S., Parikh, N., Chu, E., Peleato, B., Eckstein, J.: Distributed optimization and statistical learning via the alternating direction method of multipliers. Found. Trends Mach. Learn. 3(1), 1–122 (2011)

8. Brasco, C., Eklund, N., Shah, M., Marthaler, D.: Predictive modeling of high-bypass turbofan engine deterioration. In: Proceedings of the Annual Conference of the Prognostics and Health Management Society (PHM 2013), vol. 4. PHM Society (2013). http://www.phmsociety.org/node/1104

9. Bui, N., Zorzi, M.: Health care applications: a solution based on the internet of things. In: Proceedings of the 4th International Symposium on Applied Sciences in Biomedical and Communication Technologies, ISABEL '11, pp. 131:1–131:5. ACM, New York, NY, USA (2011). http://doi.acm.org/10.1145/2093698.2093829

10. Byrnes, N.: Cities find rewards in cheap technologies. MIT Technology Review, November 2014. http://www.technologyreview.com/news/532466/cities-find-rewards-in-cheap-technologies/

11. Chui, M., Löffler, M., Roberts, R.: The internet of things. McKinsey Quarterly 2, 1–9 (2010). http://www.mckinsey.com/insights/high_tech_telecoms_internet/the_internet_of_things

12. Cognizant Report: Reaping the benefits of the internet of things. Technical Report, May 2014. http://www.cognizant.com/InsightsWhitepapers/Reaping-the-Benefits-of-the-Internet-of-Things.pdf

13. Crankshaw, D., Bailis, P., Gonzalez, J.E., Li, H., Zhang, Z., Franklin, M.J., Ghodsi, A., Jordan, M.I.: The missing piece in complex analytics: low latency, scalable model management and serving with velox. In: Conference on Innovative Data Systems Research (CIDR). Asilomar, CA (2014)

14. Dean, J., Corrado, G., Monga, R., Chen, K., Devin, M., Mao, M., Ranzato, M., Senior, A., Tucker, P., Yang, K., Le, Q.V., Ng, A.Y.: Large scale distributed deep networks. In: Pereira, F., Burges, C.J.C., Bottou, L., Weinberger, K.Q. (eds.) Advances in Neural Information Processing Systems 25, pp. 1223–1231. Curran Associates, Inc. (2012). http://papers.nips.cc/paper/4687-large-scale-distributed-deep-networks.pdf

15. Deb, B., Shah, M., Evans, S., Mehta, M., Gargulak, A., Lasky, T.: Towards systems level prognostics in the cloud. In: Proceedings of the IEEE Conference on Prognostics and Health Management (PHM), pp. 1–6. IEEE (2013). ISBN:978-1-4673-5722-7

16. Dohr, A., Modre-Opsrian, R., Drobics, M., Hayn, D., Schreier, G.: The internet of things for ambient assisted living. In: Seventh International Conference on Information Technology: New Generations (ITNG), 2010, pp. 804–809. IEEE (2010)

17. Doukas, C., Maglogiannis, I.: Bringing IoT and cloud computing towards pervasive healthcare. In: 2012 Sixth International Conference on Innovative Mobile and Internet Services in Ubiquitous Computing (IMIS), pp. 922–926, July 2012. doi:10.1109/IMIS.2012.26

18. Feblowitz, J.: The big deal about big data in upstream oil and gas. IDC Energy Insights, October 2012

19. Feigelson, E.D., Babu, G.J.: Big data in astronomy. Significance 9(4), 22–25 (2012)

20. Fung, B.C.M., Wang, K., Chen, R., Yu, P.S.: Privacy-preserving data publishing: A survey of recent developments. ACM Comput. Surv. 42(4), 14:1–14:53, June 2010. doi:10.1145/1749603.1749605; ISSN:0360-0300; http://doi.acm.org/10.1145/1749603.1749605

21. Garcia, A.B., Bentes, C., de Melo, R.C., Zadrozny, B., Penna, T.J.P.: Sensor data analysis for equipment monitoring. Knowled. Inform. Syst. 28(2), 333–364 (2011). doi:10.1007/s10115-010-0365-1; ISSN:0219-1377; http://dx.doi.org/10.1007/s10115-010-0365-1

22. Ghose, A., Bhaumik, C., Das, D., Agrawal, A.K.: Mobile healthcare infrastructure for home and small clinic. In: Proceedings of the 2nd ACM International Workshop on Pervasive Wireless Healthcare, MobileHealth '12, pp. 15–20. ACM, New York, NY, USA (2012). doi:10.1145/2248341.2248347; ISBN:978-1-4503-1292-9; http://doi.acm.org/10.1145/2248341.2248347

23. Glas, B., Guajardo, J., Hacioglu, H., Ihle, M., Wehefritz, K., Yavuz, A.: Signal-based automotive communication security and its interplay with safety requirements. In: Proceedings of Embedded Security in Cars Conference, November 2012

24. Gonzalez, J.E., Xin, R.S., Dave, A., Crankshaw, D., Franklin, M.J., Stoica, I.: Graphx: graph processing in a distributed dataflow framework. In: 11th USENIX Symposium on Operating Systems Design and Implementation (OSDI 14), pp. 599–613. USENIX Association, Broomfield, CO, October 2014. ISBN:978-1-931971-16-4; https://www.usenix.org/conference/osdi14/technical-sessions/presentation/gonzalez

25. Gubbi, J., Buyya, R., Marusic, S., Palaniswami, M.: Internet of things (IoT): a vision, architectural elements, and future directions. Future Gen. Comput. Syst. **29**, 1645–1660 (2013)
26. Hems, A., Soofi, A., Perez, E.: Drilling for new business value: how innovative oil and gas companies are using big data to outmaneuver the competition. A Microsoft White Pater, May 2013
27. Hesla, L.: Particle physics tames big data. Symmetry **1** (2012)
28. IBM White Paper: Predictive maintenance for manufacturing. IBM (2011)
29. Japkowicz, N., Shah, M.: Evaluating Learning Algorithms: A classification perspective. Cambridge University Press (2011)
30. Jeske, M., Grüner, M., Weiß, F.: Big data in logistics: A DHL perspective on how to move beyond the hype. DHL Customer Solutions and Innovation, December 2013. http://www.delivering-tomorrow.com/wp-content/uploads/2014/02/CSI_Studie_BIG_DATA_FINAL-ONLINE.pdf
31. Joint DHL Bosch KIT Report: Self-driving vehicles in logistics: A DHL perspective on implications and use cases for the logistics industry. Technical report (2014). http://www.delivering-tomorrow.com/wp-content/uploads/2014/12/dhl_self_driving_vehicles.pdf
32. Kleiner, A., Talwalkar, A., Sarkar, P., Jordan, M.I.: A scalable bootstrap for massive data. J. Royal Statis. Soc. **76**, 795–816 (2013)
33. Krizhevsky, A., Sutskever, I., Hinton, G.E.: Imagenet classification with deep convolutional neural networks. In: Pereira, F., Burges, C.J.C., Bottou, L., Weinberger, K.Q. (eds.) Advances in Neural Information Processing Systems 25, pp. 1097–1105. Curran Associates, Inc. (2012). http://papers.nips.cc/paper/4824-imagenet-classification-with-deep-convolutional-neural-networks.pdf
34. Kurtz, J., Hoy, P., McHargue, L., Ward, J.: Improving operational and financial results through predictive maintenance. IBM Smarter Analytics Leadership Summit, Feb 2013
35. Lawson, S.: IoT groups are like an orchestra tuning up: the music starts in 2016. Computer World, Dec 2014. http://www.computerworld.com/article/2863498/networking-hardware/IoT-groups-are-like-an-orchestra-tuning-up-the-music-starts-in-2016.html
36. Le, Q.V., Monga, R., Devin, M., Chen, K., Corrado, G.S., Dean, J., Ng, A.Y.: Building high-level features using large scale unsupervised learning. In: International Conference on Machine Learning (2012)
37. Lee, J., Lapira, E., Bagheri, B., Kao, H.: Recent advances and trends in predictive manufacturing systems in big data environment. Manuf. Lett. **1**, 38–41 (2013)
38. Lee, J., Kao, H., Yang, S.: Service innovation and smart analytics for industry 4.0 and big data environment. Procedia CIRP **16**, 3–8 (2014)
39. Leuth, K.L.: IoT market segments biggest opportunities in industrial manufacturing. IoT-Analytics (2014). http://IoT-analytics.com/IoT-market-segments-analysis/
40. Lin, J., Keogh, E., Lonardi, S., Chiu, B.: A symbolic representation of time series, with implications for streaming algorithms. In: Proceedings of the 8th ACM SIGMOD workshop on Research issues in data mining and knowledge discovery, pp. 2–11. ACM (2003)
41. Mackey, L., Talwalkar, A., Jordan, M.I.: Distributed matrix completion and robust factorization. J. Mach. Learn. Res. (2014)
42. Markkanen, A., Shey, D.: The intersection of analytics and the internet of things. IEEE Internet of Things Newsletter, Nov 2014. http://IoT.ieee.org/newsletter/november-2014/the-intersection-of-analytics-and-the-internet-of-things.html
43. Marz, N., Warren, J.: Big data: principles and best practices of scalable realtime data systems. Manning Publications Co. (2015)
44. Matwin, S.: Privacy-preserving data mining techniques: survey and challenges. In: Discrimination and Privacy in the Information Society, pp. 209–221. Springer (2013)
45. McKinsey Study: Connected car, automotive value chain unbound. Technical report (2014)
46. Metz, R.: Ces 2015: Wearables everywhere. MIT Technology Review, January 2015. http://www.technologyreview.com/news/533916/ces-2015-wearables-everywhere/
47. Middleton, P., Kjeldsen, P., Tully, J.: Forecast: The Internet of Things, worldwide, 2013. Gartner, November 2013

48. Mind Commerce LLC Report: Big data in extraction and natural resource industries: Mining, water, timber, oil and gas 2014–2019. Technical report, July 2014. http://www.researchandmarkets.com/research/3qpj9t/big_data_in
49. MIT Business Report: Cities get smarter. Technical report (2015)
50. Nambiar, R., Bhardwaj, R., Sethi, A., Vargheese, R.: A look at challenges and opportunities of big data analytics in healthcare. In: 2013 IEEE International Conference on Big Data, pp. 17–22. IEEE (2013)
51. Navarro-Arribas, G., Torra, V.: Advanced Research in Data Privacy (2014)
52. Nicholson, R.: Big data in the oil and gas industry. IDC Energy Insights, September 2012
53. NIST Report: Workshop report on foundations for innovation in cyber-physical systems. Technical report, Jan 2013. http://www.nist.gov/el/upload/CPS-WorkshopReport-1-30-13-Final.pdf
54. Orts, E., Spigonardo, J.: Sustainability in the age of big data. Special Report, Initiative for Global Environmental Leadership (IGEL), Knowledge at Wharton, September 2014. http://knowledge.wharton.upenn.edu/article/the-big-data-and-energy-synergy/
55. Páez, D., Aparicio, F., de Buenaga, M., Ascanio, J.R.: Big data and IoT for chronic patients monitoring. In: Ubiquitous Computing and Ambient Intelligence. Personalisation and User Adapted Services, pp. 416–423. Springer (2014)
56. Pan, X., Jegelka, S., Gonzalez, J., Bradley, J.K., Jordan, M.: Parallel double greedy submodular maximization. In: Advances in Neural Information Processing Systems 22, (2014)
57. Poulymenopoulou, M., Malamateniou, F., Vassilacopoulos, G.: Machine learning for knowledge extraction from phr big data. Stud. Health Technol. Inform. 202, 36–39 (2013)
58. Reddy, A.S.: Reaping the benefits of the internet of things. Cognizant Reports, May 2014
59. Salakhutdinov, R.: Learning deep generative models. Ph.D. thesis, University of Toronto, Toronto, Canada (2009)
60. Seshadri, M.: Big data science challenging the oil industry. Energyworld (2013). http://web.idg.no/app/web/online/event/energyworld/2013/emc.pdf
61. Socher, R., Pennington, J., Huang, E.H., Ng, A.Y., Manning, C.D.: Semi-supervised recursive autoencoders for predicting sentiment distributions. In: Proceedings of the Conference on Empirical Methods in Natural Language Processing, EMNLP '11, pp. 151–161. Association for Computational Linguistics, Stroudsburg, PA, USA, 2011. ISBN:978-1-937284-11-4. http://dl.acm.org/citation.cfm?id=2145432.2145450
62. Sowe, S.K., Kimata, T., Mianxiong, D., Zettsu, K.: Managing heterogeneous sensor data on a big data platform: IoT services for data-intensive science. In: 2014 IEEE 38th International Computer Software and Applications Conference Workshops (COMPSACW), pp. 295–300, July 2014. doi:10.1109/COMPSACW.2014.52
63. Tracey, D., Sreenan, C.: A holistic architecture for the internet of things, sensing services and big data. In: 2013 13th IEEE/ACM International Symposium on Cluster, Cloud and Grid Computing (CCGrid), pp. 546–553, May 2013. doi:10.1109/CCGrid.2013.100
64. Turner, V., Gantz, J.F., Reinsel, D., Minton, S.: The digital universe of opportunities: rich data and the increasing value of the internet of things. IDC White Paper, April 2014. http://idcdocserv.com/1678
65. Vandermerwe, S., Rada, J.: Servitization of business: adding value by adding services. Eur. Manage J. 6(6), 314–324 (1989)
66. Vermesan, O., Friess, P.: Internet of Things: Converging Technologies for Smart Environments and Integrated Ecosystems. River Publishers (2013)
67. Waller, M.A., Fawcett, S.E.: Data science, predictive analytics, and big data: a revolution that will transform supply chain design and management. J. Bus. Logist. 34(2), 77–84 (2013)
68. Wang, Y., Bai, H., Stanton, M., Chen, W., Chang, E.Y.: Plda: parallel latent dirichlet allocation for large-scale applications. In: Proceedings of the 5th International Conference on Algorithmic Aspects in Information and Management, AAIM '09, pp. 301–314. Springer-Verlag, Berlin, Heidelberg (2009). doi:10.1007/978-3-642-02158-9_26; ISBN:978-3-642-02157-2; http://dx.doi.org/10.1007/978-3-642-02158-9_26

69. Witten, B.: Top 10 IoT security mishaps 2014. In: Industrial Internet Consortium Web blog post. IIC (2014). http://blog.iiconsortium.org/2014/12/top-10-IoT-security-mishaps-2014-.html
70. Yashiro, T., Kobayashi, S., Koshizuka, N., Sakamura, K.: An internet of things (IoT) architecture for embedded appliances. In: Humanitarian Technology Conference (R10-HTC), 2013 IEEE. Region, vol. 10, pp. 314–319 (2013). doi:10.1109/R10-HTC.2013.6669062
71. Yavuz, A.A.: Practical immutable signature bouquets (pisb) for authentication and integrity in outsourced databases. In: Data and Applications Security and Privacy XXVI, pp. 179–194. Springer (2013)
72. Zaki, M., Neely, A.: Optimising asset management within complex service networks: the role of data. Cambridge Service Alliance, working paper:1–11 (2014)
73. Zanella, A., Bui, N., Castellani, A., Vangelista, L., Zorzi, M.: Internet of things for smart cities. IoT J IEEE, 1(1):22–32 (2014). doi:10.1109/JIoT.2014.2306328; ISSN:2327-4662
74. Zaslavsky, A, Perera, C., Georgakopoulos, D.: Sensing as a service and big data. arXiv:1301.0159 (2013)
75. Zhai, K., Boyd-Graber, J., Asadi, N., Alkhouja, M.L.: Mr. lda: A flexible large scale topic modeling package using variational inference in mapreduce. In: Proceedings of the 21st International Conference on World Wide Web, WWW '12, pp. 879–888, ACM, New York, NY, USA (2012). doi:10.1145/2187836.2187955; ISBN:978-1-4503-1229-5; http://doi.acm.org/10.1145/2187836.2187955
76. Zhou, Z., Chawla, N., Jin, Y., Williams, G.: Big data opportunities and challenges: discussions from data analytics perspectives [discussion forum]. IEEE Comput. Intell. Magaz. 9(4), 62–74 (2014)
77. Zicari, R.V., Akerkar, R. (ed.): Big data computing. In: Big Data: Challenges and Opportunities, pp. 103–128. Chapman and Hall/CRC (2013)

Social Network Analysis in Streaming Call Graphs

Rui Sarmento, Márcia Oliveira, Mário Cordeiro, Shazia Tabassum and João Gama

Abstract Mobile phones are powerful tools to connect people. The streams of Call Detail Records (CDR's) generating from these devices provide a powerful abstraction of social interactions between individuals, representing social structures. Call graphs can be deduced from these CDRs, where nodes represent subscribers and edges represent the phone calls made. These graphs may easily reach millions of nodes and billions of edges. Besides being large-scale and generated in real-time, the underlying social networks are inherently complex and, thus, difficult to analyze. Conventional data analysis performed by telecom operators is slow, done by request and implies heavy costs in data warehouses. In face of these challenges, real-time streaming analysis becomes an ever increasing need to mobile operators, since it enables them to quickly detect important network events and optimize business operations. Sampling, together with visualization techniques, are required for online exploratory data analysis and event detection in such networks. In this chapter, we report the burgeoning body of research in network sampling, visualization of streaming social networks, stream analysis and the solutions proposed so far.

Keywords Call graphs · Network sampling · Network visualization · Streaming networks

1 Introduction

Technological advances in computer processing power, disk storage and databases have enabled the collection of a wealth of data on social interactions by mobile telecom operators. The communication among individuals through mobile devices has been used as a proxy of their social relationships and is captured in the form of Call Detail Records (CDRs). CDRs provide detailed information regarding each

R. Sarmento · M. Oliveira · M. Cordeiro · S. Tabassum · J. Gama (✉)
LIAAD/INESC TEC, University of Porto, Porto, Portugal
e-mail: jgama@fep.up.pt

R. Sarmento · M. Oliveira · J. Gama
FEP, School of Economics and Management, University of Porto, Porto, Portugal

© Springer International Publishing Switzerland 2016
N. Japkowicz and J. Stefanowski (eds.), *Big Data Analysis: New Algorithms for a New Society*, Studies in Big Data 16, DOI 10.1007/978-3-319-26989-4_10

phone call made between two individuals, such as time, call duration, source number and destination number. These data implicitly define a call graph, where nodes represent the individuals (or mobile operator subscribers or phone users) and there is an edge between two individuals if they called each other. This graph is typically weighted and directed. Weighted because the edges of the graph are assigned a weight indicating the frequency of phone calls, and directed because it includes information about who initiated the phone call. Since call graphs model the communication among individuals, these represent *social networks* and its analysis falls within the scope of Social Network Analysis (SNA). SNA is an interdisciplinary methodology concerned with the discovery of patterns in the structure of social networks. The focus of SNA is on the link structure of the network rather than on the attributes of the nodes. Important SNA tasks include topological analysis, centrality analysis and community detection. Topological analysis aims at discovering structural properties that characterize the overall topology of the network (e.g., degree distribution, average path length, effective diameter, clustering coefficient, connected components), whereas centrality analysis is focused on finding the key nodes in the network based on the position they occupy in the network structure. The importance of nodes is typically measured in terms of their centrality in the network, which can be quantified by, for instance, the degree, closeness and betweenness centralities. On a different level, the goal of community detection is to discover implicit communities comprised in the network. These communities are groups of nodes that interact more often with each other than with other nodes in the networks and share stronger ties among them.

The analysis of the social networks underlying call graphs, using the SNA methodology, can deliver valuable business insights to mobile telecom operators that can support relevant business tasks. Pinheiro [34] addresses these tasks by combining areas such as social network analysis, analytical studies and marketing expertise to propose improvements in customer service for telecommunications networks. Kayastha et al. [23] reveal another work quite relevant, presenting a compilation of applications, architecture and issues associated with protocols designed for social networking obtained with mobile communications. Examples of applications enabled by SNA methodology include churn prediction, identification of influencers and most active callers, fraud detection and design of targeted marketing campaigns to increase subscribers' loyalty.

However, the network data collected by mobile network operators has specific features that introduces complexity in the design of new methods. CDRs are being continuously generated by the communication activity among subscribers. In addition to the large volume, this data arrives at high rates. Thus, the developed methods should be able to cope with data speed and volume and operate under the one-pass constraint of data streams. Besides scalability and computational efficiency issues, it is also desirable that the outputs of the developed algorithms are comprehensible, in order to foster their real-world deployment. Resorting to appropriate visualization techniques eases the understanding of patterns in the data, especially for non-experts. Since visualization plays an important role in the presentation of results and proves useful in supporting business decisions by the operators' managers, methods for streaming network analysis should be coupled with visualization techniques. Hitherto, few research work has been done to tackle these challenges.

The analysis of data collected by mobile telecom operators is usually performed offline and heavily relies on batch processing. Business Intelligence techniques and tools are typically used to transform these large volumes of raw data into useful business information, mostly by means of querying and reporting. This information serves several purposes, such as the identification of trends and the extraction of patterns from both users and equipment events. The useful knowledge obtained from the data analysis process is then used to support a wide range of business decisions. Nevertheless, the prevailing *modus operandi* for analysing data by the telecommunication providers is slow, done by request and requires high costs in data warehouses. These characteristics, coupled with an increasing need to quickly react to events (or even anticipate them), avoid customer churn and improve customer service, places real-time streaming analysis in the forefront of the analytics solutions of telecom operators. The development and application of streaming methods, specifically tailored to network mining, grants telecom operators the ability to adapt to changes in the evolution of the network and detect key events in an efficient and timely manner. This ability to react and quickly adapt to real-time events brings benefits to the operators by means of increased revenue and cost reduction. In short, a streaming solution would provide the operators with the means to operate with little or even no latency, therefore being able to automatically respond to events, in a shorter time.

In this chapter, we generically cover the solutions proposed so far on sampling, visualization, community detection and centrality measures for streaming social networks.

The chapter starts with Sect. 2, where we describe general structural properties of call graphs. In Sect. 3, we address the sampling process for both static and streaming social networks. Section 4 is devoted to an introduction, followed by a critical discussion, of windowing data models to capture different kind of network events in a streaming environment. In Sect. 5, we also present network centrality measures that were developed for the analysis of dynamic and large-scale social networks. In Sect. 6, we discuss community detection algorithms. Finally, in Sect. 7 we summarize this chapter and discuss open challenges.

2 The Case Study

All the experiments and empirical results presented in this chapter are supported by a case study conducted with real-world data streams collected, and made available, by a mobile telecom operator. The characteristics of the network data will be presented in this section.

The communication among mobile users generates huge amounts of data that arrives at high rates. To conduct the case study we had access to 135 days of anonymized CDRs retrieved from equipment distributed geographically. These CDRs implicitly define a directed weighted call graph, which is stored in the form of a sparse weighted edge list. This call graph depicts the communication among

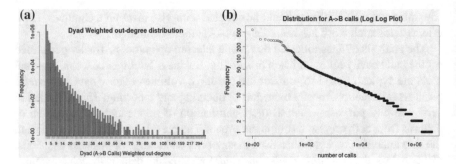

Fig. 1 **a** Distribution of the A→B phone calls and **b** the corresponding log-log plot

the operator's subscribers, by modelling the subscribers as nodes, and the phone calls made among them as edges. These edges are weighted by the number of phone calls made. This call graph has, on average, 10 million phone calls per day made by approximately 6 million subscribers. Each edge represents a private phone call between source number A and destination number B. For each edge/call, there is information regarding the date and time (seconds) when the call was initiated, as well as its duration. The volume of data speed ranges from 10 up to 280 calls per second, usually around mid-night and mid-day time, respectively.

To study the distribution of the available data, we aggregate the data in two different ways:

1. **Dyad weighted out-degree distribution** Count the number of calls, per day, from source number A to destination number B (A→B);
2. **Out-degree distribution** Count the number of calls, per day, made by each subscriber.

After the previous operation we observed the distribution of the aggregated data and there is some evidence that the tail of these distributions follows a power-law [3], as can be ascertained in Figs. 1a and 2a. The analysis of these figures suggest that, for a 1-day period, it is expected a high amount of single phone calls between some phone numbers A→B, and a low amount of many phone calls between a few phone numbers A→B. Therefore, it is expected a low amount of highly active subscribers and a large amount of low activity subscribers. We also plotted the distribution of the daily aggregated data in a log-log scale (see Figs. 1b and 2b). These plots show a monomial approximation that suggests that this data is derived from a power-law distributions.

The power-law hypothesis was tested by following the guidelines described in [9] and using the *poweRlaw* R package. Figure 3 illustrates the hypothesis test for the power-law distribution presenting the mean estimate of parameters x_{min}, α and the *p-value*, being x_{min} the lower bound of the power-law distribution. Estimation parameter α is the scaling parameter ("Par 1" in Fig. 3) and $\alpha > 1$. The dashed-lines give approximate 95 % confidence intervals. The observed *p-value* when testing the

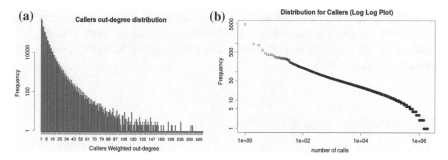

Fig. 2 a Out-degree distribution of the subscribers and **b** the corresponding log-log plot

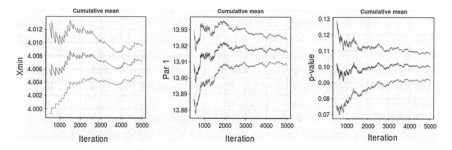

Fig. 3 Original network—hypothesis test for the caller power-law distribution

null hypothesis H_0 that the original data is generated from a power-law distribution is 0.1. Since we set the significance level to be 0.05, H_0 cannot be rejected because the *p-value* is higher than 0.05. Given that the tail of the distribution of the phone calls follows a power-law, we can expect that the use of sampling techniques to extract the most active subscribers is feasible and desirable.

3 Sampling

The analysis of large streaming networks poses processing or memory issues when using conventional hardware or software. Even if the available computational resources are able to perform the analysis of a network comprised of millions of nodes, it is difficult for the analyst to gather valuable knowledge from the outcome. Thus, in this section, we introduce several sampling methods, as well as effective visualizations, of large streaming networks in order to address the above-mentioned problems. More specifically, we introduce the *top-K* sampling method that focuses on extracting a sample of the most active nodes in the network. This approach is suitable for networks exhibiting a power-law behaviour and is able to preserve the same distribution of the original network and the global community structure. Besides, it is highly efficient, either for visualization or analysis purposes, because it relies on the *Space-Saving algorithm*.

3.1 Sampling Large Static Networks

Large-scale network sampling has recently become a hot topic in network analysis and only a few methods have been proposed so far. The most common approaches for static network sampling are the *random sampling* and the *snowball sampling*.

In the *snowball sampling* [15], a starting node is chosen. Then, the connections in the 1st, 2nd, to *n*, neighborhood-order of this starting node are extracted until the network reaches the desirable size. This approach, while easy to implement, has known problems: it is biased towards the region of the network to where the starting node belongs to, and potentially misses important network properties. Yet, it is one of the most common sampling approaches.

On the other hand, *random sampling* [16], randomly selects a user-defined percentage of nodes and keeps all the corresponding edges. Alternatively, the sampling can be performed on the edges, by randomly selecting a user-defined percentage of edges and keeping the corresponding nodes. The main problem with this approach is that random edge sampling is biased towards high-degree nodes, whereas the random node sampling may be unable to generate a representative sample, since the structural properties of the sample may not reflect the ones observed on the original network. Despite these drawbacks, random sampling is easy to understand and implement.

The task, therefore, must be to generate a sample in such a way that the sampled network is representative of the original one in terms of structural properties. A primary question is related with the definition of *representative sample*. Existing work considers measures such as similarity in degree distributions and clustering coefficients [20, 28].

Leskovec et al. [28] introduce a great variety of graph sampling algorithms. They conclude that methods combining random node selection and some vicinity exploration generate the best network samples. They show that a 15 % sample is usually enough to match the properties of the original graph and that no list of network properties serving as basis for sampling evaluation will ever be perfect.

3.2 Sampling Large Streaming Networks

Papagelis et al. [33] introduced sampling-based algorithms that quickly obtains a near-uniform random sample of nodes in its neighborhood, given a selected node in the social network. The authors also introduce and analyze variants of these basic sampling schemes, aiming the minimization of the total number of nodes in the visited network, by exploring correlations across samples.

Several approaches have been proposed to gather information from streaming graphs. Typical SNA problems, such as triangle counting, centrality analysis and community detection, have already been implemented in streaming settings. We will delve deeper on these topics further in the chapter.

Network sampling of streaming graphs is still a promising area for future research since, to the best of our knowledge, only few stream-based sampling methods were proposed so far. Ahmed et al. [1] present a novel approach to graph streaming sampling. According to the authors, there was no previous contribution in this topic. The authors propose a novel sampling algorithm, dubbed PIES, based on edge sampling and partial induction by selecting the edges that connect sampled nodes.

3.3 Top-K Sampling with Top-K Itemsets

Researchers have been trying to achieve efficient ways of analyzing data streams and performing graph summarization. The exact solution implies the knowledge of the frequency of all nodes and edges, which might be impossible to obtain in large-scale networks.

The problem of finding the most frequent items in a data stream S of size N is basically how to discover the elements e_i whose relative frequency f_i is higher than a user-defined support ϕN, with $0 \leq \phi \leq 1$ [14]. Given the space requirements that exact algorithms addressing this problem would need [8], several algorithms were already proposed to find the top-k frequent elements, being roughly classified into *counter-based* and *sketch-based* [30]. *Counter-based* techniques keep counters for each individual element in the monitored set, which is usually a lot smaller than the entire set of elements. When an element is identified as not currently being monitored, various algorithms take different actions to adapt the monitored set accordingly. *Sketch-based* techniques provide less rigid guarantees, but they do not monitor a subset of elements, providing frequency estimators for the entire set.

Simple *counter-based* algorithms that process the stream in compressed size, such as *Sticky Sampling* and *Lossy Counting*, were proposed by Manku et al. in [29]. Yet, these have the disadvantage of keeping a large amount of irrelevant counters. *Frequent* [11], by Demaine et al., keeps only k counters for monitoring k elements, incrementing each element counter when it is observed, and decrementing all counters when an unmonitored element is observed. Zeroed-counted elements are replaced by new unmonitored element. This strategy is similar to the one applied by the *Space-Saving algorithm*, proposed by Metwally et al. [30], which give guarantees for the *top-m* most frequent elements. *Sketch-based* algorithms usually focus on families of hash functions which project the counters into a new space, keeping frequency estimators for all elements. The guarantees are less strict but all elements are monitored. The *CountSketch* algorithm [8], by Charikar et al., solves the problem with a given success probability, estimating the frequency of the element by finding the median of its representative counters, which implies sorting the counters. Also, *GroupTest* method [10] proposed by Cormode et al., employs expensive probabilistic calculations to keep the majority elements within a given probability of error. Despite the fact of being generally accurate, its space requirements are large and no information is given about frequencies or ranking.

Algorithm 1 represents the proposed *top-K* Method application using the *Space-Saving algorithm.*

This type of application is based on a landmark window model [14], which implies a growing number of inspected events in the accumulating time window. This landmark application is useful also in other contexts, e.g., when the network is relatively small and the user wants to check all events in it.

Experiments using the landmark window model showed that this model suffers from the problems we would like to avoid, such as exceeding memory limits. This happens when the number of nodes and edges exceeds dozens of thousands of nodes. The *top-K* algorithm, based on a landmark window model, is an efficient approach for large-scale data. It focus on the most active nodes and discards the least active ones, which are the most frequent according to the power-law distribution. The alternative option to the landmark window model, i.e., the sliding window model [14], would not be appropriate for the *top-K* approach, since it may remove less recent nodes. Those nodes may yet be included in the *top-K* list we want to maintain.

In our scenario, the *top-K* representation of data streams implies knowing the K elements of the simulated data stream from the database. Network nodes that have higher frequency of outgoing connections, incoming connections, or even specific connections between any node A and B, may be included in the graph, as well as their connections.

For this application, the user can insert as input a start date and hour and also the maximum number of *top-K* nodes to be represented (the K parameter), along with their connections.

With the inserted start date and hour, the *top-K* application is expected to return the evolving network of the *top-K* nodes. Functions *getTopKNodes* and *updateTopNodesList* in **Algorithm 1** implement the *Space-Saving algorithm.* As the network evolves over time, new *top-K* nodes are added to the graph. Nodes that exit the *top-K* list of numbers are removed from the *top-K* list and, thus, removed from the graph along with their connections.

Figure 4 represents the network induced by the top-100 subscribers with the highest number of phone calls, since the midnight of the first day of July 2012, until 00h44m33s. The algorithm shows the 100 most active phone numbers in that period. Figure 5 depicts a similar network but after running the layout algorithm. This time, the output considers results until 01h09m45s.

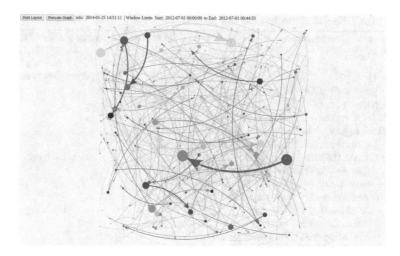

Fig. 4 Network induced by the top-100 subscribers with the highest number of phone calls and corresponding direct connections. This network was generated without running the layout algorithm

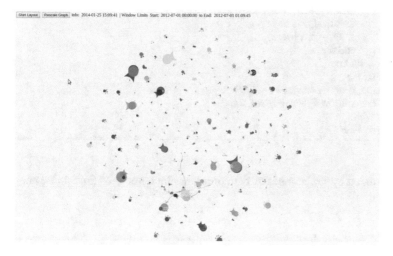

Fig. 5 Network induced by the top-100 subscribers with the highest number of phone calls and corresponding direct connections. This network was generated after running the layout algorithm

4 Window-Based Visualization

Resorting to time window models is an useful strategy to limit the amount of data available for analysis, since it is based on setting a fixed point in time (the so-called *landmark*) from which the data starts being observed. A disadvantage of this method is that the amount of data inside the window quickly grows to a prohibitive size. Other way of limiting data is by using a fixed sliding window model. These windows

---

Algorithm 1 Top-K algorithm for call graphs

Input: $start, k\_param, tinc$ ▷ start timestamp, k parameter and time increment
Output: $edges$
1: $R \leftarrow \{\}$ ▷ data rows
2: $E \leftarrow \{\}$ ▷ edges currently in the graph
3: $R \leftarrow$ getRowsFromDB $(start)$
4: $new\_time \leftarrow start$
5: **while** $(R <> 0)$ **do**
6:     **for all** $edge \in R$ **do**
7:         $before \leftarrow$ GETTOPKNODES(k_param)
8:         UPDATETOPNODESLIST$(edge)$ ▷ update node list counters
9:         $after \leftarrow$ GETTOPKNODES(k_param)
10:        $maintained \leftarrow before \bigcap after$
11:        $removed \leftarrow before \setminus maintained$
12:        **for all** $node \in after$ **do**   ▷ add top-k edges
13:            **if** $node \subset edge$ **then**
14:                ADDEDGETOGRAPH$(edge)$
15:                $E \leftarrow E \bigcup \{edge\}$
16:            **end if**
17:        **end for**
18:        **for all** $node \in removed$ **do**   ▷ remove non top-k nodes and edges
19:            REMOVENODEFROMGRAPH$(node)$
20:            **for all** $edge \in node$ **do**
21:                $E \leftarrow E \setminus \{edge\}$
22:            **end for**
23:        **end for**
24:    **end for**
25:    $new\_time \leftarrow new\_time + tinc$
26:    $R \leftarrow$ getRowsFromDB (new_time)
27: **end while**
28: $edges \leftarrow E$

---

are bounded by the number of data points or the number of time units, being both constant.

4.1 Landmark Windows

Algorithm 2, regarding streaming landmark window models, provides the representation of all the events (e.g., edge and node addition or removal) that occur in the network, starting at a specific time stamp, for example, 01h48m09s of 1st of January 2012.

This type of window model is not very useful in a streaming scenario, because it implies a growing number of events outputted on the screen and the comprehensibility lowers as this number reaches, or exceeds, a few thousands of events. This landmark application is however useful in other contexts, for example, if the network is relatively small and the analyst is interested in checking all events in the

---

Algorithm 2 Algorithm based on a landmark window model [37]

Input: *start, wsize, tinc*   ▷ start timestamp, window size and time increment
Output: *edges*
1: $R \leftarrow \{\}$ ▷ data rows
2: $E \leftarrow \{\}$ ▷ edges currently in the graph
3: $R \leftarrow$ getRowsFromDB (*start*)
4: *new_time* $\leftarrow$ *start*
5: **while** $(R <> 0)$ **do**
6:     **for all** *edge* $\in R$ **do**
7:         ADDEDGETOGRAPH(*edge*)
8:         $E \leftarrow E \bigcup \{edge\}$
9:     **end for**
10:     *new_time* $\leftarrow$ *new_time* + *tinc*
11:     $R \leftarrow$ getRowsFromDB (*new_time*)
12: **end while**
13: *edges* $\leftarrow E$

---

evolution of the network. Nevertheless, if the analyst wants to follow the evolution of a large streaming network, the application described in the next subsection is more appropriate.

4.2 Sliding Windows

For this large data stream, we generate a dynamic sample representation of the data by using a sliding window model. This sliding window is defined as a data structure with fixed number of registered events. Each event is a phone call between pairs of subscribers. Since these events are annotated with time stamps, the time period between the first call and the last call in the window is easily computed. The input parameters of this algorithm are (i) starting date and time, and the (ii) maximum number of events/calls the sliding window can have. The SNA model used in this application is a full weighted directed network, since all the nodes and edges in the network are outputted for a particular instance of the sliding window [19].

An example of the obtained results is provided in Fig. 6. Nodes are sized according to their weighted degree. Thus, larger nodes correspond to more active subscribers, i.e., subscribers associated with a higher number of phone calls (either received or made). This is the representation of a window with 1000 events/calls, for a time period starting at 00h01m52s and ending at 00h02m40s. The evolution of the network is visually and immediately conclusive. There are three nodes with the largest number of connections/phone calls in the network.

From the figure, we can also see the connection established between two of these three largest nodes. Figure 6 also displays the average data speed in the window (approximately 22 calls per second). This average data speed is computed by counting the number events (i.e., phone calls) inside the window, which comprises all the events observed during the time period associated with the window length. Under

Algorithm 3 Algorithm based on a sliding window model [37]

Input: *start, wsize, tinc*   ▷ start timestamp, window size and time increment
Output: *edges*
1: $R \leftarrow \{\}$ ▷ data rows
2: $E \leftarrow \{\}$ ▷ edges currently in the graph
3: $V \leftarrow \{\}$ ▷ buffer to manage removal of old edges
4: $R \leftarrow$ getRowsFromDB *(start)*
5: *new_time* ← *start*
6: $p \leftarrow \{\}$
7: **while** $(R <> 0)$ **do**
8:    **for all** *edge* ∈ R **do**
9:        ADDEDGETOGRAPH(*edge*)
10:       $E \leftarrow E \bigcup \{edge\}$
11:       $k \leftarrow 1 + (p \bmod wsize)$
12:       *old_edge* ← $V[k]$
13:       REMOVEEDGEFROMGRAPH(*old_edge*)
14:       $E \leftarrow E \setminus \{old\_edge\}$
15:       $V[k] \leftarrow edge$
16:       $p \leftarrow p + 1$
17:   **end for**
18:   *new_time* ← *new_time* + *tinc*
19:   $R \leftarrow$ getRowsFromDB *(new_time)*
20: **end while**
21: *edges* ← E

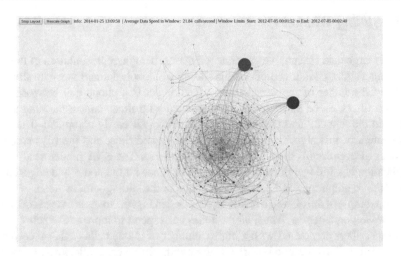

Fig. 6 Visualization of the call graph using a sliding window approach [37]

different experimental conditions, namely when analysing the window obtained for
mid-day, the data speed increases, with more phone calls per second. After several
experiments with different window sizes, and considering that the data speed changes,
we concluded that this speed should not be smaller than approximately 100 events
and also not larger than approximately 1000 events. With the minimum data speed

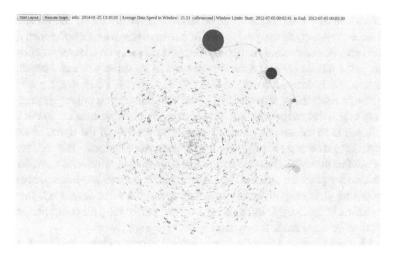

Fig. 7 Visualization of the call graph using a sliding window approach (2nd version) [37]

conditions, 100 events represents a window period of around 10 s of events. With the maximum data speed and a window of 1000 events, it represents around 5 s of data. Using this data, less than 100 events represents changes in the window that are too fast to be visually comprehensible, and more than 1000 events represents too many events, reducing the visual comprehensibility of the output.

Figure 7 represents the next window instance, starting at 00h02m41s and ending at 00h03m30s. Considering the previous Fig. 6, we can visually observe the evolution of the network and observe that there is a new smaller node establishing connection to the most active nodes identified before, in this window of 1000 events.

5 Centrality Analysis

One of the most relevant tasks in SNA is the computation and interpretation of centrality measures. Centrality is often used for finding important nodes by analyzing their position in the network. The concept of *importance* is multidimensional. Each relevant dimension, such as reachability, embeddedness, influence, support, ability to span structural holes and control of information flow, is captured by different centrality measures. Examples of classical measures are *degree*, *betweenness*, *closeness* and *eigenvector centrality*. The first three were proposed by Freeman [13] and the last one by Bonacich [5]. There are two fundamentally different classes of centrality measures: node-level measures (e.g., degree) and network-level measures (e.g., density). The former evaluates the centrality of each node/vertex (or edge/link) in the network, whereas the latter evaluates the centrality of the whole network.

While some of these measures can be straightforwardly computed for streaming graphs, since they only require the update of counters (e.g., degree and density), other measures are computationally expensive because they rely on the computation of the shortest paths across all nodes in the network (e.g., closeness and betweenness). The computation of the all pairs shortest paths is feasible when dealing with small networks of a few tens of thousands of nodes and links, but it quickly becomes prohibitively expensive as the network grows larger. A possible solution to circumvent this problem is to calculate these measures on snapshots of the dynamic network. However, since streaming networks typically change at a fast pace, this solution is frequently not fast enough to provide accurate and up-to-date information, which makes it unfeasible for practical applications. Given this limitation, incremental algorithms for computing shortest-path-based centrality measures were developed. Incremental algorithms keep analytic information without performing all computations from scratch, thus being suitable for dynamically changing networks.

Since betweenness and closeness are among the most widely used centrality measures, are expensive to compute in streaming environments and are essential to identify important subscribers in a call graph (e.g., mobile users with high reachability or users who control the information flow between communities), here we report the advances on algorithms for computing these two measures. Note that these algorithms assume a landmark window model when updating the centralities.

5.1 Betweenness

Betweenness is a classical centrality measure that quantifies the importance of a network element (node or edge) based on the frequency of its occurrence in shortest paths between all possible pairs of nodes in the network. The intuitive idea behind this measure is to identify graph elements that act as bridges, i.e., which connect dense regions of the network and without which the information would not pass from one of these regions to the other. An edge with high betweenness is likely to act as a bridge between dense graph regions and, thus, occurs in many shortest paths. Nodes with high betweenness are usually located at the ends of these edges. These nodes occupy a strategic position in the network, which allows them to control the information transfer between different network regions, either by blocking the information between them or by accessing it before other nodes belonging to their region.

Node (or edge) betweenness is computed as the fraction of shortest paths that go through a given node (or edge) among all shortest paths in the network. This is a global centrality measure since it requires complete information about the network in order to compute all pairs shortest paths. Since it is based on the computation of shortest paths for the whole network, betweenness is computationally demanding. The best known algorithm for computing this measure in static unweighted graphs was proposed by Brandes [7]. This algorithm runs in $O(nm)$ time, being n the number of nodes/vertices and m the number of links/edges, and has a space complexity of

$O(n^2 + nm)$, which is prohibitive for networks with millions of nodes and billions of links. Given this, it is necessary to develop new algorithms that avoid the full recomputation of betweenness every time a new edge or node is added to, or removed from, the network. Incremental algorithms offer a solution to this problem, since they can handle large-scale data and can adapt to incremental changes in evolving networks. This is achieved by performing early pruning and by updating only the regions of the network affected by the changes.

Recently, several solutions were proposed to compute node and/or edge betweenness centrality in streaming networks. Lee et al. [27] introduced QUBE, an incremental algorithm that relies on the decomposition of the graph into biconnected components. The performance of the algorithm is strongly associated with the size of the components, which is usually very large in real-world applications. Consequently, it suffers from scalability and efficiency problems, which are of utmost importance in streaming settings. In the same year, Green et al. [17] proposed an exact algorithm, which extends Brandes's approach [7], to compute node betweenness in unweighted streaming graphs. The idea is to preserve information from prior computations of betweenness values and the needed data structures and update only the values and data structures directly affected by the changes in the network. However, the algorithm has some drawbacks: it only supports one type of change (insertion of new edges), and has the same space complexity of Brandes's static algorithm, which is $O(n^2 + nm)$. Hence, the algorithm is not fully suited to handle large-scale and dynamic networks. Kas et al. [22] propose a slightly better solution. They present an incremental algorithm for dynamic maintenance of node betweenness centrality values in rapidly growing networks, using as a building block the dynamic all pairs shortest path algorithm introduced by Ramalingam and Reps [36]. Similarly to [17], the technique of Kas et al. is also based on keeping in memory information from previous computations but using data structures that are faster to update (e.g., shortest distances and number of shortest paths). Although the computational complexity can be lower than the one obtained by [17], the space complexity is similar, turning it prohibitive for very large graphs. More recently, Kourtellis et al. [26] proposed an incremental and scalable algorithm for online computing of both node and edge betweenness centralities in very large dynamic networks. Besides being adapted to fully dynamic networks, where nodes and links are added and removed over time, these authors propose an algorithmic framework that can be efficiently used for real-world deployment (e.g., for identifying strategic subscribers in call graphs) since the algorithm allows for out-of-core implementation and is tailored for modern parallel stream processing engines.

A different approach from the above mentioned was introduced by Kim and Anderson [25]. They presented the time-ordered graph model, which converts a dynamic network into a static network with directed flows, and propose temporal centrality measures to extract information from the graph. These measures are simple extensions of the classical measures to this specific type of dynamic model. However, this method was devised for dynamic networks which evolve in non-streaming scenarios, thus not being fully suitable to the online analysis of call graphs.

5.2 *Closeness*

Closeness is another classical SNA measure that quantifies the importance of a node based on its ability to reach other nodes in a network through shortest paths. The higher the closeness, the less the cost for a node to reach the rest of the network. For instance, in a call graph, the subscriber with highest closeness can quickly spread information to other nodes of the network, as long as the nodes belong to the same connected component.

Closeness is computed as the inverse of the sum of the shortest path distances between a node i and the remaining $n - 1$ nodes in a network of size n. Similarly to betweenness, a few incremental algorithms were proposed to compute the closeness centrality in large dynamic graphs. Kas et al. [21] proposed an algorithm for the fast computation of closeness in large-scale networks. Their technique supports efficient computation of all pairs shortest paths and is suited to dynamic networks, since it handles addition, removal and modification of nodes and links. This algorithm is similar to the one proposed by the same authors in [22]. A more recent work by Khopkar et al. [24] presents a partially dynamic and incremental closeness algorithm that runs in $O(n^2)$.

6 Community Detection

In a social network, a community represents individuals that form a group distinguishable by its properties or characteristics. In other words, when we say we encountered a community it might be, for example, a group of friends, family, work colleagues or other group of individuals sharing the same role, characteristics or label in the context of a network. Detection of communities on a network has many applications, for example, clients that have the same interests and are geographically close to each other might benefit with the implementation of mirror servers. These servers provide faster services on the World Wide Web. The identification of retail clients with similar interests in products enables the retailer to give better recommendation services and therefore increases the probability of increasing profits and the service quality. In telecommunications and computer networks, the community structure of nodes may help the improvement of the compactness of routing tables, maintaining efficient choice of communication paths. Regarding community structure, several areas consider important if the node is located in the center of a community or if it lies on the boundaries of the community. In the first case, the node might have an important control and stability function within the community. In the second case, the node might have functions enabling information exchange between communities. The identification of central and peripheral nodes is of high importance, for example, in social and metabolic networks [12]. One of the most efficient methods used for community detection, either in static environments or in dynamic ones, is the Louvain method [4].

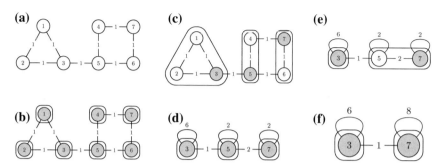

Fig. 8 The original Louvain algorithm steps. **a** Original network. **b** Initial communities. **c** Step 1 of 1st iteration. **d** Step 2 of 1st iteration. **e** Step 1 of 2nd iteration. **f** Step 2 of 2nd iteration

Figure 8 briefly explains how the *Louvain* algorithm works, by illustrating the sequential steps that the algorithm performs for identifying communities in the network. The Louvain method is non-deterministic and performs a greedy optimization to maximize the modularity [31] of all the network partitions. A two-step optimization is performed at each iteration. In step 1, the algorithm seeks for small communities by locally optimizing the modularity. Only local changes of communities are allowed. In step 2, nodes belonging to the same community are aggregated in a single node representing that community in order to build a new aggregated network of communities. Steps are repeated iteratively until no increase of modularity is possible and a hierarchy of communities is generated. Figure 8a represents the initial network; Fig. 8b represents initial individual node communities; Fig. 8c represents local modularity optimization after the first step; Fig. 8d represents the community aggregation results and the new initial communities; Fig. 8e, f, are the two Louvain steps, where the local modularity optimization and community aggregation for the second iteration are presented; The algorithm stops at the 2nd iteration, once increasing modularity is no longer possible. Despite the fast convergence property of the algorithm, which allows the identification of communities in very large networks, its non-deterministic behaviour and internal network structure make it only suitable for static networks.

Methods for detecting communities in dynamic networks were already proposed in the literature. Shang et al. [39] propose the addition of new edges in a two-step approach by using the Louvain Method in the first step and then applying incremental updating strategies on the detected communities in the second step. Thus, the obtained results with this algorithm are dependent on the community structure at its starting point. Another extension of the Louvain method is the one introduced by Nguyen et al. [32]. The proposed QCA algorithm for the efficient detection of communities, starts by detecting the initial communities and then adapts itself to the communities changes by doing addition and removals of nodes and edges within and across communities. Bansal et al. [2], propose a fast community detection algorithm that uses community information from previous time steps. In their experiments, these authors achieve execution time improvements of as much as 30 % over the static methods and maintaining the quality of the community partitions. More recently, the use of spectral clustering became one of the most used methods for community detection. There are several examples of its use. In [43], Yun et al., introduce community detection with partial information. With their method, only a sample of the nodes is observed. Again, the developed algorithm developed is spectral. They extract the clusters only when it is possible. Thus, the authors address the memory limit problem for community detection from a streaming point of view, and achieve asymptotic reconstruction of the clusters with a memory requirement which is sublinear with the size of the network. The memory requirement of the algorithm is non-increasing. Bouchachia et al. [6] present an algorithm that performs enhanced spectral incremental clustering. This algorithm does not need to calculate the clustering for the whole network each time new nodes or edges appear in the stream. Several Label propagation algorithms such as, for example, LPA [35, 41], COPRA [18] and SLPA [42] were also proposed. Results shown that they perform well in static networks, however when those algorithms are applied in different snapshots of an evolving network they produce different communities at each run. When tracking communities in dynamic networks the instability of the partitions is undesirable. LabelRankT, a new label propagation technique proposed by [40] was designed to overcome this problem.

6.1   Top-K Communities

Top-K communities are defined as groups of densely connected nodes in top-K networks. As the name implies, in this work these communities are detected considering only the top-K nodes and their 1st and 2nd neighborhood-order connections. This method samples the original network in a way that preserves its structural properties and the community structure of the original network. We apply top-K sampling to obtain the nodes that belong to the top-K group. To retrieve their network we query the database so as to collect all connections/edges representing the network with the neighbors of the top-K nodes. After generating the sampled networks, the Louvain method [4] is applied to find the communities. Figure 9 represents the matching between the community membership obtained for the top-10000 network and the community membership of the original network retrieved by the Louvain method. This task was performed for an entire day of streaming data. The matching of communities for the two scenarios (sampled network and original network) is performed by retrieving the percentage of matching community members between any top-k network community and the original network communities.

Further analysis of Fig. 9 shows the matching of the 100 largest communities for the sampled network and the 20 largest communities in the original network. The proportion of community member matching varies with a color gradient between 0 (light grey) and 1 (black). There is considerable matching of the top-10000 sampling communities and the 20 largest communities of the original network. These highly active callers and the communities they belong to are therefore represented in the top-K sampling, as expected.

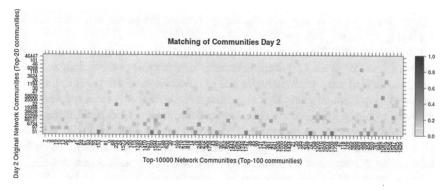

Fig. 9 Matching of community members in the community structure of the sampled network and the community structure of the original network

Experiments were also conducted using other days of the dataset. The results are very similar and consistent throughout full day data comparisons and for the complete dataset of more than 100 days. In all comparisons it is visible that larger original dataset communities are matched by communities retrieved with the proposed *top-K* sampling method.

7 Conclusions and Open Issues

Sampling of large social networks is still in its infancy and there are important open research issues and unsolved problems not yet satisfactorily addressed by the scientific community. Firstly, it is necessary to achieve consensus on the definition of *representative sample*. Based on previous research, a representative sample is a subgraph of the original network that matches its structural properties. However, it is not yet clear which structural properties (e.g., degree distribution, global clustering coefficient, motifs, community structure) should be preserved in the sample. Another matter of concern is the sample size. Despite the promising results of empirical studies [28], a rigorous formal study of the most appropriate sample size according to the size and characteristics of the original network still has to be done.

Regarding visualization, the proposed solutions for streaming networks are still quite simple and leave much space for improvement. A straightforward extension of the presented techniques would be to include additional information by means of visual cues, such as color, size and shape. An idea would be to integrate the visualization techniques with the incremental algorithms used to compute computationally demanding centrality measures (e.g., closeness and betweenness) and then include this node-level information on the visual output. On the other hand, it is necessary to explore new types of representations, that go beyond the graph model, for visually displaying interesting patterns in streaming networks.

Community detection on large dynamic social networks faces many of the challenges of static community detection, namely in what regards the lack of a consensual definition of *network community* and the evaluation of the network partitions produced by community detection algorithms. A possible way to circumvent these problems would be to support the definition of community on the specific domain and develop evaluation measures specifically tailored for the application. For instance, telecommunication providers might be interested in finding communities defined not only by the connections induced by the mobile communications, but also by business variables (e.g., revenue generated by each user), geographical position and demographic attributes. Depending on the purpose of the community detection task, telecommunication companies could, for instance, perform the evaluation based on the similarity of the users' response to the marketing campaigns targeted to the community they were assigned to. Another important issue, especially for telecommunication providers, is to create models and procedures to characterize and define profiles of communities.

The telecommunication networks graphs represent the pattern of interactions from a social space, the analysis of such graphs provides useful insights into the social behaviors that would have a positive social impact. These behavioral patterns apart from providing significant gains to telecom service providers from maximizing profits by customer segmentation, profiling, acquisition, retention, churn and fraud detection etc., it also provides consecutive benefits to society in terms of users or subscribers. Users could gain better services with low prices. Improved and timely support to service issues. Enhanced user satisfaction with optimized value added services and service schemes. Overall, the major challenge is to devise a system that integrates all the relevant steps involved in the process of extracting useful knowledge from large streaming social networks. This system should be built upon rigorous methods and appropriate algorithms, while being user-friendly, in order to encourage its use by business managers and decision makers.

Acknowledgments This work was supported by Sibila and Smartgrids research projects (NORTE-07-0124-FEDER-000056/59), financed by North Portugal Regional Operational Programme (ON.2 O Novo Norte), under the National Strategic Reference Framework (NSRF), through the Development Fund (ERDF), and by national funds, through the Portuguese funding agency, Fundação para a Ciência e a Tecnologia (FCT), and by European Commission through the project MAESTRA (Grant number ICT-2013-612944). The authors also acknowledge the financial support given by the project number 18450 through the "SI I&DT Individual" program by QREN and delivered to WeDo Business Assurance. Márcia Oliveira gratefully acknowledges funding from FCT, through Ph.D. grant SFRH/BD/81339/2011.

References

1. Ahmed, N.K., Neville, J., Kompella, R.: Space-efficient sampling from social activity streams. In: Proceedings of the 1st International Workshop on Big Data, Streams and Heterogeneous Source Mining: Algorithms, Systems, Programming Models and Applications (BigMine 2012), pp. 53–60. ACM (2012)
2. Bansal, S., Bhowmick, S., Paymal, P.: Fast community detection for dynamic complex networks. In: da, L., Costa, F., Evsukoff, A., Mangioni, G., Menezes, R. (eds.) Complex Networks, Communications in Computer and Information Science, vol. 116, pp. 196–207. Springer, Berlin (2011)
3. Barabasi, A.L.: The origin of bursts and heavy tails in human dynamics. Nature **435**(7039), 207–211 (2005)
4. Blondel, V.D., Guillaume, J.L., Lambiotte, R., Lefebvre, E.: Fast unfolding of communities in large networks. J. Statis. Mech.: Theory Exper. **2008**(10), P10,008 (2008)
5. Bonacich, P.: Power and centrality: a family of measures. Am. J. Sociol. **92**(5), 1170–1182 (1987)
6. Bouchachia, A., Prossegger, M.: Incremental spectral clustering. In: Sayed-Mouchaweh, M., Lughofer, E. (eds.) Learning in Non-Stationary Environments, pp. 77–99. Springer, New York (2012)
7. Brandes, U.: A faster algorithm for betweenness centrality. J. Math. Sociol. **25**(2), 163–177 (2001)
8. Charikar, M., Chen, K., Farach-Colton, M.: Finding frequent items in data streams. In: Proceedings of the 29th International Colloquium on Automata, Languages and Programming (ICALP 2002), pp. 693–703. Springer (2002)
9. Clauset, A., Shalizi, C.R., Newman, M.E.: Power-law distributions in empirical data. SIAM Rev. **51**(4), 661–703 (2009)
10. Cormode, G., Muthukrishnan, S.: What's hot and what's not: tracking most frequent items dynamically. ACM Trans. Database Syst. **30**(1), 249–278 (2005)
11. Demaine, E.D., López-Ortiz, A., Munro, J.I.: Frequency estimation of internet packet streams with limited space. In: Mohring, R., Raman, R. (eds.) Algorithms-ESA 2002, Lecture Notes in Computer Science, vol. 2461, pp. 348–360. Springer, Berlin (2002)
12. Fortunato, S.: Community detection in graphs. Phys. Rep. **486**(3–5), 75–174 (2010)
13. Freeman, L.C.: Centrality in social networks conceptual clarification. Soc. Netw. **1**(3), 215–239 (1979)
14. Gama, J.: Knowledge Discovery from Data Streams, 1st edn. Chapman & Hall/CRC (2010)
15. Goodman, L.A.: Snowball sampling. Ann. Math. Statis. **32**(1), 148–170 (1961)
16. Granovetter, M.: Network sampling: some first steps. Am. J. Sociol. **81**(6), 1267–1303 (1976)
17. Green, O., McColl, R., Bader, D.A.: A fast algorithm for streaming betweenness centrality. In: Proceedings of the 2012 International Conference on Privacy, Security, Risk and Trust (PASSAT 2012) and 2012 International Conference on Social Computing (SocialCom 2012), pp. 11–20. IEEE Computer Society (2012)
18. Gregory, S.: Finding overlapping communities in networks by label propagation. New J. Phys. **12**(10), 103,018 (2010)
19. Hanneman, R.A., Riddle, M.: Introduction to Social Network Methods. University of California, Riverside, Riverside, CA, USA (2005). http://www.faculty.ucr.edu/~hanneman/nettext/index.html
20. Hubler, C., Kriegel, H.P., Borgwardt, K., Ghahramani, Z.: Metropolis algorithms for representative subgraph sampling. In: Proceedings of the 8th IEEE International Conference on Data Mining (ICDM 2008), pp. 283–292. IEEE Computer Society (2008)
21. Kas, M., Carley, K.M., Carley, L.R.: Incremental closeness centrality for dynamically changing social networks. In: Proceedings of the 2013 IEEE/ACM International Conference on Advances in Social Networks Analysis and Mining (ASONAM 2013), pp. 1250–1258. IEEE Computer Society (2013)

22. Kas, M., Wachs, M., Carley, K.M., Carley, L.R.: Incremental algorithm for updating betweenness centrality in dynamically growing networks. In: Proceedings of the 2013 IEEE/ACM International Conference on Advances in Social Networks Analysis and Mining (ASONAM 2013), pp. 33–40. IEEE Computer Society (2013)
23. Kayastha, N., Niyato, D., Wang, P., Hossain, E.: Applications, architectures, and protocol design issues for mobile social networks: a survey. Proc. IEEE **99**(12), 2130–2158 (2011)
24. Khopkar, S.S., Nagi, R., Nikolaev, A.G., Bhembre, V.: Efficient algorithms for incremental all pairs shortest paths, closeness and betweenness in social network analysis. Soc. Netw. Anal. Min. **4**(1), 1–20 (2014)
25. Kim, H., Anderson, R.: Temporal node centrality in complex networks. Phys. Rev. E **85**(2), 026,107 (2012)
26. Kourtellis, N., Morales, G.D.F., Bonchi, F.: Scalable online betweenness centrality in evolving graphs. arXiv:1401.6981 (2014)
27. Lee, M.J., Lee, J., Park, J.Y., Choi, R.H., Chung, C.W.: QUBE: a quick algorithm for updating betweenness centrality. In: Proceedings of the 21st International Conference on World Wide Web, pp. 351–360. ACM (2012)
28. Leskovec, J., Faloutsos, C.: Sampling from large graphs. In: Proceedings of the 12th ACM SIGKDD International Conference on Knowledge Discovery and Data Mining (KDD 2006), pp. 631–636. ACM (2006)
29. Manku, G.S., Motwani, R.: Approximate frequency counts over data streams. In: Proceedings of the 28th International Conference on Very Large Data Bases (VLDB 2002), pp. 346–357. VLDB Endowment (2002)
30. Metwally, A., Agrawal, D., El Abbadi, A.: Efficient computation of frequent and top-k elements in data streams. In: Proceedings of the 10th International Conference on Database Theory (ICDT 2005), pp. 398–412. Springer (2005)
31. Newman, M.E., Girvan, M.: Finding and evaluating community structure in networks. Phys. Rev. E **69**(2), 026,113 (2004)
32. Nguyen, N.P., Dinh, T.N., Xuan, Y., Thai, M.T.: Adaptive algorithms for detecting community structure in dynamic social networks. In: Proceedings of the 2011 IEEE International Conference on Computer Communications (INFOCOM 2011), pp. 2282–2290. IEEE Computer Society (2011)
33. Papagelis, M., Das, G., Koudas, N.: Sampling online social networks. IEEE Trans. Knowled. Data Eng. **25**(3), 662–676 (2013)
34. Pinheiro, C.A.R.: Social network analysis in telecommunications, vol. 37. Wiley (2011)
35. Raghavan, U.N., Albert, R., Kumara, S.: Near linear time algorithm to detect community structures in large-scale networks. Phys. Rev. E **76**(3), 036,106 (2007)
36. Ramalingam, G., Reps, T.: On the computational complexity of dynamic graph problems. Theoret. Comput. Sci. **158**(1), 233–277 (1996)
37. Sarmento, R., Cordeiro, M., Gama, J.: Visualization for streaming networks. In: Proceedings of the 3rd Workshop on New Frontiers in Mining Complex Patterns (NFMCP 2014), pp. 62–74 (2014)
38. Sarmento, R., Cordeiro, M., Gama, J.: Streaming networks sampling using top-k networks. In: Proceedings of the 17th International Conference on Enterprise Information Systems (ICEIS 2015), p. to appear. INSTICC (2015)
39. Shang, J., Liu, L., Xie, F., Chen, Z., Miao, J., Fang, X., Wu, C.: A real-time detecting algorithm for tracking community structure of dynamic networks. In: Proceedings of the 6th SNA-KDD Workshop (SNA-KDD 2012), pp. 1–9. ACM (2012)
40. Xie, J., Chen, M., Szymanski, B.K.: Labelrankt: Incremental community detection in dynamic networks via label propagation. In: Proceedings of the Workshop on Dynamic Networks Management and Mining (DyNetMM 2013), pp. 25–32. ACM (2013)
41. Xie, J., Szymanski, B.K.: Community detection using a neighborhood strength driven label propagation algorithm. In: Proceedings of the IEEE Network Science Workshop (NSW 2011), pp. 188–195. IEEE Computer Society (2011)

42. Xie, J., Szymanski, B.K.: Towards linear time overlapping community detection in social networks. In: Tan, P.N., Chawla, S., Ho, C., Bailey, J. (eds.) Advances in Knowledge Discovery and Data Mining. Lecture Notes in Computer Science, vol. 7302, pp. 25–36. Springer, Berlin (2012)

43. Yun, S.Y., Lelarge, M., Proutiere, A.: Streaming, memory limited algorithms for community detection. In: Advances in Neural Information Processing Systems (NIPS 2014), pp. 3167–3175 (2014)

Scalable Cloud-Based Data Analysis Software Systems for Big Data from Next Generation Sequencing

Monika Szczerba, Marek S. Wiewiórka, Michał J. Okoniewski and Henryk Rybiński

Abstract Next generation sequencing (NGS) technology has become a serious computational challenge since its commercial introduction in 2008. Currently, thousands of machines worldwide produce daily billions of sequenced nucleotide base pairs of data. Due to continuous development of faster and economical sequencing technologies, processing the large amounts of data produced by high throughput sequencing technologies became the main challenge in bioinformatics. It can be solved by the new generation of software tools based on the paradigms and principles developed within the Hadoop ecosystem. This chapter presents the overall perspective for data analysis software for genomics and prospects for the emerging applications. To show genomic big data analysis in practice, a case study of the SparkSeq system that delivers tool for biological sequence analysis is presented.

Keywords Genomics · Big data · RNA · DNA · Next-generation sequencing · Biobanking

1 Introduction

Since the advent of the commercial *next generation sequencing* technology (NGS) in 2008 [1], the output generated by the technology can be considered as *big data*. Currently, every day thousands of machines worldwide generate about billions of sequenced nucleotide base pairs of data. Typically Illumina HiSeq sequencer outputs 1 T base pairs in 100 bp short reads within a 10-day run. The sequences may come from various molecular biology laboratory technologies. Those are mainly DNA genotyping [2, 3], RNA expression profiling [4, 5], genome methylation searches

M. Szczerba · M. Wiewiórka · M. Rybiński
Institute of Computer Science, Warsaw University of Technology, Warsaw, Poland

M.J. Okoniewski (✉)
Research Informatics, Scientific IT Services, ETH Zürich, Zurich, Switzerland
e-mail: michalo@id.ethz.ch

© Springer International Publishing Switzerland 2016
N. Japkowicz and J. Stefanowski (eds.), *Big Data Analysis: New Algorithms for a New Society*, Studies in Big Data 16, DOI 10.1007/978-3-319-26989-4_11

263

[6, 7], but there are also various other types of biological sequences libraries that can be sequenced with NGS.

Due to the decrease in prices of the sequencing machines, the capacity of biological data keeps increasing. NGS has become the working horse technology of molecular biology and we can observe its application more widely in the medical applications, where sequencing of DNA and RNA can be used in many stages of diagnosing and treatment procedures, especially since the importance of the genomics (DNA) and transcriptomics (RNA) knowledge will grow with the improvements of the databases of translational and personalized medicine, as well as, functional knowledge bases. The area of knowledge that stores and annotates clinical sequencing data that emerged recently is called *biobanking*.

The abundance and diversity of sequencing data becomes hard to address with the first generation tools of sequencing bioinformatics. Some of them, like BLAST [8] or early genome aligners [9] have been developed for the data from Sanger sequencing [10], before the NGS era. Other, like genome aligners [11], de-novo genome assemblers [12, 13], feature extraction and statistical software [14–17] have been developed specifically for the new sequencing technologies. They are often not efficient enough in the case of the new big genomic datasets, which might reveal the new knowledge behind them, if analyzed for many biological samples, and with fine-grained nucleotide precision (the human genome has over 3 billion of nucleotide base pairs).

Fortunately, in the recent years there was a new wave of big data techniques and software, particularly scalable cloud-based solutions, which can be used to address the data storage and analytic challenges in modern genomics and transcriptomics. In this chapter we would like to present the early outlook of the combination of those two dynamically developing research areas, and outline the possible consequences for the future biological and medical research, along with the opportunities and needs from the computer science side.

The chapter presents an overview of cloud big data analytic tools that are currently used, tested or may possibly be adapted for the genomic data analysis. Those tools, coming mainly from the Hadoop ecosystem, have been already experiencing its fast-paced progress. In particular the presentation will include the existing and possible applications tools such as map-reduce frameworks, columnar databases or SQL-like solutions and distributed machine learning solutions. The discussion will also include the ability of the computational infrastructure to run those novel tools.

This will be supplemented by presenting a detailed case study of SparkSeq—the dedicated genomics big data processing system which has been already applied in a number of biological sequencing analysis projects on a number of infrastructures. Through the example of SparkSeq, the perspectives for similar system applications in biology and medicine will be presented.

2 Typical Software Tools for Analysing Data from Sequenced DNA and RNA

NGS data analysis is currently done with a great variety of tools. The overall goal is to transform the millions of short sequences into the useful information that can be combined and added up to the existing knowledge in molecular biology and medicine. Most of the software tools are not scalable, as they were often designed as academic prototypes that were running on relatively small datasets. Currently, widespread applications of NGS generate datasets, which are often by orders of magnitude bigger. Many of NGS tools are reaching their limits when processing increasingly larger sets of files and broader experimental schemes.

The role of bioinformaticians is often also redefined. From a researcher that adjusted the local experimental schema and analyzed the data from the local sequencer, the bioinformatician of today needs to get much of the data from global repositories (GEO [18], SRA [19], ENA [20]). Also the genome atlases, such as, e.g., TCGA cancer atlas, are expected to be in even wider use in coming years, as it is easier to get the information that already exists in silico, rather than to run the whole new sequencing experiment. This is why we describe here several of the typical processing pipelines (see Table 1) and look at them from the perspective of possible cloud scalability.

2.1 Search for Genome Variants with DNA Analysis

In the DNA analysis, often the goal is to find out the differences between the genomic code of one organism and the reference genome assembly. The reads of a particular biological sample are aligned to the reference genome in a way that allows mismatches on single nucleotides [21], or small differences (several nucleotides), but also mutations of various sizes [22]. This is achieved by genome mappers, most typical example is BWA [23] (an implementation of the Burrows-Wheeler algorithm [24]). The product of the alignment is typically an alignment file in the SAM or BAM format [25]. The subsequent step after mapping is the variant calling [26], which uses the information on mismatches to select the differences between the sequenced sample and the reference genome, or between cancer and germline [27] of the same patient. For the cases of species without a defined reference genome, it is possible to produce a de novo assembly from the DNA short reads [13].

2.2 RNA Expression Profiling

In the case of sequencing of RNA (sequenced in fact as cDNA recoded from RNA [28]) the primary data analysis includes mainly the alignment to the reference

Table 1 Typical functionality needed in the phases of DNA and RNA sequencing data analysis

Type of analysis	DNA	RNA
Primary	Alignment with mismatches	Alignment with exon junctions
Secondary	Variant calling	Counting reads in genes statistical tests
Functional	Linkage	Expression profiling

genome. Due to the splicing of mRNA [29], the exons are interlaced on the reference genome with introns, thus splice-aware aligners are splitting the reads and may be aligning them in several subsequent exon locations [30, 31]. The secondary analysis includes typically the counting of reads aligned to the genes or exons, finding genome junctions [32] and performing statistical analyses to find differentially expressed genes between the treatment groups [5].

Other techniques of RNA-seq analysis may include mapping to ready transcriptomes in order to find out the isoform deconvolution of the expressed genes [33]. De novo methodologies are also applied [34] in order to discover complete transcript sequences, also those that may not be present in the reference genome.

The amount of data for a single sample exome-DNA or mRNA sequencing experiment can be counted as several gigabytes of FASTQ files, similarly—several gigabytes of alignment file after mapping to the reference genome and megabyte-size summary tables. Typical genomic experiment may consist of several biological samples and currently there are projects that have hundreds of them, soon this number will reach thousands in specific cases. The emerging applications will produce even more data, e.g. diagnostic sequencing in a single hospital may produce thousand of patient's samples every year.

3 Cloud-Based Data Analysis Software and the Potential of Genomic Applications

In parallel to the introduction of the NGS technologies for molecular biology and medicine, in data mining a fast development of the new parallel data analysis techniques can be observed. They are focused on the works done for the projects belonging to the so called Hadoop ecosystem [35]. Many of the technologies in that area has been already tested in genomic applications or have the potential to be usefully applied there. In the sequel we will discuss two frameworks that are potentially applicable for genomic computations, namely Apache Hadoop MapReduce and Apache Spark.[1]

---

[1] Apache Spark is another state-of-the-art case of a distributed computing framework, it is an open source software developed by AMPLab at UC Berkeley [36]. Apache Spark has been widely adopted as a successor to Apache Hadoop MapReduce, and can be deployed on single machines, small clusters of a few nodes up to large cluster deployment with thousands of nodes.

3.1 Distributed Computing Frameworks and Analytics Engines

Analytics engine is a piece of software intended for distributed execution of a series of complex jobs represented internally by a graph of tasks. In general the major challenge for contemporary analytics engines is to ensure scalability and reliability, so that such engine is able to supports high-throughput processing of large amounts of data simultaneously. Otherwise, there are number of limitations on availability and latency which may affect end users directly [37, 38].

As cloud computing becomes a fast emerging research area, and the data processing needs in genomics are growing, the distributed computing technologies being deployed in the cloud become a significant issue. Some proposals in this respect refer to the recent cloud technologies, offering the ability to increase the workload on its current hardware resources [39, 40]. As a result, the benefits of such approach include scalability, reliability and redundancy of the hardware, making possible to add resources dynamically to a running system in order to handle processing demand. This in turn makes the modern system cost-efficient, flexible and robust with regard to various data and applications [41].

Yet another advantage of the modern system is a fault-tolerant architecture, satisfying a reliable data processing by means of replicating data on different computing nodes or across cluster, so that in the event of a failure, the system continues to re-process the data [42]. Because complex systems are most likely to have more failure points therefore they require high levels of availability. It can be met by using redundancy which is overcome high level of common cause failures. Storing data redundantly also increases fault tolerance [43]. The popularity of cloud computing is growing also due to many open source projects that are being deployed, many of them belonging to the Hadoop ecosystem [35].

Usually, the analytics engine is equipped with an API. The API provides a definition of tasks to be performed. For the case of Apache Hadoop MapReduce API is in a form of low level *map* and *reduce* tasks, whereas for Apache Spark it is a higher-level form of transformations and actions. In particular, the transformations for Apache Spark can create a new dataset from an existing one, and actions would run a calculation and return a value when completed [44].

Data locality is an important factor for the Apache Hadoop MapReduce performance. Hadoop balances the load by distributing data to multiple nodes based on disk space availability [45]. When data is written in HDFS, a copy is written locally, second copy is written to another node in a different rack and a third copy—written to another node within the same rack. Because data is copied across a public cloud, it means that the infrastructure is shared with other customers. As a result, when running VMs there is a limited control over which server the data is being spun up, and if data are to be located on the same physical server. There is no rack awareness that one has access to, and can configure in the name node [46]. Apache Hadoop MapReduce has also a high degree of fault tolerance, and it achieves it through restarting tasks.

In the case of jobs that experience a high failure rate, Apache Hadoop MapReduce can assure the task gets completed.

A goal of Apache Spark was to reach much higher memory efficiency than Apache Hadoop MapReduce, mainly by using optimized ways of parallel and in-memory processing. The central paradigm of memory optimization is that Apache Spark attempts to minimize storing intermediate files on the disk, and thus has low runtime overhead. Apache Spark job (application) creates a DAG of task stages and performs them on the cluster. Compared to Apache Hadoop MapReduce, which creates a DAG with two predefined stages—Map and Reduce, DAG created by Apache Spark can contain any number of stages. This allows some jobs to be completed faster than with Apache Hadoop MapReduce (with simple jobs completing after just one stage, and more complex tasks completing in a single run of many stages), rather than having to be split into multiple jobs.

Apache Spark jobs perform work on Resilient Distributed Datasets (RDDs), an abstraction for a collection of elements that can be operated on in parallel. When running Apache Spark in a Hadoop cluster, RDDs are created from files in the distributed file system in any format supported by Apache Hadoop, such as text files SequenceFiles, or anything else supported by Apache Hadoop InputFormat. Once data is read into RDD object in Apache Spark, a variety of operations can be performed calling abstract Apache Spark API. Two major types of operations available are:

1. Transformations: they return a new, modified RDD based on the original one. Several transformations are available through the Apache Spark API, including $map()$, $filter()$, $sample()$, and $union()$.
2. Actions: they return a value based on some computation being performed on an RDD. Some examples of actions supported by the Apache Spark API include $reduce()$, $count()$, $first()$, and $foreach()$.

Some Apache Spark jobs will require several actions or transformations to be performed on a particular data set, making it highly desirable to hold RDDs in memory for rapid access. Apache Spark exposes a simple API to do this— $cache() : RDD.this.type$ which persist this RDD with the default storage level (MEMORY_ONLY). Also $persist(newLevel : StorageLevel) : RDD.this.type$ which sets this RDD's storage level to persist its values across operations after the first time it is computed. After performing $cache$ or $persist$, an action called *lazy evaluation* is required . Once this API is called on an RDD, future operations called on the RDD will return in a fraction of the time they would if retrieved from the disk [44].

Apache Spark is written in Scala and has its own version of Scala interpreter. Yet another advantage of Apache Spark framework is spark-REPL (read-evaluate-print-loop), which is an interactive shell execution engine for Scala expressions.

Apache Spark can run in Hadoop clusters through Hadoop YARN, Apache Spark's standalone mode, Apache Mesos and its APIs allow to access data in HDFS, HBase, Cassandra, Hive and any Apache Hadoop InputFormat. It is designed to perform both batch processing (similar to MapReduce) and new workloads, like streaming, interactive queries, and machine learning [44].

Apache Spark has number of advantages over Apache Hadoop MapReduce. The ability to run computation in-memory, such as extracting a working set, cache and query it repeatedly, results in running the Apache Spark applications up to 100 times faster than MapReduce.

The Apache Spark framework provides a number of internal components. For example Apache Spark SQL, providing a SQL-like interface to query the data from Apache Spark RDDs or sources such as Hive tables, Parquet files or JSON files. Apache Spark supports queries written in SQL-like languages such as HiveQL, which can be converted to Apache Spark jobs [47]. It has its own API in Scala, Java and Python and interactive command line interface (in Scala or Python) for low-latency, ad-hoc data exploration on a cluster.

3.2 Data Serialization System and In-memory Data Structures

Data serialization is an important functionality, as it focuses on performance of an application. Some formats are slow or consume a large number of bytes, which slows down the computation. Apache Avro is a data serialization system developed by Apache Software Foundation (ASF). It defines data types, protocols and schema with JSON, and serializes data into a binary format. Avro includes experimental higher level language, called Avro interface description language (IDL), which instead of defining schema, as it is in Avro, can be used to specify the object types instead.

Kryo is fast and efficient serialization library based on object graph serialization framework. Kryo is faster than Java serialization and by default Apache Spark includes Kryo serializers for generally used core Scala classes.

Google's Protocol Buffers is developed by Google, used for serializing structured data in a way that data is defined according to a structure, and generated code can be used to write and read structured data faster from data streams.

3.3 Data Storage and Representation

Hadoop ecosystem includes also several database engines, many of them SQL-like. Examples of them are Apache Spark SQL, Hive-QL, PIQL, SQLLine, Optiq. Parquet is a file format that has advantages of compressed, efficient columnar data representation for Hadoop [48].

RCFile (Record Columnar Format) is a data format that stores relational tables on cluster. RCFile is a collaborative effort from Facebook and Ohio State University, Institute of Computing Technology and Chinese Academy of Sciences. The

RCFile structure characterize the following components: data storage format, data compression approach, and optimization techniques for data reading. File format fulfills requirements such as fast data loading, fast query processing and highly efficient storage space utilization [49].

ORCFile optimizes row columnar (ORC) file format, introduced in Hive 0.11. Data stored in ORCFile can be read or written through a Map/Reduce process. In addition to this, ORCFile offers high level compression algorithms: no compression, ZLIB and SNAPPY.

3.4 Data Warehouse Infrastructure

Among them are data-warehousing solutions based on traditional Massively Parallel Processor (MPP) architectures (such as Amazon Redshift), systems which MPP-like execution engines integrates with Hadoop ecosystem such as distributed storage, security (Impala, HAWQ), and systems which optimise MapReduce to improve performance on analytical workloads (Tez). The comparison is summarised in the table which shows following features: UDF Support—User Defined Functions support, Mid-query fault tolerance—a recovery from mid-query failures, Open source—open to the community, license free software, User API—program interface for user interaction and HDFS Compatible—compatible with Hadoop Distributed File System.

3.5 Current and Potential Uses of Hadoop Ecosystem Tools in Genomics

Currently, the modern cloud-based software described above have found only partially direct applications in genomics, still there is a great potential of using it in specific applications and stages of NGS data analysis. The genomic-specific implementations and tools related that use the software described before are: HadoopBAM [50], Seq-Pig [51] SparkSeq [52] BioPig [53] Biodoop [54]. The ADAM project [55] is an initiative to create a genomics processing engine and specialised file format built using Apache Avro, Apache Spark and Parquet. ADAM has currently several side-projects, that address the RNA sequencing (RNAadam), DNA variant calling (avocado) and other types of utilities for parallel processing the genomic data (Table 2).

Many of the software classes mentioned above have not been yet adapted to processing and storing of genomic data. Still many of them have a potential to be used after specific developers' adaptation. The data warehousing systems with SQL-based languages may be possibly used to store heterogeneous genomic data combined with experimental and clinical metadata. This can be driven by the need for specific needs in the large-scale personalised, genome-based healthcare which is currently a matter of hospital implementation, as the sequencing techniques are mature enough to be

Table 2 Summary of SQL-like systems in the ecosystem (MqFT stands for Mid-query fault tolerace)

Engine	User API	UDF Sup	MqFT	HDFS compatible
Apache Spark SQL	HiveQL (HQL), SQL	Yes	Yes	Yes
Hive on Tez	HiveQL (HQL)	Yes	Yes	Yes
Impala	HQL + extensions	Java/C++	No	Yes
Amazon Redshift	Full SQL 92	No	No	No

applied in the clinic. The same drive for novel applications can be expected from the areas of pharmacogenomics or genome-based nutrition biotechnology. The amount of biological samples that are currently available, and are about to be sequenced in near future will convince developers and decision-makers to use the modern scalable solutions.

The SparkSeq system described below is an example of productive prototype of a scalable information system that can perform many genomic data operations with the use of many of the software classes described above.

4   SparkSeq Case Study

SparkSeq [52] is one of the first cloud-ready solutions for interactive analysis of next generation sequencing data. The particular development principle was putting the emphasis on single-nucleotide resolution of data processing and results. It has been developed as a prototype that addresses many of the challenges present in parallel and distributed processing of NGS data. SparkSeq takes advantage of Apache Spark computing framework as well as some other Hadoop ecosystem components. Besides, it adapts and implements many of the techniques and technologies that are gradually becoming de facto standards in big data ecosystems. To some extent it can be treated as an attempt to design a reference architecture, providing hints for cloud-based NGS data studies, and addressing challenges and many novel approaches.

4.1   Data Storage and Parallel Access

NGS-experiments tend to generate huge amounts of data. However, most of them do not follow well-known and generally supported formats like flat files (e.g. DSV—delimiter separated values) nor any kind of hierarchically structured files (e.g. XML or JSON). This is why storing and efficient manipulation of the NGS data is not directly available in HDFS and requires additional adaptation efforts.

While analyzing NGS data the most time is spent on processing alignment files (BAM file format [25]). They can be decomposed into basic operations, like filtering, summarizing, pileup coverage function of mapped short reads using their genome coordinates and record properties. For instance, in RNA-seq studies it is very often necessary to filter out records that do not possess descent mapping quality scores, calculate base coverage or calculate genomic regions coverage. Some of these operations can be very easily parallelized, since they can be computed independently and directly on short reads level (e.g. counting, reads flags or quality filtering). Other, such as coverage function (pileup), or genomic regions counts, can be efficiently parallelized at data partitions level, i.e. the data partitioned by genomic coordinates. Taking into consideration that each BAM file contains typically millions of reads mapped to the genomic positions across the whole genome, both kinds of operations exhibit a very high degree of parallelism.

Unfortunately, the BAM file format is not well suited for parallel and distributed computing [50, 55]. This is because of using a centralised file header and implementing gzip-compatible compression method that is not records aware in that sense it allows records to be split between blocks. To overcome these shortcomings a few approaches have been proposes so far. The most naïve one is to use the SAM format instead, or convert the BAM file to any other text-based format that can be easily handled by the HDFS built-in libraries. This approach however results in several times higher disk requirements and interconnect network loads. Text files can be stored compressed in HDFS, still then only block compressions like bzip2, Snappy or LZO allow files to be splittable and eligible for being processed in parallel. Such an approach would require an additional, very time-consuming, step of transcoding the BAM files to compressed SAM files.

Another idea was proposed in [50]—it consists in implementing custom HDFS InputFormats that transparently takes care of accessing short reads kept in BAM files and stored in the HDFS clusters. This method is very convenient, as it only requires copying alignment files to HDFS storage without any need of data preprocessing. On the other hand, it suffers from the same drawbacks of the BAM format design as mentioned above, which in some cases may deteriorate scalability, due to file header segment contention. This data access method was initially implemented in SparkSeq, as it currently seems to be the best trade-off between convenience of use and performance.

The most recent approach [55] aims at introducing a completely new data format for storing alignment data. The basic idea is to keep reading data as self-contained records using the Avro serialization. The serialized records are then stored on disk using the columnar compressed file format Apache Parquet that is designed for the distribution across multiple computers (like in HDFS). Thanks to these optimizations, two appealing goals have been achieved: elimination of the centralized header and better scalability characteristics, as well as, disk occupancy reduction: the ADAM files are up to 25 % smaller than BAM without losing any information. The application of the ADAM format in data processing may be unfortunately still problematic due to the current lack of support for this file format in mapping software. BAM files

Fig. 1 The impact of the HDFS block size selection

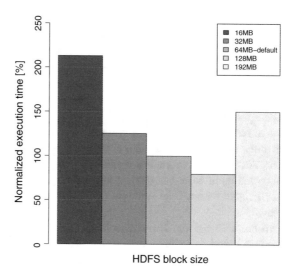

need to be transcoded to the ADAM format beforehand—this step may be sometimes more time-consuming than running analysis using BAM files.

HDFS read performance HDFS read performance is in many cases a key factor for a robust NGS data analysis, as many of the algorithms are more IO- than CPU bound. Many of those operations require fast sequential read access to data stored in the HDFS cluster. It has been shown in [52] that the proper HDFS block selection can cut processing time by more than 20%. As a rule of thumb, increasing the block size from the default value of 64 MB is advisable. However, it is a parameter, the value of which should be adjusted in line with throughput of the underlying storage performance. Increasing it too much would lead to a serious performance degradation, as shown in Fig. 1.

Another possible optimisation is to take advantage of the feature that has been introduced in HDFS 2.1, called "Short-Circuit Local Reads ". This option allows DFSClient, e.g. Apache Spark worker, located on the same machine as DataNode (which very often is the case) to read the locally stored BAM files splits directly from the local disk, rather than over the TCP socket and DataTransferProtocol. In such a way, yet another gain of 15–20% in the read throughput can be achieved [56].

Data movement A very frequent need is to transfer BAM alignment files to the HDFS cluster efficiently, since the alignment and secondary analysis are done using different systems. In order to facilitate data transfer between mapping and analytics environment one can use a specialized gateway, since the direct copy operation requires that source host has access over the network to all DataNodes in the HDFS cluster. Two popular solutions are HTTPFs that enables data requests to be done over the simple http REST protocol and HDFS NFS gateway. The latter one has been introduced in the Hadoop 2.6 release, and it makes possible mounting HDFS using the NFSv3 distributed file system protocol [57].

Data security Data security is one the main concerns in sequencing data analyses performed in the cloud [58, 59], as the cost of data production is still high, and the market value of the state-of-the art biological data or patients' medical data is inestimable. This is why providing a solid security level for both data in motion and at rest seems to be a crucial factor for cloud-based NGS analytics adoption.

A data in motion encryption has been for a long time available in Hadoop. It encompasses, inter alia, Hadoop RPC, HDFS Data transfer and WebUIs. For the data at rest in Hadoop, the encryption is available since the recent version 2.6. For the transparent encryption a new abstraction to HDFS was introduced: the encryption zone. An encryption zone is a special directory which contents will be transparently encrypted upon write and transparently decrypted upon read. Each encryption zone is associated with a single encryption zone key which is specified when the zone is created.

In order to enforce security policies at any level (i.e. files, folders, etc., stored in HDFS), and to manage fine-grained access control to higher level objects like the Hive, HBase tables or columns one can take advantage of Apache Ranger [60]. Finally, the easiest way to protect access to the services provided by various Hadoop ecosystem components (such as, e.g., WebHDFS, Stargate—HBase REST gateway or Hive JDBC) is to use the Apache Knox solution [61].

4.2   Storing Results

Once primary and secondary analysis has been accomplished, the end- and inter-mediate results should be conveniently stored in order to run further data mining algorithms or serve them to end users [62]. Basing on the predominant data query patterns, one can choose between

- non-relational distributed databases, like Apache HBase or Cassandra in the case of random read/write data access, or
- data warehousing solutions like Apache Hive, when large data scans and running complex data queries are the main analysis goals.

In many real-world scenarios the combination of both approaches can be optimal. Users that require very fast access to only a small fraction of genomic data—e.g. to study nucleotide-level phenomena across samples at a given position can turn to an HBase-based solution, whereas analyzing the whole-genome with many samples at once could be addressed by an Hive-based application.

4.3   Programming Paradigms and Interfaces

Apache Spark with its support for programming in the languages like Python or Java that are popular among bioinformaticians [63, 64] offers a shallow learning curve.

Since the alignment records are stored in a tabular format, scientists skilled in SQL can easily query them using the Apache Spark SQL component. Bioinformaticians that are not only interested in querying NGS datasets but also in running more sophisticated analyses on them, would like to achieve the highest job performance and to get more concise code should definitely turn to the Scala programming language. Its object-functional syntax and alignment records represented as Apache Spark's RDD, as well as the Mlib machine learning library, seem to be a perfect fit for many kinds of NGS data processing steps, ranging from custom quality control to running the most sophisticated secondary analyses. SparkSeq with its built-in specialized routines for many common filtering options and reads pileup calculations can make them even easier to implement.

4.4 Scalability and Performance

As it has been shown in [50–52] all big data solutions using Hadoop-BAM library as the data access layer exhibit similar scalability characteristics. It is almost ideal up to 7–8 worker nodes, after which scaling worsens. Besides, more computationally intensive operations show better scalability than those mainly limited by the overall IO-throughput (both storage system and network interconnects). The difference between these tools is however in performance measured. According to [52] SparkSeq, which is Apache Spark-based solution, clearly outperforms its Hadoop MapReduce competitor—SeqPig by the order of magnitude.

4.5 Caching Strategies

Cache experiments conducted in [52] show clearly that using fast data serializer (like KryoSerializer) instead of Java built-in together with LZF or the Snappy compression can speed up multi-pass data querying up 80–110 times, and reduce memory consumption approximately 13 times (Fig. 2). SparkSeq can use the same storage level settings as Apache Spark, by default it is MEMORY_ONLY as it seems to be the most CPU-efficient one.

4.6 Cluster and Resource Management

The state-of-art big data architectures are becoming constantly more complex and getting beyond the initial ecosystems consisting of just distributed storage and computing engines. This is why there is a constant need to facilitate, automate their management and deployment processes. Apache Ambari [65] is one of the first solutions providing an integrated web-based user interface that helps one to manage and

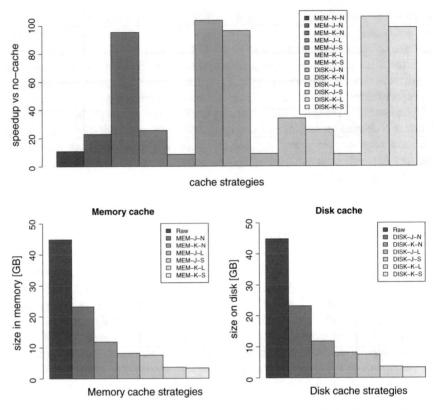

Fig. 2 Apache Spark cache strategies comparison. *Raw* Size of Scala RDD without serialization or data compression. *MEM* Memory. *J* JavaSerializer. *K* KryoSerializer. *L* LZF compression. *S* Snappy compression. *N* No serialization or no compression, e.g. *MEM-K-N* means in memory data cache, objects serialized with KryoSerializer and without any compression

monitor many components of the Hadoop ecosystem (using the Nagios and Gandlia tools). It also offers built-in wizards that make it possible to install, as well as add new hosts to the existing cluster very easily.

A proper resource management is crucial for making different types of bioinformatics processing run smoothly. Some tasks, like alignment, quality checks or variant calling, are clearly batch processes, which do not suffer much from latencies, whereas other, like interactive sequencing data browsing or analytical queries, surely do. The resource managers e.g. Hadoop YARN [66] or Apache Mesos [67, 68] abstract computer resources (like CPU, memory, etc.) from physical or virtual machines, and help one to allocate them dynamically in an efficient way. Thanks to this abstraction layer, the applications running on top of them are no longer strictly assigned to specific hosts—they can be launched on any machines that belong to the cluster governed by the resource manager. This way, the fault-tolerance as well as dynamic process reallocation can be achieved. On the other hand, static resources reservation is still possible—such as the "coarse-grained" mode helps one meeting low-latency requirements in the case of interactive queries or handling web requests.

Furthermore, in order to reduce hardware virtualization overheads (especially in the case of the IO operations) and at the same time minimize software installation efforts, the software container frameworks, like Docker [69], can be used. They provide resources isolation basing on the operating system-level virtualization, which is complementary to the already mentioned resource managers. Besides, software containers can be combined with the Hadoop management software, such as Apache Ambari, to make cluster deployments even easier in the cloud environments that support this technology.

4.7 Integration with R Environment

R-project [70] is one of the main analytical tools that is commonly used in many bioinformatics studies. The R language is currently the common language in many sub-areas of the data science. Its lightweight syntax together with the great data visualization features and the BioConductor package repository for genomic applications [71]—make it often essential for NGS data analysis. Unfortunately, the R internal architecture does not provide feats for big data mining in distributed environments. In the case of SparkSeq it has been decided to combine the computation power of Apache Spark and ease of high-level data analytics of R. Thus RSparkSeq [52] has been developed, so that it makes possible to call SparkSeq routines directly in the R environment by means of the rJvmr package. The initial experiments show that it is possible to integrate these two computing solutions in order to find in the future the optimum balance for their use in genomic data analysis.

4.8 NGS Big Data Reference Architecture Considerations

Figure 3 depicts the next generation sequencing big data architecture that summarizes the above discussion. The bottom layer presents computing nodes the resources of which can be virtualized with either one of hardware or operating system-level solutions. A two-step approach (i.e. first hardware, than os virtualization strategy is also possible and beneficial in cases when a different version of kernel or operating system then installed on the host is required). The next layer constitutes the abstraction of resources: storage (as the distributed filesystem—HDFS), computing resources (Apache YARN) and other resources (e.g. network interfaces, operating system—Docker). One layer above is dedicated to the NGS data access components and file formats: Hadoop-BAM and ADAM file formats family. Two upper layers are computing/storage engines and user interfaces. The architecture layout is accompanied by a data exchange layer (i.e. NFS and HTTPFs gateways) at the bottom together with management and security components to the right of the diagram.

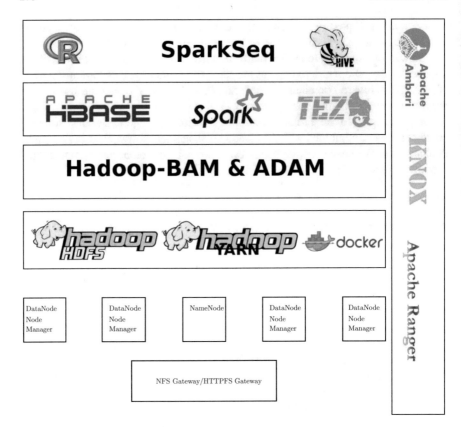

Fig. 3 Scalable cloud ready reference architecture for big data from next generation sequencing

This is by no means a complete picture of all the components that are involved but the most important ones are highlighted. In case of many of the components there is more than option to choose from: e.g. instead of HBase, one can opt for Cassandra, instead of Hadoop YARN Apache Mesos can be used, etc.

5 Open Challenges and Near Future Direction for Cloud-Based Genomics

Currently, the systems and APIs such as SparkSeq, SeqPig or ADAM standards are mostly prototypical. Still the experience from it have shown that only scaling the data processing into multiple machines in cluster and computational clouds may bring the answer to the unmet need of more efficient and more precise data analysis of genomic data. The cost of sequencing is significant, however the cost and effort on data analysis is even a more important issue to solve. The emerging areas of applications of NGS, such as personalized medicine, but also pharmacogenomics, biomaterials, genomics of plants and animals in agriculture, or biosafety are waiting for the solutions in efficient data analysis.

The scalable solutions that can be run in a reliable way on single machines, on clusters or in the proprietary or public clouds should be the future of genomic data analysis which is shaped by the growing size of datasets but also by the need for precise and confidential processing of highly valuable patients' data. In order to successfully apply existing cloud-based software in genome bioinformatics still many open issues need to be addressed. The bioinformatic software scene is rich with various types of software, but at the moment they do not have many common points with Hadoop ecosystem and cloud-based technologies. To make this two technology worlds meet, many of the algorithms must be re-implemented with the use of modern computing frameworks, and bioinformatic academic and commercial software needs to switch to emerging NGS big data formats, such as the ADAM formats family. In order to successfully apply existing cloud-based software in genome bioinformatics still many open issues need to be addressed.

In this chapter there were discussed many challenges at various stages of the NGS data processing. They may be addressed with various tools and solutions. The cloud-based solutions are among the novel and promising ones, still for a large part of genome bioinformatics the working-horse techniques remain at the moment those rooted in classic high-performance computing. The software for quality control, alignment and genomic features extraction is mainly run in multi-threaded way under control of distributed resource management tools such as SGE or LSF. The complexity of genomic data processing pipelines and multi-node scalability can be achieved using bioinformatic-oriented workflow management systems such as SnakeMake [72] or Big Data Script [73]. It is likely, and will be interesting to see in the near future the use of those or similar tools to integrate the data processing in both HPC and cloud environments.

It is also important, that at the level of human skills there is a need for really cross-disciplinary expertise. Bioinformaticians should become aware of the cloud technologies, biological and medical experts should know the full potential of the informational value of the big genomic datasets and medicine and life sciences should educate a new generation of physicians and researchers being able to formulate the requirements for the data scientists. All this processes have been already initiated and they are directed towards the synergy between big data, cloud computing and genomics, which will in turn give boost to the novel medical and biotechnology applications.

References

1. Shendure, J., Ji, H. (eds.): Next-generation DNA sequencing. In: Shendure, J., Ji, H., (eds.) Nature Biotechnology, vol. 26. Nature Publishing Group (2008)
2. DePristo, M.A., Banks, E., Poplin, R., Garimella, K.V., Maguire, J.R., Hartl, C., Philippakis, A.A., del Angel, G., Rivas, M.A., Hanna, M., McKenna, A., Fennell, T.J., Kernytsky, A.M., Sivachenko, A.Y., Cibulskis, K., Gabriel, S.B., Altshuler, D., Daly, M.J.: A framework for variation discovery and genotyping using next-generation DNA sequencing data. Nat Genet **43**, 491–498 (2011)
3. Duitama, J., Quintero, J.C., Cruz, D.F., Quintero, C., Hubmann, G., Foulquié-Moreno, M.R., Verstrepen, K.J., Thevelein, J.M., Tohme, J.: An integrated framework for discovery and genotyping of genomic variants from high-throughput sequencing experiments. Nucleic Acids Res. **42**, e44 (2014)
4. Ozsolak, F., Milos, P.M.: RNA sequencing: advances, challenges and opportunities. Nat. Rev. Genet. **12**, 87–98 (2011)
5. Anders, S., McCarthy, D.J., Chen, Y., Okoniewski, M., Smyth, G.K., Huber, W., Robinson, M.D.: Count-based differential expression analysis of RNA sequencing data using R and Bioconductor. Nat. Protoc. **8**, 1765–1786 (2013)
6. Bird, A.P.: Cpg-rich islands and the function of dna methylation. Nature **321**, 209–213 (1985)
7. Suzuki, M.M., Bird, A.: Dna methylation landscapes: provocative insights from epigenomics. Nat. Rev. Genet. **9**, 465–476 (2008)
8. Tatusova, T.A., Madden, T.L.: BLAST 2 Sequences, a new tool for comparing protein and nucleotide sequences. FEMS Microbiol. Lett. **174**, 247–250 (1999)
9. Pearson, W.R., Lipman, D.J.: Improved tools for biological sequence comparison. In: Proceedings of the National Academy of Sciences of the United States of America (1988)
10. DNA sequencing with chain-terminating inhibitors. In: Proceedings of the National Academy of Sciences of the United States of America, National Academy of Sciences of the United States of America (1977)
11. Langmead, B., Salzberg, S.L.: Fast gapped-read alignment with bowtie 2. Nat. Methods **9**, 357–359 (2012)
12. Li, R., Zhu, H., Ruan, J., Qian, W., Fang, X., Shi, Z., Li, Y., Li, S., Shan, G., Kristiansen, K., et al.: De novo assembly of human genomes with massively parallel short read sequencing. Genome Res. **20**, 265–272 (2010)
13. Zerbino, D.R., Birney, E.: Velvet: algorithms for de novo short read assembly using de Bruijn graphs. Genome Res. **18**, 821–829 (2008)
14. Frazee, A.C., Sabunciyan, S., Hansen, K.D., Irizarry, R.A., Leek, J.T.: Differential expression analysis of RNA-seq data at single-base resolution. Biostatistics (Oxford, England) (2014)
15. Anders, S., Huber, W.: Differential expression analysis for sequence count data. Nature Precedings (2010)
16. Robinson, M.D., McCarthy, D.J., Smyth, G.K.: edger: a bioconductor package for differential expression analysis of digital gene expression data. Bioinformatics **26**, 139–140 (2010)
17. Anders, S., McCarthy, D.J., Chen, Y., Okoniewski, M., Smyth, G.K., Huber, W., Robinson, M.D.: Count-based differential expression analysis of RNA sequencing data using R and Bioconductor (2013)
18. Barrett, T., Troup, D.B., Wilhite, S.E., Ledoux, P., Evangelista, C., Kim, I.F., Tomashevsky, M., Marshall, K.A., Phillippy, K.H., Sherman, P.M., Muertter, R.N., Holko, M., Ayanbule, O., Yefanov, A., Soboleva, A.: NCBI GEO: archive for functional genomics data sets-10 years on. Nucleic Acids Res. **39**, D1005–D1010 (2011)
19. Kodama, Y., Shumway, M., Leinonen, R.: The sequence read archive: explosive growth of sequencing data. Nucleic Acids Res. **40**, D54–D56 (2012)
20. Cochrane, G., Akhtar, R., Bonfield, J., Bower, L., Demiralp, F., Faruque, N., Gibson, R., Hoad, G., Hubbard, T., Hunter, C., Jang, M., Juhos, S., Leinonen, R., Leonard, S., Lin, Q., Lopez, R., Lorenc, D., McWilliam, H., Mukherjee, G., Plaister, S., Radhakrishnan, R., Robinson, S.,

Sobhany, S., Hoopen, P.T., Vaughan, R., Zalunin, V., Birney, E.: Petabyte-scale innovations at the European Nucleotide Archive. Nucleic Acids Res. **37**, D19–25 (2009)

21. Kwok, P.Y.: Single Nucleotide Polymorphisms. Humana, Totowa, NJ (2003)
22. Okoniewski, M.J., Meienberg, J., Patrignani, A., Szabelska, A., Mátyás, G., Schlapbach, R.: Precise breakpoint localization of large genomic deletions using PacBio and Illumina next-generation sequencers. BioTechniques **54**, 98–100 (2013)
23. Langmead, B., Trapnell, C., Pop, M., Salzberg, S.L., et al.: Ultrafast and memory-efficient alignment of short dna sequences to the human genome. Genome Biol. **10**, R25 (2009)
24. Li, H., Durbin, R.: Fast and accurate short read alignment with burrows-wheeler transform. Bioinformatics **25**, 1754–1760 (2009)
25. Li, H., Handsaker, B., Wysoker, A., Fennell, T., Ruan, J., Homer, N., Marth, G., Abecasis, G., Durbin, R.: 1000 Genome Project Data Processing Subgroup: The Sequence Alignment/Map format and SAMtools. Bioinformatics (Oxford, England) **25**, 2078–2079 (2009)
26. Li, H., Ruan, J., Durbin, R.: Mapping short DNA sequencing reads and calling variants using mapping quality scores. Genome Res. **18**, 1851–1858 (2008)
27. Saunders, C.T., Wong, W.S.W., Swamy, S., Becq, J., Murray, L.J., Cheetham, R.K.: Strelka: accurate somatic small-variant calling from sequenced tumor-normal sample pairs. Bioinformatics (Oxford, England) **28**, 1811–1817 (2012)
28. Thomas, M.F., Ansel, K.M.: Construction of small RNA cDNA libraries for deep sequencing. Methods Mol. Biol. (Clifton, N.J.) **667**, 93–111 (2010)
29. Kornblihtt, A.R., Schor, I.E., Allo, M., Dujardin, G., Petrillo, E., Muñoz, M.J.: Alternative splicing: a pivotal step between eukaryotic transcription and translation. Nat. Rev. Mol. Cell Biol. **14**, 153–165 (2013)
30. Kim, D., Pertea, G., Trapnell, C., Pimentel, H., Kelley, R., Salzberg, S.L.: TopHat2: accurate alignment of transcriptomes in the presence of insertions, deletions and gene fusions. Genome Biol. **14**, R36 (2013)
31. Dobin, A., Davis, C.A., Schlesinger, F., Drenkow, J., Zaleski, C., Jha, S., Batut, P., Chaisson, M., Gingeras, T.R.: STAR: ultrafast universal RNA-seq aligner. Bioinformatics (Oxford, England) **29**, 15–21 (2013)
32. Trapnell, C., Roberts, A., Goff, L., Pertea, G., Kim, D., Kelley, D.R., Pimentel, H., Salzberg, S.L., Rinn, J.L., Pachter, L.: Differential gene and transcript expression analysis of RNA-seq experiments with TopHat and Cufflinks. Natu. Protoc. **7**, 562–578 (2012)
33. Li, B., Dewey, C.N.: Rsem: accurate transcript quantification from rna-seq data with or without a reference genome. BMC Bioinf. **12**, 323 (2011)
34. Grabherr, M.G., Haas, B.J., Yassour, M., Levin, J.Z., Thompson, D.A., Amit, I., Adiconis, X., Fan, L., Raychowdhury, R., Zeng, Q., Chen, Z., Mauceli, E., Hacohen, N., Gnirke, A., Rhind, N., di Palma, F., Birren, B.W., Nusbaum, C., Lindblad-Toh, K., Friedman, N., Regev, A.: Full-length transcriptome assembly from RNA-Seq data without a reference genome. Nat. Biotechnol. **29**, 644–652 (2011)
35. White, T.: Hadoop: The Definitive Guide. O'Reilly Media, Inc. (2012)
36. Franklin, M.: Spark Becomes Top Level Apache Project
37. Ousterhout, K., Wendell, P., Zaharia, M., Stoica, I.: Sparrow: Distributed, low latency scheduling. In: Proceedings of the Twenty-Fourth ACM Symposium on Operating Systems Principles, pp. 69–84. SOSP '13, New York, NY, USA, ACM (2013)
38. Bykov, S., Geller, A., Kliot, G., Larus, J., Pandya, R., Thelin, J.: Orleans: Cloud computing for everyone. In: ACM Symposium on Cloud Computing (SOCC 2011), ACM (2011)
39. O'Driscoll, A., Daugelaite, J., Sleator, R.D.: 'big data', hadoop and cloud computing in genomics. J. Biomed. Inf. 774–781 (2013)
40. Dove, E.S., Joly, Y., Tassé, A.M.: Genomic cloud computing: legal and ethical points to consider. Eur. J. Hum. Genet. (2014)
41. Kuo, A.M.H.: Opportunities and challenges of cloud computing to improve health care services. J. Med. Internet Res. **13** (2011)
42. Dai, L., Gao, X., Guo, Y., Xiao, J., Zhang, Z.: Bioinformatics clouds for big data manipulation. J. Med. Internet Res. **36**(6), 4031–4036 (2012)

43. Jimerson, B.: Software Architecture for High Availability in the Cloud
44. Apache: Spark programming guide (2014)
45. Xie, J., Yin, S., Ruan, X., Ding, Z., Tian, Y., Majors, J., Manzanares, A., Qin, X.: Improving mapreduce performance through data placement in heterogeneous hadoop clusters. In: 2010 IEEE International Symposium on Parallel Distributed Processing, Workshops and Phd Forum (IPDPSW), pp. 1–9 (2010)
46. Kumar, V.: Running Hadoop in the Cloud
47. Apache: Spark sql programming guide (2014)
48. Apache: Parquet (2014)
49. He, Y., Lee, R., Huai, Y., Shao, Z., Jain, N., Zhang, X., Xu, Z.: Rcfile: A fast and space-efficient data placement structure in mapreduce-based warehouse systems. In: 2011 IEEE 27th International Conference on Data Engineering (ICDE), pp. 1199–1208 (2011)
50. Niemenmaa, M., Kallio, A., Schumacher, A., Klemelä, P., Korpelainen, E., Heljanko, K.: Hadoop-bam: directly manipulating next generation sequencing data in the cloud. Bioinformatics **28**, 876–877 (2012)
51. Schumacher, A., Pireddu, L., Niemenmaa, M., Kallio, A., Korpelainen, E., Zanetti, G., Heljanko, K.: Seqpig: simple and scalable scripting for large sequencing data sets in hadoop. Bioinformatics **30**, 119–120 (2014)
52. Wiewiórka, M.S., Messina, A., Pacholewska, A., Maffioletti, S., Gawrysiak, P., Okoniewski, M.J.: Sparkseq: fast, scalable, cloud-ready tool for the interactive genomic data analysis with nucleotide precision. Bioinformatics 2652–2653 (2014)
53. Nordberg, H., Bhatia, K., Wang, K., Wang, Z.: Biopig: a hadoop-based analytic toolkit for large-scale sequence data. Bioinformatics **29**, 3014–3019 (2013)
54. Leo, S., Santoni, F., Zanetti, G.: Biodoop: bioinformatics on hadoop. In: IEEE International Conference on Parallel Processing Workshops, 2009. ICPPW'09, pp. 415–422 (2009)
55. Massie, M., Nothaft, F., Hartl, C., Kozanitis, C., Schumacher, A., Joseph, A.D., Patterson, D.A.: Adam: Genomics formats and processing patterns for cloud scale computing. Technical Report UCB/EECS-2013-207, EECS Department, University of California, Berkeley (2013)
56. McCabe, C.: How Improved Short-Circuit Local Reads Bring Better Performance and Security to Hadoop. http://blog.cloudera.com/blog/2013/08/how-improved-short-circuit-local-reads-bring-better-performance-and-security-to-hadoop/ (2013)
57. Callaghan, B., Pawlowski, B., Staubach, P.: Nfs version 3 protocol specification. Technical report, RFC 1813, Network Working Group (1995)
58. Dove, E.S., Joly, Y., Tassé, A.M., Burton, P., Chisholm, R., Fortier, I., Goodwin, P., Harris, J., Hveem, K., Kaye, J., et al.: Genomic cloud computing: legal and ethical points to consider. Eur. J. Hum. Genet. (2014)
59. Beck, M., Haupt, V.J., Roy, J., Moennich, J., Jäkel, R., Schroeder, M., Isik, Z.: Genecloud: Secure cloud computing for biomedical research. In: Trusted Cloud Computing, pp. 3–14. Springer (2014)
60. Hortonworks. Manage Security Policy for Hive & HBase with Knox & Ranger. http://hortonworks.com/hadoop-tutorial/manage-security-policy-hive-hbase-knox-ranger/ (2014)
61. Sharma, P.P., Navdeti, C.P.: Securing big data hadoop: a review of security issues, threats and solution. Int. J. Comput. Sci. Inf. Technol. **5** (2014)
62. Merelli, I., Pérez-Sánchez, H., Gesing, S., D'Agostino, D.: Managing, analysing, and integrating big data in medical bioinformatics: open problems and future perspectives. BioMed Res. Int. **2014** (2014)
63. Cock, P.J., Antao, T., Chang, J.T., Chapman, B.A., Cox, C.J., Dalke, A., Friedberg, I., Hamelryck, T., Kauff, F., Wilczynski, B., et al.: Biopython: freely available python tools for computational molecular biology and bioinformatics. Bioinformatics **25**, 1422–1423 (2009)
64. Holland, R.C., Down, T.A., Pocock, M., Prlić, A., Huen, D., James, K., Foisy, S., Dräger, A., Yates, A., Heuer, M., et al.: Biojava: an open-source framework for bioinformatics. Bioinformatics **24**, 2096–2097 (2008)
65. Wadkar, S., Siddalingaiah, M.: Apache ambari. In: Pro Apache Hadoop, pp. 399–401. Springer (2014)

66. Vavilapalli, V.K., Murthy, A.C., Douglas, C., Agarwal, S., Konar, M., Evans, R., Graves, T., Lowe, J., Shah, H., Seth, S., et al.: Apache hadoop yarn: Yet another resource negotiator. In: Proceedings of the 4th Annual Symposium on Cloud Computing, p. 5. ACM (2013)
67. Franklin, M.: The berkeley data analytics stack: Present and future. In: 2013 IEEE International Conference on Big Data, pp. 2–3 (2013)
68. Xiao, W., Ji, C.L., Li, J.D.: Design and implementation of massive data retrieving based on cloud computing platform. Appl. Mech. Mater. **303**, 2235–2240 (2013)
69. Turnbull, J.: The Docker Book: Containerization is the new virtualization. James Turnbull (2014)
70. Team, R.C., et al.: R: A language and environment for statistical computing (2012)
71. Gentleman, R.C., Carey, V.J., Bates, D.M., Bolstad, B., Dettling, M., Dudoit, S., Ellis, B., Gautier, L., Ge, Y., Gentry, J., et al.: Bioconductor: open software development for computational biology and bioinformatics. Genome Biol. **5**, R80 (2004)
72. Kaster, J., Rahmann, S.: Snakemake—a scalable bioinformatics workflow engine. Bioinformatics **28**, 2520–2522 (2012)
73. Cingolani, P., Sladek, R., Blanchette, M.: Bigdatascript: a scripting language for data pipelines. Bioinformatics **31**, 10–16 (2015)

Discovering Networks of Interdependent Features in High-Dimensional Problems

Michał Dramiński, Michał J. Dąbrowski, Klev Diamanti,
Jacek Koronacki and Jan Komorowski

Abstract The availability of very large data sets in Life Sciences provided earlier by the technological breakthroughs such as microarrays and more recently by various forms of sequencing has created both challenges in analyzing these data as well as new opportunities. A promising, yet underdeveloped approach to Big Data, not limited to Life Sciences, is the use of feature selection and classification to discover interdependent features. Traditionally, classifiers have been developed for the best quality of supervised classification. In our experience, more often than not, rather than obtaining the best possible supervised classifier, the Life Scientist needs to know which features contribute best to classifying observations (objects, samples) into distinct classes and what the interdependencies between the features that describe the observation. Our underlying hypothesis is that the interdependent features and rule networks do not only reflect some syntactical properties of the data and classifiers but also may convey meaningful clues about true interactions in the modeled biological system. In this chapter we develop further our method of Monte Carlo Feature Selection and Interdependency Discovery (MCFS and MCFS-ID, respectively), which are particularly well suited for high-dimensional problems, i.e., those

---

We thank the reviewer for providing valuable and detailed comments.

M. Dramiński · M.J. Dąbrowski · J. Koronacki
Institute of Computer Science, Polish Acad. Sci, Ordona 21, Warsaw, Poland
e-mail: Michal.Draminski@ipipan.waw.pl

M.J. Dąbrowski
e-mail: Michal.Dabrowski@ipipan.waw.pl

J. Koronacki
e-mail: Jacek.Koronacki@ipipan.waw.pl

K. Diamanti
Department of Cell and Molecular Biology, Uppsala University,
Box 596, Uppsala, Sweden
e-mail: Klev.Diamanti@icm.uu.se

J. Komorowski (✉)
Department of Cell and Molecular Biology, Uppsala University and Institute
of Computer Science, Polish Acad. Sci, Uppsala, Sweden
e-mail: Jan.Komorowski@icm.uu.se

© Springer International Publishing Switzerland 2016
N. Japkowicz and J. Stefanowski (eds.), *Big Data Analysis: New Algorithms for a New Society*, Studies in Big Data 16, DOI 10.1007/978-3-319-26989-4_12

where each observation is described by very many features, often many more features than the number of observations. Such problems are abundant in Life Science applications. Specifically, we define Inter-Dependency Graphs (termed, somewhat confusingly, ID Graphs) that are directed graphs of interactions between features extracted by aggregation of information from the classification trees constructed by the MCFS algorithm. We then proceed with modeling interactions on a finer level with rule networks. We discuss some of the properties of the ID graphs and make a first attempt at validating our hypothesis on a large gene expression data set for CD4$^+$ T-cells. The MCFS-ID and ROSETTA including the Ciruvis approach offer a new methodology for analyzing Big Data from feature selection, through identification of feature interdependencies, to classification with rules according to decision classes, to construction of rule networks. Our preliminary results confirm that MCFS-ID is applicable to the identification of interacting features that are functionally relevant while rule networks offer a complementary picture with finer resolution of the interdependencies on the level of feature-value pairs.

Keywords MCFS-ID · ROSETTA · Ciruvis · High-dimensional problems · Gene expression data

1 Introduction

Technical developments of the last decades enabling researchers to deal with Big Data allow one to look much deeper into the workings of complex systems, in particular those from Life Sciences. Specifically, within genomic studies Encyclopedia of DNA Elements (ENCODE, cf. [1, 2]), Genome-wide association study (GWAS; cf. [3]), NIH Roadmap project (cf. [4]) and 1000 Genomes project (A Deep Catalog of Human Genetic Variation; cf. [5]) are the ongoing projects that offer continuously updated information about transcribed and non-transcribed genomic regions, epigenetic marks, RNA-seq, SNPs, the biological traits they are associated with, and integrated maps of genetic variation from 1092 human genomes. Combining, processing and analyzing these data provides a conceptual bridge to investigate features and functions of the human genome. Associating the genetic background of a specific disease with, for instance, histone modifications, transcription factors, chromatin states and mutations has become one of the major streams of the investigations in genomics today. However, it requires a deep understanding of the mechanisms and the development of complex computational techniques for managing and merging data sources so as to enable biological interpretation of the results.

Another major challenge in the analysis of such biological data is due to their sizes: a small number of objects (records, samples) versus several orders of magnitude greater number of attributes or features for each record. Such problems are usually referred to as "small n large p problems", which we renamed to "small n large d problems" (where n stands for the number of records and p or d as that of features).

By far, it is not only in Life Sciences, where problems of this type appear and have to be dealt with. Indeed, in our own work, we met challenging problems of commercial origin, including transactional data from a major multinational fast-moving consumer goods company and geological data from oil wells operated by a major oil company.

Independently of whether the data are to explain a quantitative—as in regression—or categorical—as in classification—trait, such problems are quite different from typical data mining ones in which the number of features is much smaller than the number of samples. In a sense, these are ill-posed problems. This fact is immediately clear in the case of linear regression fitted by ordinary least-squares, where one gets a few linear equations with many more unknowns.

For two-class classification, at least from the geometrical point of view, the task is trivial, since in a d-dimensional space, as many as $d + 1$ points can be divided into two arbitrary and disjoint subsets by some hyperplane, provided that these points do not lie in a proper subspace of the d-dimensional space. This is a well-known result on the Vapnik-Chervonenkis dimension for the class of halfspaces in R^d. It is another matter that the found hyperplane, or any other classification rule, should have the ability to generalize.

In conclusion, since it is rather a rule than an exception that most features in the data are not informative, but are essentially a noise or are redundant, it is of utmost importance to select the few ones that are informative and that may form a basis for class prediction, or for a proper regression model. Accordingly, before building a classifier or a regression model, or while building any of them, we would like to find out which features are specifically linked to the problem at hand and should be included in the solution.

Mathematically, properly formulated sparsity constraints should be included when seeking a solution. This requirement can be fulfilled by regularization or randomization. In the later sections of this exposition, we shall confine ourselves to the latter approach.

Regarding classification, and from now on we shall deal with classification only, one more important issue should be emphasized. More often than not, rather than obtaining the best possible classifier, the Life Scientist and, we claim, any other user of classifiers need to know which features contribute best to classifying observations (samples) into distinct classes and what are the *interdependencies* between such informative features.

In the area of feature ranking and selection, very significant progress has been achieved. For a brief account, up to 2002, see [6] and for an extensive survey and somewhat later developments see [7].

Without coming to details let us note that feature selection can be *wrapped* around the classifier construction or directly built (*embedded*) into the classifier construction, and not performed prior to addressing the classification task per se by *filtering* out noisy features first and keeping only informative ones for building a classifier. An early and successful method with embedded feature selection included, not mentioned by [7], was developed by Tibshirani et al. (see [8, 9]). More recently and within non-filter approaches, a Bayesian technique of automatic relevance determination, the use of support vector machines, and the use of ensembles of classifiers, all these either alone or in combination, have proved promising. For further details see [10–12] and the literature there.

Moreover, the last developments by the late Leo Breiman deserve special attention. In his Random Forests (RFs), he proposed to make use of the so-called variable (i.e. feature) importance for feature selection. Determination of the importance of the variable is not necessary for random forest construction, but it is a subroutine performed in parallel to building the forest; cf. [13]. Ranking features by variable importance can thus be considered to be a by-product of building the classifier.

At the same time, nothing prevents one from using such variable importances within, say, the embedded approach; cf., e.g., [14]. In any case, feature selection by measuring variable importance in random forests should be seen as a very promising method, albeit under one proviso. Namely, the problem with variable importance as originally defined is that it is biased towards variables with many categories; cf. [15–17]. Accordingly, proper debiasing is needed, in order to obtain true ranking of features; cf. [18]. And, however sound such debiasing may be, it incurs much additional computational cost.

Most recently, much work has been done to give embedded feature selection procedures, in particular those used within RFs (whether biased or unbiased), a clear statistical meaning; cf. [19] (see also [20] and the literature there).

On the other hand, one potential advantage of the filter approach is that it constructs a group of features that contribute the most to the classification task, and therefore are informative or "relatively important" for this given task, *regardless* of the classifier that will be used. Of course, for this to be the case, a filter method used for feature selection should be capable of incorporating interdependencies between the features. Indeed, the fact that a feature may prove informative only in conjunction with some other features, but not alone, needs be taken into account. It should be noticed here that the aforementioned algorithms for measuring variable importance in RFs possess this last capability.

In 2005, a novel, effective and reliable filter method for ranking features according to their importance for a given supervised classification task has been introduced by [21]. The method, which relies on Monte Carlo approach to select informative features and was fully developed in [22] as Monte Carlo Feature Selection Algorithm or MCFS, is capable of incorporating interdependencies between features. It bears some remote similarity to the RF methodology, but differs entirely in the way features ranking is performed. Specifically, our method does not require debiasing (cf. [23]) and is conceptually simpler. A more important and newer result is that it provides explicit information about interdependencies among features (cf. [24]; see also [25, 26]).

In this chapter, we substantially expand the ideas from [24] by adding to our MCFS the functionality of discovering interdependencies between features, i.e., on extending the MCFS to Monte Carlo Feature Selection and Interdependency Discovery Algorithm or MCFS-ID. Within our approach, discovering interdependencies builds on identifying features which "cooperate" in determining that some samples belong to one class, others samples to another class, still another to still another class and so on.

It is worthwhile to emphasize that this is completely different from the usual approach which aims at finding features that are similar in some sense (cf. the first paragraph of Sect. 3.1). Instead, it can be said that our way to discover interdependencies between features amounts to determining multidimensional dependency between the classes and sets or sequences of features. In this sense, we are in fact interested in *contextual interdependencies between features*, since the dependency in question requires the context of class information.

In [24], we presented a way to provide undirected networks of interdependent features with only a few features in one network. Here, we replace undirected networks by a directed graph with as many features (as the graph nodes) as needed to fully describe the interdependencies involved. The graph does not tell the differences between the classes, i.e., it does tell what interdependencies make the samples belong to different classes but does not give rules which determine any given class. Accordingly and separately, a way to construct rule networks is also provided, where the networks are constructed from IF-THEN rules with one network per each decision class. Our approach to construct rule networks is borrowed from [27, 28].

It is nearly impossible to overemphasize—and let us repeat it once again—that we do not aim at classification. While in our approach we heavily rely on using classifiers, we do not use them for the classification task per se. Indeed, we use classifiers to: (i) rank features according to their importance with respect to their discriminative power to distinguish between classes; (ii) discover interdependencies between features; (iii) and find sets of rules which provide additional information on the interdependencies mentioned. To put it otherwise, given the top features found in step (i), one can later use them for classification by any classifier, but this is neither required nor of our interest. Even more clearly, steps (ii) and (iii) are aimed at something vastly different from sheer solving the classification task.

The procedure from [22] for Monte Carlo feature selection is briefly recapitulated in Sect. 2. In Sect. 3, our new way to discover interdependencies between features is provided. A short description of finding interdependencies in rule sets as generated by ROSETTA is given there, too. In Sect. 4 application of the method is illustrated on a real-life example with genes level expression measured by microarray of fresh $CD4^+$ T cells response to activation from healthy individuals. Interpretation of the obtained results is provided in Sect. 4.2. We close with concluding remarks in Sect. 5.

2 Monte Carlo Feature Selection or the MCFS Algorithm

We begin with a brief recapitulation of our MCFS; see [22], which can be consulted for details and rationale for our approach to feature selection.

We consider a particular feature to be important, or informative, if it is likely to take part in the process of classifying samples into classes "more often than not". This "readiness" of a feature to take part in the classification process, termed

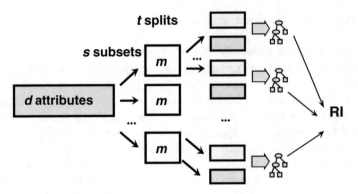

Fig. 1 Block diagram of the main step of the MCFS procedure. From [22]

relative importance of a feature, is measured via intensive use of classification trees. In the main step of the procedure, we estimate relative importance of features by constructing thousands of trees for randomly selected subsets of features.

More precisely, out of all d features, s subsets of m features are randomly selected, m being fixed and $m \ll d$, and for each subset of features, t trees are constructed and their performance is assessed (one can easily see that the procedure is essentially the same as that of the Random Subspace Method, the fact the authors were not aware of at the time they wrote their proposal; cf [29]).

Each of the t trees in the inner loop is trained and evaluated on a different, randomly selected training and test sets that come from a split of the full set of training data into two subsets: each time, out of all n samples, 2/3 of the samples are drawn at random for training in such a way as to preserve proportions of classes from the full set of training data, and the remaining samples are used for testing. See Fig. 1 for a block diagram of the procedure.

The relative importance of feature g_k, RI_{g_k}, is defined as

$$\mathrm{RI}_{g_k} = \sum_{\tau=1}^{s \cdot t} \mathrm{wAcc}_\tau^u \sum_{n_{g_k}(\tau)} \mathrm{IG}(n_{g_k}(\tau)) \left(\frac{\text{no. in } n_{g_k}(\tau)}{\text{no. in } \tau} \right)^v, \tag{1}$$

where summation is over all $s \cdot t$ trees and, within each τth tree, over all nodes $n_{g_k}(\tau)$ of that tree on which the split is made on feature g_k, wAcc_τ stands for the weighted accuracy of the τ's tree, $\mathrm{IG}(n_{g_k}(\tau))$ denotes information gain for node $n_{g_k}(\tau)$, (no. in $n_{g_k}(\tau)$) denotes the number of samples in node $n_{g_k}(\tau)$, (no. in τ) denotes the number of samples in the root of the τth tree, and u and v are fixed positive reals (now set to 1 by default; cf. [24]). The normalizing factor (no. in τ), which has the same value for all τ, has been included mainly for computational reasons.

With u and v set to 1, there are three parameters, m, s and t to be set by an experimenter. Note that, overall, $s \cdot t$ trees are constructed and evaluated in the main step of the procedure. Both s and t should be sufficiently large, so that each feature has a chance to appear in many different subsets of features and randomness due to inherent variability in the data is properly accounted for. The choice of subset size m of features selected for each series of t experiments should take into account the trade-off between the need to prevent informative features from being masked too severely by the relatively most important ones and the natural requirement that s be not too large. Indeed, the smaller m, the smaller the chance of masking the occurrence of a feature. However, a larger s is then needed, since all features should have a high chance of being selected into many subsets of the features. For classification problems of dimension d ranging from several thousands to hundreds of thousands, we have found that taking m equal to a few hundreds (say, $m = 300-500$) and t equal to maximum 20 (even $t = 5$ usually suffices) is a good choice in terms of reliability and overall computational cost of the procedure. Finally, for a given m, s can be made a running parameter of the procedure, and the procedure executed for increasing s until the rankings for successive values of the s of top scoring $p\%$ features prove (almost) the same.

The above provides one with a ranking of features. We skip discussion on how to find a cut-off between informative and noninformative featrues, and refer the reader to [24].

3 Discovering Feature Interdependencies

Once features are ranked by the MCFS procedure, a natural issue to be raised concerns possible interdependencies among the features. In Sect. 3.1, we introduce a way to present such interdependencies in the form of a directed graph. The graph provides information about the interdependencies which make the samples belong to different classes but does not give rules which determine any given class. Thus, in Sect. 3.2, a way to construct rule networks is provided, with one network per each decision class.

3.1 Interdependency Discovery or the ID Part of the MCFS-ID Algorithm

The interdependencies among features are often modeled using interactions, similarly as in experimental design and analysis of variance. Perhaps the most widely used approach to recognizing interdependencies is finding correlations between features or finding groups of features that behave in some sense similarly across samples. A classical bioinformatics example of this problem is finding co-regulated features, most often genes or, rather more precisely, their expression profiles. Searching groups of similar features is usually done with the help of various clustering techniques, frequently specially tailored to a task at hand. See [7, 30–32] and the literature there.

As has already been mentioned, our approach to interdependency discovery (abbreviated ID, with MCFS-ID used as an abbreviation for the whole procedure) is significantly different in that we focus on identifying features that "cooperate" in determining that a sample belongs to a particular class. Put otherwise, in the sense just described, our aim is to discover *contextual* interdependencies between the features.

To be more specific, consider a single classification tree. Now, for each class, a set of decision rules defining this class is provided by the tree. Each decision rule is produced as a conjunction of conditions imposed on the particular features. But this simply means that producing decision rules amounts to pointing to interdependencies between the features appearing in the conditions.

Given a single rule-based classifier, our trust in the decision rules that are learned and thus in the discovered interdependencies, is naturally limited by the predictive ability of that classifier. Even more importantly, the classifier is trained on just one training set. Therefore, our conclusions are necessarily dependent on the classifier and are conditional upon the training set, since these conclusions follow from just one solution of the classification problem. In the case of classification trees, the problem is aggravated by their high variance, i.e., their tendency to provide varying results even for slightly different training sets. It should now be obvious, however, that the way out of the trouble is through an aggregation of the information provided by all the $s \cdot t$ trees (cf. Fig. 1), which anyhow are built within the MCFS part of the MCFS-ID algorithm.

Generally, the idea presented here is similar to that of [24]. The main difference is that we propose a new interdependency measure that allows creating a directed graph of features' interdependencies. The graph will be referred to as ID Graph, ID standing this time for Inter-Dependency (we apologize for using the same abbreviation for two slightly different terms). In order to describe how an ID Graph is built, let us recall that the MCFS-ID algorithm is based on building a multitude of classification trees, where each node in a tree represents a feature on which a split is made. Now, for each node in each classification tree its all antecedent nodes can be taken into account along the path to which the node belongs; note that each node in a tree has only one parent and thus for the given node we simply consider its parent, then the parent of the parent and so on. In practice, the maximum possible depth of such analysis, i.e. the number of antecedents considered, if available before the tree's root is attained, is set to some predetermined value, which is the procedure's parameter (its default value being 5). For each pair [*antecedent node* → *given node*] we add one directed edge to our ID Graph from *antecedent node* to *given node*. Let us emphasize again that a node is equated with the feature it represents and thus any directed edge found is in fact an edge joining two uniquely determined features in a directed way. To put it otherwise, while the edges are found as directed pairs of nodes appearing along the paths in all the $s \cdot t$ MCFS-ID trees, they represent directed pairs of features which are uniquely determined for each pair. In particular, the same edge can appear more than once even in a single tree.

The strength of the interdependence between two nodes, actually two features, connected by a directed edge, termed ID weight of a given edge, or ID weight for short, is equal to the gain ratio (GR) in the given node multiplied by the fraction of objects in the given node and the antecedent node. Thus, for node $n_k(\tau)$ in τth tree, $\tau = 1, \ldots, s \cdot t$, and its antecedent node $n_i(\tau)$, ID weight of the directed edge from $n_i(\tau)$ to $n_k(\tau)$, denoted $w[n_i(\tau) \rightarrow n_k(\tau)]$, is equal to

$$w[n_i(\tau) \rightarrow n_k(\tau)] = \mathrm{GR}(n_k(\tau)) \left(\frac{\text{no. in } n_k(\tau)}{\text{no. in } n_i(\tau)} \right), \tag{2}$$

where $\mathrm{GR}(n_k(\tau))$ stands for gain ratio for node $n_k(\tau)$, (no. in $n_k(\tau)$) denotes the number of samples in node $n_k(\tau)$ and (no. in $n_i(\tau)$ denotes the number of samples in node $n_i(\tau)$.

The final ID Graph is based on the sums of all ID weights for each pair [*antecedent node* → *given node*]; i.e. for each directed edge found, its ID weights are summed over all occurrences of this edge in all paths of all MCFS classification trees. For a given edge, it is this sum of ID weights which becomes the ID weight of this edge in the final ID Graph. Algorithm 1.1, where T denotes the set of all $s \cdot t$ trees and $D = \{1, 2, \ldots, depth\}$ with *depth* being the predetermined number of antecedents considered, gives the pseudo code that describes the calculation.

In the ID Graphs, as seen in Fig. 2, some additional information is conveyed with the help of suitable graphical means. The color intensity of a node is proportional to the corresponding feature's relative importance RI. The size of a node is proportional to the number of edges related to this node. The width and level of darkness of an edge is proportional to the ID weight of this edge.

Algorithm 1.1 ID graph building procedure

$w[n_\delta \rightarrow n] = 0$
for $\tau_k \in T$ **do**
  for $n \in \tau_k$ **do**
    for $\delta \in D$ **do**
      $n_\delta = \delta$-th antecedent of n
      $w[n_\delta \rightarrow n] = w[n_\delta \rightarrow n] + \mathrm{GR}(n) \left(\frac{\text{no. in } n}{\text{no. in } n_\delta} \right)$
    end for
  end for
end for

The ID Graph is a way to present interdependencies that follow from all of the MCFS classification trees. Each path in a tree represents a decision rule and by analyzing all tree paths we in fact analyze decision rules to find the most frequently observed features that along with other features form good decision rules. The ID Graph thus presents some patterns that frequently occur in thousands of classification trees built by the MCFS procedure.

Note that an edge $n_i \rightarrow n_k$ from node n_i to node n_k is directed as is the edge (if found) from n_k to n_i, $(n_k \rightarrow n_i)$. Interestingly, in most cases of ID Graphs, we find that one of such two edges is dominating, i.e. has a much larger ID weight than the other. Whenever it happens, it means that not only n_i and n_k form a sound partial decision (a part of a conjunction rule) but also that their succession in the directed rule is not random.

In sum, an ID Graph provides a general roadmap that not only shows all the most variable attributes that allow for efficient classification of the objects but, moreover, it points to possible interdependencies between the attributes and, in particular, to a hierarchy between pairs of attributes. High differentiation of the values of ID weights in the ID Graph gives strong evidence that some interdependencies between some features are much stronger than others and that they create some patterns/paths calling for biological interpretation.

3.2 Rule Networks

A rule-based classifier consists of a set of IF-THEN rules that describe the relations in the training data. Rough sets [33], which we use here, stand out among many other rule-based approaches by their firm mathematical foundations in Boolean reasoning and the much appreciated property of finding minimal sets of features that discern between classes or objects. These minimal sets of features are termed *reducts*.

As the first approximation of feature interaction in rules one can think of the co-occurrence of features in rules. Since rules contain usually several conjuncts, interpretation of such interactions and their visualization is non-trivial. In [27] the authors presented a novel concept called rule networks and interaction detection together with a visualization paradigm collectively called Ciruvis.

We present here a brief summary of that work. Rules that have more than one condition are considered by Ciruvis. Using the visualization paradigm for circular graphs introduced by the Circos software [34] each condition that has at least one connection to another condition is placed as a node on the outer ring of the circle in an alphabetical order. Two conditions are connected inside the circle by an edge if they co-occur in some rules. The score of the connection between two conditions, x and y, is defined as

$$connection(x, y) = \sum_{r \in R(x,y)} support(r) \cdot accuracy(r), \qquad (3)$$

where R(x, y) is the set of all rules in which x and y co-occur. The connections are shown as edges between the nodes. The width and color of the edges are related to the connection score (low = yellow and thin, high = red and thick). The inner ring shows the color of the condition on the other side of the connection. The width of a node is the sum of all connections to it, scaled so that all nodes together cover the whole circle. Clearly, the connections are ordered by the *connection* values. For an illustration see Fig. 3. Supplementary material detailing the Ciruvis examples can be found at http://bioinf.icm.uu.se/~ciruvis/mcfsid_chapter/.

To find out how well the rule networks from Ciruvis could detect feature interactions the approach was tested on simulated data and real data sets. The simulated data set contained features that had a varying degree of singular correlation of features to the decision and pairs of features correlated to the decision, also to a varying degree. Somewhat surprisingly, the authors found out that a higher level of interaction (i.e. pairwise correlation) increased or at least retained the classification quality, whereas a higher correlation of a single feature to decision sometimes decreased the quality. This suggested that the rule generation algorithm was biased towards finding rules containing features correlated to the decision. When the correlated features were not present, then the combinatorial rules of higher quality were more likely to be found. The identified masking was investigated in detail and an approach to alleviating this problem was developed. It is interesting to notice that similarly to feature shadowing that may occur in feature selection of strongly correlated features interaction detection can be obscured by the strongest interaction pairs. Removing the strongest interactions rectifies the problem.

Application of Ciruvis to real data sets produced interesting results. For example, the reader is referred to [27] for a detailed description of novel interactions found in the well-known California Housing data set and the other experiment of interaction detection for various forms of leukemia and lymphoma. The Ciruvis tool is publicly available and can be used for any rule set, not necessarily rough sets.

4 Biological Validation Study

The value of a tool like MCFS-ID, or the Ciruvis one, will eventually be proven empirically, through a large number of experiments that are validated by domain experts, alternatively by a comparison with a golden standard provided its existence. However, we need to start this process and hence we propose to evaluate MCFS-ID on a substantial biological dataset and to compare this approach with a related one of discovering interactions with rule networks as implemented by Ciruvis.

4.1 Experimental Setup and Results

The earlier version of our MCFS-ID, as has already been stated capable of discovering undirected networks of interdependent features, was validated on several real data sets (cf. [22, 24–26]). Its current version, proposed in this chapter, was validated on a large, fairly complex real data set from [35]. Those authors, inter alia, have collected gene expression levels of 236 genes in $CD4^+$ T cells activated in unbiased conditions and measured after 4 and 48 h, or in biased conditions toward T helper 17 (TH17), or with addition of IFN-β. The $CD4^+$ T cells were sampled from human blood from 348 healthy patients who were of three different ancestries: European, Asian, and African-American. The authors have reported the impact of ancestry on 94 of 229

Table 1 20 attributes with the highest RI score

Rank	Attribute	Rank	Attribute	Rank	Attribute
1	UTS2_Th17_48	8	HDGFRP3_Activated_48	15	OLR1_Activated_4
2	UTS2 Activated_48	9	Weight (kg)	16	LYZ_Th17_48
3	UTS2_Activated_4	10	FGL2_Activated_4	17	IFITM3_Th17_48
4	UTS2_Unstim_4	11	NPCDR1_IFNb_4	18	Age (years)
5	UTS2_IFNb_4	12	FGL2_Unstim_4	19	CYBB_Activated_48
6	MXRA7_Th17_48	13	IFITM3_Activated_48	20	CCL2_Activated_4
7	MXRA7_Activated_48	14	IFIT2_IFNb_4		

genes. Differentially responsive genes included key indicators of TH phenotype, IL17 family cytokines and IFNG.

The main aim of our validation was to obtain a better understanding of the ancestry influence on human immune system development and current genes expression levels using nonlinear methods in contrast to [35] who examined T cell responses in different populations using a linear model. In our study, we retained observations removed from the study by [35]. This resulted in a decision table with 365 observations (objects), each with 1259 attributes. The attributes of the decision system were all gene expression features and the following donor's personal data: *age, height.cm, weight.kg, bmi, systolic, diastolic and sex*. The decision attribute (class) was chosen to be donor's ancestry. The parameters of the MCFS-ID algorithm were set to their default values: $s = 5000$, $t = 5$ and $m = 0.05d$ (i.e., m = 63 in our case).

In sum, while our results are similar to those of [35], there are also differences which are a consequence of the generality with which we take into account interdependencies between the features. For instance, two features excluded by [35] due to high false discovery rate, namely genes OLR1 and CCL2 activated in unbiased conditions and measured after 4 h, were returned in our study within 20 topmost features.

MCFS-ID returned the features ranked according to their RI score and the ID Graph that shows interdependencies between the features (see Table 1 and Fig. 2). We verified that the top ranked features are truly informative by using the first 50 of them to build several popular classifiers. In particular, using 10-fold cross-validation we obtained 78.0 % classification accuracy for KNN(5) and 82.4 % accuracy for SVM with polynomial kernel.

Secondly, for the top 20 and 100 features with the highest RI scores returned by MCFS-ID we built respectively two sets of classifiers with the help of ROSETTA, which in turn were used as input to the Ciruvis tool. We used the same decision table as for MCFS-ID, although reduced to 20 and 100 top features, respectively. The feature values were discretized using Equal Frequency Binning with 3 levels. In our decision system there were three decision classes that were slightly unbalanced: Caucasian with 190 objects; African-Americans with 99 objects; and Asian with 76 objects.

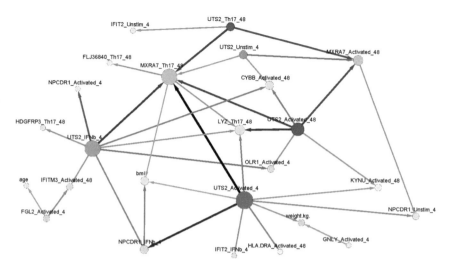

Fig. 2 ID Graph created for 50 attributes with highest RI score and ID weights ≥ 6

To reduce the impact of data quality on a classifier we subsampled the datasets. For each of the cases, viz. 20 and 100 features, we thus obtained 100 subsampled datasets where the number of objects in each decision class was equal. (Actually, balancing the classes was not needed and we performed subsampling twice, without and with balancing, to obtain corresponding results; we report only those for the balanced data, as they are more transparent to interpret.) Accuracy was obtained by taking the average of the 10-fold cross-validation performed on 100 replicates returned from the subsampling for each of the datasets. The reducts were computed using the JohnsonReducer algorithm. To avoid over-fitting, the rules with support lower than five were not included.

The mean accuracy of the returned models based on 100 attributes was 0.691 (SD = 0.091) and based on 20 attributes was 0.671 (SD = 0.091). Rather surprisingly, the 80 % reduction of the number of attributes (from 100 to 20) caused only a minor accuracy reduction of just 3 %. The expected classifier accuracy from random guessing for three decision classes would be 33 %, which means that the obtained accuracies were over two times higher than those achievable by chance. We obtained almost 70 % likelihood of predicting correct ancestry for unseen patients, thus confirming both proper choice of the attributes and high performance of our rule-based classifier.

Note that it is not the best possible rule generating classifier which is sought here. We need one that can be considered reliable and, most importantly, the one that provides possibly simple and clear-cut rule networks, easy to interpret for domain experts. This is why we have decided to use a rule generating classifier which requires discretization of data. By means of an example, while discretization of blood pressure to just three crude levels of low, medium and high cannot result in the best possible classification results, it is not only the most popular discretization used at large, but it has proved sufficient for bringing reasonable and reliable results.

All rules from 100 replicates were taken together and filtered to remove all duplicates and rules that were supersets of more significant rules. The rules p-values were calculated using the hypergeometric distribution; cf. [28]. Finally, the rules were ranked based on the p-values, which represented the probability that a random selection of the same number of objects equal to the rule support would contain an equally large or larger fraction of objects assigned to a certain decision class as in the rule accuracy. Finally, we built rule networks using Ciruvis with default settings.

4.2 Discussion

The first interpretation of the results is made on the basis of the RI ranking. For brevity of the discussion, we combine it with the interpretation of the ID Graph. By the construction, the intensity of the coloring in the ID Graph (Fig. 2) follows the RI index.

The first five features represent gene UTS2 in all five activation states. According to the number of activation states, NPCDR1 comes next being present in three activation states, MXRA7 and IFIT2in two activation states and finally LYZ, CYBB, OLR1, IFITM3 in one activation state (cf. Fig. 2). Moreover, taking into account not only the genes names but also the time at which the measurement was made, one gets that the direction of significant connections is consistent with that of the lapse of time. That is, the gene expressions measured after 4 h point to gene expressions measured at the same hour or later. It is also worth a mention that in all cases the arrows have only one direction when 50 attributes are taken into account. In contrast, in the ID Graph for 100 top ranked attributes there are three cases with edges directed both ways, namely the edges between IFITM3 and SRD5A3, MXRA7 and LYZ, MXRA7 and FLJ36840, suggesting possible biological feedback (data not shown due to to very high level of detail). However, in all three cases, one of the connections is much weaker than the other.

It is now interesting to notice that the UTS2 gene nodes have the largest number of outgoing edges. These findings are interestingly in agreement with the biological role of the gene. Regarding biological meaning of the observed interdependencies between attributes, one can easily see that genes coding proteins that have a more general function point to those that encode proteins with more specific functions. Indeed, the UTS2 gene encodes a mature peptide that is an active cyclic heptapeptide absolutely conserved from lamprey to human. The urotensin-2 protein, i.e. the product of UTS2 is a potent vasoconstrictor and agonist for the orphan receptor GPR14 (cf. [36]). Activation of the urotensin-2 receptor leads to increase in Ca(2^+), activation of phospholipase A(2) (PLA(2)) and increase in arachidonic acid (cf. [37]). Human urotensin-2 is found within both vascular and cardiac tissue (including coronary atheroma) and effectively constricts isolated arteries from non-human primates. The potency of vasoconstriction of urotensin-2 is one order of magnitude greater than that of endothelin-1, making human urotensin-2 the most potent mammalian vasoconstrictor identified so far. Furthermore, as urotensin-2 immunoreactivity is also found within central nervous system and endocrine tissues, it may have additional activities. In the ID Graph, the arrows anchored at UTS2 are directed toward MXRA7 gene. Its proteins are expected to localize in various compartments mitochondrion, nucleus, and are integral to membrane, which suggests its higher specificity than of the UTS2 gene product.

Continuing this line of reasoning, we find that UTS2 is also linked with LYZ, OLR1, CYBB, NPCDR1. These genes respectively encode: (1) Lysozyme, one of the anti-microbial agents. (2) Oxidized low-density lipoprotein receptor 1 which is the receptor protein that belongs to the C-type lectin superfamily. Proteins that contain C-type lectin domains have a diverse range of functions including cell-cell adhesion, immune response to pathogens and apoptosis. (3) Beta chain of cytochrome b (-245). It has been proposed as a primary component of the microbicidal oxidase system of phagocytes. CYBB deficiency is one of five described biochemical defects associated with chronic granulomatous disease (CGD). (4) Nasopharyngeal carcinoma, down-regulated gene protein 1.

These examples show that although the ID Graph may not be translated directly into biological pathways, it gives reasonable hypotheses for further studies. Moreover, it appears that under some circumstances the ID Graph might return results with a direct biological function.

The ID graph analysis was followed by a development of rule-based classifiers with ROSETTA and creating rule networks with Ciruvis. Examples of rules, one per each class is given below. The numbers after each rule are, respectively, accuracy, support and p-value of the rule.

- IF UTS2_Activated_4=low and CYBB_Activated_48=high THEN African-American 0.579; 57; 7.73E-08
- IF UTS2_Activated_4=high and IFITM3_Th17_48=high THEN Asian 0.763; 59; 6.19E-25
- IF UTS2_Activated_4=low and IFIT2_IFNb_4=high THEN European 0.872; 78; 2.84E-13

ROSETTA was applied to data sets reduced to 20 and 100 topmost features, respectively. Next, for the classifier built on 20 attributes, we investigated the top 20 rules related to each particular decision. The obtained frequency of each attribute condition among various ancestries is given in Table 2. There are two phenotypic attributes followed by 18 attributes describing the gene, its activation pattern and time point of gene expression measurement. With the time point and activation pattern skipped, one finds 11 genes (FGL2, UTS2, CCL2, OLR1, IFIT2, NPCDR1, CYBB, HDGFRP3, IFITM3, MXRA7, LYZ) which are the most significant for the classification process and, hence, suggestive of importance to biological characterization of the differences between the classes.

Rules provide a refined view that may be used with advantage in interpreting biological results. For instance, one may notice that for the African-American and Asian classes (Table 2) attributes in all but one case have exactly one of the three possible values; it is only IFIT2_IFNb_4 that takes two values. In the European class

Table 2 Frequencies of the attributes' conditions within 20 top rules sorted by *p*-values and related to the ancestry decision class

Attribute name	African-American			Asian			European		
	Low	Medium	High	Low	Medium	High	Low	Medium	High
Age			0.5					0.05	
Weight.kg.			0.2					0.05	
FGL2_Unstim_4	0.05					0.05			
UTS2_Unstim_4				0.15		0.2			
CCL2_Activated_4							0.05		0.05
FGL2_Activated_4									0.05
OLR1_Activated_4							0.05		
UTS2_Activated_4	0.05					0.2	0.05		0.05
IFIT2_IFNb_4	0.05	0.1							0.95
NPCDR1_IFNb_4				0.3					0.05
UTS2_IFNb_4						0.15	0.05		
CYBB_Activated_48			0.25				0.05		
HDGFRP3_Activated_48			0.05	0.15					
IFITM3_Activated_48			0.05			0.15	0.05		
MXRA7_Activated_48			0.3				0.05		
UTS2_Activated_48	0.2					0.3	0.05		0.1
IFITM3_Th17_48						0.25			
LYZ_Th17_48			0.1				0.05		0.05
MXRA7_Th17_48			0.2				0.05		
UTS2_Th17_48	0.15					0.05	0.05		0.05

There are two phenotypic attributes, age and weight (kg), followed by 18 gene expression attributes; their names contain: gene name, type of stimulation and time point measurement

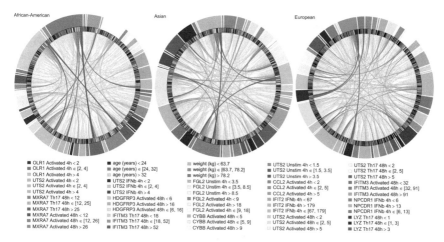

African-American   Asian   European

■ OLR1 Activated 4h < 2	■ age (years) < 24	■ weight (kg) < 63.7	■ UTS2 Unstim 4h < 1.5	UTS2 Th17 48h < 2
■ OLR1 Activated 4h ∈ [2, 4]	■ age (years) ∈ [24, 32]	■ weight (kg) ∈ [63.7, 78.2]	■ UTS2 Unstim 4h ∈ [1.5, 3.5]	UTS2 Th17 48h ∈ [2, 5]
OLR1 Activated 4h > 4	age (years) > 32	weight (kg) > 78.2	■ UTS2 Unstim 4h > 3.5	■ UTS2 Th17 48h > 5
■ UTS2 Activated 4h < 2	■ UTS2 IFNb 4h < 2	■ FGL2 Unstim 4h < 3.5	■ CCL2 Activated 4h < 2	■ IFITM3 Activated 48h < 32
■ UTS2 Activated 4h ∈ [2, 4]	UTS2 IFNb 4h ∈ [2, 4]	■ FGL2 Unstim 4h ∈ [3.5, 8.5]	■ CCL2 Activated 4h ∈ [2, 5]	■ IFITM3 Activated 48h ∈ [32, 91]
■ UTS2 Activated 4h > 4	■ UTS2 IFNb 4h > 4	FGL2 Unstim 4h > 8.5	■ CCL2 Activated 4h > 5	IFITM3 Activated 48h > 91
■ MXRA7 Th17 48h < 12	HDGFRP3 Activated 48h < 6	■ FGL2 Activated 4h < 9	■ IFIT2 IFNb 4h < 67	■ NPCDR1 IFNb 4h < 6
■ MXRA7 Th17 48h ∈ [12, 25]	■ HDGFRP3 Activated 48h > 16	■ FGL2 Activated 4h > 18	■ IFIT2 IFNb 4h > 179	■ NPCDR1 IFNb 4h > 13
■ MXRA7 Th17 48h > 25	HDGFRP3 Activated 48h ∈ [6, 16]	FGL2 Activated 4h ∈ [9, 18]	IFIT2 IFNb 4h ∈ [67, 179]	■ NPCDR1 IFNb 4h ∈ [6, 13]
■ MXRA7 Activated 48h < 12	IFITM3 Th17 48h < 18	CYBB Activated 48h < 5	■ UTS2 Activated 48h < 2	■ LYZ Th17 48h < 1
■ MXRA7 Activated 48h ∈ [12, 26]	■ IFITM3 Th17 48h ∈ [18, 52]	■ CYBB Activated 48h ∈ [5, 9]	■ UTS2 Activated 48h ∈ [2, 5]	■ LYZ Th17 48h ∈ [1, 3]
■ MXRA7 Activated 48h > 26	■ IFITM3 Th17 48h > 52	CYBB Activated 48h > 9	UTS2 Activated 48h > 5	■ LYZ Th17 48h > 3

Fig. 3 All three figures follow the color code as explained in the legend below and are the Ciruvis online tool results showing the attribute interactions for the individuals belonging to the decision classes of African-American, Asian and European ancestry. For all three we used the filtered rule-set from ROSETTA as input and we set minimum support to 5. Additionally we apply custom color code and group code

40 % of the attributes may take two values. The observed distribution of attribute values (low, medium, high) across the classes suggests that African-Americans and Asians are much more homogeneous than Europeans at least for these genes.

Finally, the rules returned by ROSETTA were filtered and input to the CIRUVIS tool, for each ancestry class (see Fig. 3), to reveal the most frequently interacting conditions of the factors. The interested reader may browse these graphs on-line [38]. The most clearly displayed pairs of the attribute conditions for the three classes are as follows:

(1) African, weight = high and MXRA7_Th17_48 = high, age = high and LYZ_Th17_48 = high, weight = high and IFIT2_IFNb_4 = medium, CCL2_Activated_4 = high and UTST_Activated_48 = medium;

(2) Asian, weight = low and NPCDR1_IFNb_4 = low, NPCDR1_IFNb_4 = low and HDGFRP3_Activated_48 = low, weight = low and HDGFRP3_Activated_48 = low, UTS2_Activated_48 = high and IFITM3_Th17_48 = high;

(3) European, weight = medium and OLR1_Activated_4 = low, weight = medium and UTS2_IFNb_4 = low, weight = high and UTS2_IFNb_4 = low.

5 Concluding Remarks

Working with large, complex data sets requires well-suited multiple approaches. We have shown how a combination of feature selection, construction of directed graphs of interactions between features extracted by aggregation of information from the classification trees, rule-based learning and rule networks may help in understanding complex relations in Big Data. Our set-up is classification. It seems to be particularly well suited to applications in Life Sciences in which two or more states such as, for instance, healthy-ill, treated-untreated, binding-non-binding, resistant-partly resistant-susceptible or benign-malignant are often studied. We propose a methodology that is able to cope with the "small n large p problems", a requirement that is unavoidable in the context of many biomedical experiments that are either prohibitively expensive or face the unavailability of large numbers of subjects. In the first step, a characterization of features that are significant in discerning between the classes is provided. The second step discovers networks of feature interdependencies. These interdependencies have a coarse granularity, but on the other hand seem to show an intriguing property of directedness that cannot be resolved on the level of correlation. The third step of rule generation results in a much finer characterization of the data since the rules are minimal combinations of features and their values associated to the classes. Since we provide several statistical properties of the rules, they may also be studied one-by-one, not necessarily as collections, i.e. full classifiers, and thus allowing for local interpretation. Finally, interactions between the features may be studied using our approach of rule networks.

We have illustrated our approach with a complex biological example and provided samples of possibilities in interpreting the results by a biologist.

MCFS-ID seems to be a competitive tool to produce significant features in the context of the "small n large p problems" as it has been proven by other applications and suggested by the sample application here. In the course of research for this paper, the ID part revealed several most interesting properties that will be studied further.

The ROSETTA system produces rules that provide a finer characterization of the experiment. They obviously use a subset of the significant features as identified by MCFS-ID. Studying the rules gives further important clues as to which features are used for classifying objects in each class, with the final advantage of specifying the values of the features. Putting it in another way, rules form combinatorial markers that predict classes for the sample at hand.

In the last step, the Ciruvis tool offers a possibility to explore the space of interactions on a detailed yet comprehensive level. The biological insights that we have presented here are meant to illustrate the potential power that may be eventually delivered by expert interpretation. A full biological study of this data will be published elsewhere.

If a deep understanding of experimental data is the goal of research, we advocate the use of transparent method such as decision trees or rules, even for the price of possibly lower accuracy of the classifiers. Learning that classification may be done at a higher accuracy, as often is the case with black box approaches (e.g. SVM, NN, etc.), does not bring a deeper understanding of the experiment unless parts and pieces are available to expert inspection. Lower accuracy is easily outweighed by the transparency and richness of available details to be revealed by interactions in the data.

References

1. Consortium, Encode Project, Bernstein et al: An integrated encyclopedia of DNA elements in the human genome. Nature **489**(7414), 57–74 (2012). doi:10.1038/nature11247
2. Birney, E., et al.: Identification and analysis of functional elements in 1% of the human genome by the ENCODE pilot project. Nature **447**(7146), 799–816 (2007)
3. Beck, T., Hastings, R.K., Gollapudi, S., Free, R.C., Brookes, A.J.: GWAS Central: a comprehensive resource for the comparison and interrogation of genome-wide association studies. Eur. J. Hum. Genet. **22**(7), 949–952 (2014). doi:10.1038/ejhg.2013.274
4. Bernstein, B.E., et al.: The NIH roadmap epigenomics mapping consortium. Nat. Biotechnol. **28**(10), 1045–1048 (2010). doi:10.1038/nbt1010-1045
5. Genomes Project, Consortium, Abecasis, G. R. et al: An integrated map of genetic variation from 1,092 human genomes. Nature **491**(7422), 56–65 (2012). doi:10.1038/nature11632
6. Dudoit, S., Fridlyand, J.: Classification in microarray experiments. In: Speed, T. (ed.) Statistical Analysis of Gene Expression Microarray Data, pp. 93–158. Chapman & Hall/CRC (2003)
7. Saeys, Y., Inza, I., Larrañaga, P.: A review of featrure selection techniques in bioinformatics. Bioinformatics **23**(19), 2507–2517 (2007)
8. Tibshirani, R., Hastie, T., Narasimhan, B., Chu, G.: Diagnosis of multiple cancer types by nearest shrunken centroids of gene exressions. Proc. Natl. Acad. Sci. USA **99**, 6567–6572 (2002)
9. Tibshirani, R., Hastie, T., Narasimhan, B., Chu, G.: Class prediction by nearest shrunken centroids, with applications to DNA microarrays. Statis. Sci. **18**, 104–117 (2003)
10. Li, Y., Campbell, C., Tipping, M.: Bayesian automatic relevance determination algorithms for classifying gene expression data. Bioinformatics **18**(10), 1332–1339 (2002)
11. Lu, C., Devos, A., Suykens, J.A., Arús, C., Van Huffel, S.: Bagging linear sparse bayesian learning models for variable selection in cancer diagnosis. IEEE Trans. Inf. Technol. Biomed. **11**, 338–347 (2007)
12. Chrysostomou, K., Chen, Sherry Y., S.Y. and Liu, X.: Combining multiple classifiers for wrapper feature selection. Int. J. Data Mining Modell. Manag. **1**, 91–102 (2008)
13. Breiman, L., Cutler, A.: Random forests—classification/clustering manual. http://www.math.usu.edu/~adele/forests/cc_home.htm (2008)
14. Diaz-Uriarte, R., de Andres, S.A.: Gene selection and classification of microarray data using random forest. BMC Bioinform. **7**(3), (2006). doi:10.1186/1471-2105-7-3
15. Strobl, C., Boulesteix, A.-L., Zeileis, A., Hothorn, T.: Bias in random forest variable importance measures: illustrations, sources, and a solution. BMC Bioinform. **8**(25), (2007). doi:10.1186/1471-2105-8-25
16. Archer, K.J., Kimes, R.V.: Empirical characterization of random forest variable importance measures. Comp. Stat. Data Anal. **52**(4), 2249–2260 (2008)
17. Nicodemus, K.K., Malley, J.D., Strobl, C., Ziegler, A.: The behaviour of random forest permutation-based variable importance measures under predictor correlation. BMC Bioinform. **11**, 110 (2010)
18. Strobl, C., Boulesteix, A.-L., Kneib, T., Augustin, T., Zeileis, A.: Conditional variable importance for random forests. BMC Bioinform. **9**(307), (2008). doi:10.1186/1471-2105-9-307
19. Paul, J., Dupont, P.: Inferring statistically significant features from random forests. Neurocomputing **150**, 471–480 (2015)
20. Huynh-Thu, V.A.A., Saeys, Y., Wehenkel, L., Geurts, P.: Statistical interpretation of machine learning-based feature importance scores for biomarker discovery. Bioinformatics **28**(13), 1766–1774 (2012)
21. Dramiński, M., Koronacki, J., Komorowski, J.: A study on Monte Carlo Gene screening. In: Intelligent Information Processing and Web Mining, pp. 349–356. Springer (2005)
22. Dramiński, M., Rada Iglesias, A., Enroth, S., Wadelius, C., Koronacki, J., Komorowski, J.: Monte Carlo feature selection for supervised classification. Bioinformatics **24**(1), 110–117 (2008)

23. Dramiński, M., Kierczak, M., Nowak-Brzezińska, A., Koronacki, J.: The Monte Carlo feature selection and interdependency discovery is practically unbiased. Control Cybern. **40**(2), 199–211 (2011)
24. Dramiński, M., Kierczak, M., Koronacki, J. and Komorowski, J.: Monte Carlo feature selection and interdependency discovery in supervised classification. In: Advances in Machine Learning, vol. 2, pp. 371–385. Springer (2010)
25. Kierczak, M., Ginalski, K., Dramiński, M., Koronacki, J., Rudnicki, W., Komorowski, J.: A rough set-based model of HIV-1 RT Resistome. Bioinformatics a. Biol. Insights **3**, 109–127 (2009)
26. Kierczak, M., Dramiński, M., Koronacki, J., Komorowski, J.: Computational analysis of local molecular interaction networks underlying change of HIV-1 resistance to selected reverse transcriptase inhibitors. Bioinformatics a. Biol. Insights **4**, 137–146 (2010)
27. Bornelöv, S., Marillet, S., Komorowski, J.: Ciruvis: a web-based tool for rule networks and interaction detection using rule-based classifiers. BMC Bioinform. **15**, 139 (2014)
28. Hvidsten, T.R., Wilczyński, B., Kryshtafovych, A., Tiuryn, J., Komorowski, J., Fidelis, K.: Discovering regulatory binding-site modules using rule-based learning. Genome Res. **15**(6), 856–866 (2005)
29. Ho, T.K.: The random subspace method for constructing decision forests. IEEE Trans. Pattern Analysis Mach. Intell. **20**(8), 832–844 (1998)
30. Gyenesei, A., Wagner, U., Barkow-Oesterreicher, S., Stolte, E., Schlapbach, R.: Mining co-regulated gene profiles for the detection of functional associations in gene expression data. Bioinformatics **23**(15), 1927–1935 (2007)
31. Hastie, T., Tibshirani, R., Botstein, D., Brown, P.: Supervised harvesting of expression trees. Genome Biol. **2**(1), research0003.1-0003.12 (2001)
32. Smyth, G.K., Yang, Y.H., Speed, T.: Statistical issues in cDNA microarray data analysis. In: Brownstein, M.J., Khodursky, A.B. (eds.) Functional Genomics: Methods and Protocols. Methods in Molecular Biology, vol. 224, pp. 111–136. Humana Press (2003)
33. Pawlak, Z.: Information systems: theoretical foundations. Inform. Syst. **6**(3), 205–218 (1981)
34. Krzywinski, M., Schein, J., Birol, İ., Connors, J., Gascoyne, R., Horsman, D., Jones, S.J., Marra, M.A.: Circos: an information aesthetic for comparative genomics. Genome Res. **19**(9), 1639–1645 (2009)
35. Ye, C.J., et al.: Intersection of population variation and autoimmunity genetics in human T cell activation. Science **345**(6202), 1254665 (2014)
36. Ames, R.S., et al.: Human urotensin-II is a potent vasoconstrictor and agonist for the orphan receptor GPR14. Nature **401**(6750), 282–6 (1999). doi:10.1038/45809
37. Lehner, U., et al.: Ligands and signaling of the G-protein-coupled receptor GPR14, expressed in human kidney cells. Cell. Physiol. Biochem. **20**(1–4), 181–192 (2007)
38. Ciruvis CD4+example. http://bioinf.icm.uu.se/~ciruvis/results/result_format_rules_TOXhXJ18/ (2014)

Final Remarks on Big Data Analysis and Its Impact on Society and Science

Jerzy Stefanowski and Nathalie Japkowicz

Abstract In this chapter, we summarize the lessons learned from the contributions to this book, add some of the important points regarding the current state of the art in Big Data Analysis that have not been discussed at length in the contributions per se, but are worth being aware of, and conclude with a discussion of the influence that Stan Matwin has had throughout the years on the successive related fields of Machine Learning, Data Mining and Big Data Analysis.

1 Introduction

Big Data is one of the most popular phrases in the current computer science literature. Researchers, specialists working on various applications, philosophers of science and journalists argue that we are living in a new era of the information technology, where Big Data analysis will play a critical role and may change our lives and society.

The rapid development of computer and electronic technology facilitates collecting and processing huge amounts of data. It should be noticed that standard data bases and business transaction systems are not the only, or even the main, sources of such data. Nowadays more and more Big Data is acquired from the Internet, and in particular from social media and other services tracking users' activities. Manufacturing systems, smart meters, sensor and network applications also produce massive volumes of data about business processes, technical conditions of device components, etc. In scientific or engineering tasks, data may have an even more complex structure than the typical business data considered in standard information systems. Big Data manifests itself further in healthcare and medical information systems, which are also a rapidly growing area which includes quite large, complex and heterogeneous data

J. Stefanowski (✉)
Institute of Computing Sciences, Poznań University of Technology, Poznań, Poland
e-mail: Jerzy.Stefanowski@cs.put.poznan.pl

N. Japkowicz
School of Electrical Engineering & Computer Science,
University of Ottawa, Ottawa, Canada
e-mail: nat@site.uottawa.ca

© Springer International Publishing Switzerland 2016
N. Japkowicz and J. Stefanowski (eds.), *Big Data Analysis: New Algorithms for a New Society*, Studies in Big Data 16, DOI 10.1007/978-3-319-26989-4_13

repositories about patients and their treatment. Furthermore, new smart-phones and other mobile devices offer multiple options for data acquisition as well as a variety of data formats, e.g., geographical localization of the phone, records of audio, photographs or other multimedia by internal cameras, the recording of human gestures or movements via accelerators and sensors [37]. This leads toward the development of new software systems, which need to operate efficiently with terabytes of data, often in real time and with tight demands for memory or other resources.

With these new data sources, the machines or measurement devices continuously produce data. Then, the data are processed and analysed by algorithms making human inspection very limited as compared to its role in traditional data analysis. While mining such data may allow us to discover new types of knowledge about people or events, it also opens the gate to new dangers or risks, such as privacy breeches, data protection failures, or ethical issues related to the use of automatic decisions made on human lives or others. These questions were considered that deeply before the advent of Big Data.

The term Big Data does not refer to the massive size or volume only. In the introduction to this book we have briefly surveyed popular definitions of Big Data and stressed the role of other properties—which are often called the "many V's" (besides volume, we also have velocity, variety, veracity, value and variability among others). Therefore, the term "Big Data" encompasses the presence of many heterogeneous data representations, various data sources linked together, complex structures as well as the need to deal with high speed of arrival, processing time requirements, evolving data characteristics, and uncertainty of the data elements. Larger, complex or non-static and, generally speaking, "more difficult" data sets pose new challenges for researchers and call for a variety of new dedicated approaches.

Changes resulting from Big Data are also visible in the new kinds of applications being considered. For example, some researchers attempt to predict an epidemic outbreak based on analyzing web search queries or users' tweets mentioning special-related keywords (see, e.g., a case study of flu prediction [18, 29]). Some other companies identify financial trends by linking many various data sources (in addition to the contents of data bases, social networks, tweets, logs of other Web users' activities and sentiments of their opinions are integrated together and explored [45]); others look for patterns of human mobility by analysing the records of mobile phone calls [16] (note that combining mobile data with additional data may, not only support controlling transport systems, but can also be used by police to predict crime sub-areas for patrols [49]); still others try to optimize the maintenance of city infrastructure or predict dangerous failures of technical devices [40]; some support preparing multi-dimensional astronomical maps by generalizing the results of exploring a huge collection of images covering the sky (see the Sloan Sky Digital Survey [3]); some apply sophisticated, powerful computer technology and natural language algorithms to understand queries and imitate human answers (see experiences with question answering systems and IBM Watson [28]). Finally, e-commerce companies often analyze customers' purchases and track their behavior in order to make more accurate product recommendations, and so on.

We refer the reader to popular books such as [36] to learn about additional successful applications of Big Data Analysis in medicine, natural sciences, engineering and many other fields. These and other case studies described by the authors of this edited book clearly show that Big Data algorithms already have an impact on our daily life and may affect society even more deeply in the near and more distant future.

Machine learning and data mining are among the core methods used in Big Data Analysis. In the introductory chapter of this book we have shown that machine learning researchers had already faced some issues related to the mining of massive and complex data. However, we have also identified several differences between earlier machine learning and present Big Data characteristics of the basic issues (see Sect. 1.2 of chapter "A Machine Learning Perspective on Big Data Analysis" in this book). Besides requirements of computational efficiency, Big Data has opened new research problems, which have never been considered, or considered only in a limited range before. We should also note that the new applications and research challenges can be viewed as multi-disciplinary problems that should be handled by teams of researchers coming from different fields such as databases, data mining, machine learning, statistics, pattern recognition and distributed / parallel computations.

In view of the above considerations and in honour of Stan Matwin who has made a significant contribution to the field over the years, we have decided to prepare this special edited volume. We have invited several well known researchers coming from machine learning and other related disciplines to present their views on how studying Big Data affects the research in their field, to discuss the most interesting new research directions that emerged from their work and to express their opinions and the lessons they have gathered from their experience. Furthermore, we have encouraged them to discuss the impact of Big Data on society as well as the possible dangers or risks that such research could cause.

In the next section we summarize the main problems raised by our authors and describe some of the lessons learned from the case studies discussed in their chapters. Then, in Sect. 3, we briefly survey some other Big Data opportunities and challenges that were not considered in great detail by our authors, but in our opinion, are important to consider given their influence on Big Data analysis research and on society in general. The final section of this chapter concludes the book by presenting a short summary of Stan Matwin's influence on the field of Big Data Analytics.

2 Lessons on Big Data Opportunities and Challenges

In this section, we will describe several issues or problems raised by our authors and conclude with some promising research directions they pointed out. Please note that this section is organized by themes which are conveyed by our subsection titles. Many of our authors' contributions raise several of these themes simultaneously and will, consequently, be discussed in several of our subsections.

2.1 Relation Between Traditional Data Analysis and Big Data Research Paradigms

Many popular texts on Big Data talk about changes in the way scientific research will increasingly be carried out. They suggest that sophisticated statistical models will be replaced by more data intensive methods applied to huge amounts of data. Similarly, controversial opinions are expressed concerning claims that discovering correlations in big data sets reduces the needs for discovering causality and attempts at understanding the data. The question of whether Big Data offers a new less theoretical methodology has also been raised in a few chapters of our book.

R. Sparks, A. Ickowicz and H. Lenz discuss such questions in chapter "An Insight on Big Data Analytics", within the historical context of modern science and in particular, statistical analysis. They overview the historical traditions of statistics, which was often based on constructing models to better understand the data. However, they also explain that statisticians use empirical models to approximate "real data models", and integrate this with mathematical theory to understand processes and construct knowledge. Although traditional statistical methods are based on strong theoretical assumptions and intuitive models—due to the small size of data samples and the strict requirements imposed on them—statisticians recognize that some of these theoretical frameworks are too restrictive. They say that such a realization is a useful step in the right direction for finding new ideas to solve problems without making unrealistic assumptions. On the one hand, R. Sparks, A. Ickowicz and H. Lenz cite works of philosophers of science, such as I. Kant or C. Popper that suggest it is impossible "to start with pure observations alone, without anything in the nature of a theory". On the other hand, they also conclude that "good models shape the data in trying to best fit it, and that the data also shapes the model in that it helps to use models with the appropriate assumptions".

In their chapter R. Sparks, A. Ickowicz and H. Lenz also nicely describe the tension between the traditional statistics and data mining communities. Data mining is usually seen as a methodology that favours data-exploration at the expense of theory. Moreover, in the authors' opinion, non-statistically trained data-miners quite often drop theoretical considerations and test a lot of methods. This, they believe, is an unsound approach. The authors thus disagree that Big Data is going to drive knowledge in the complete absence of a theoretical framework or models.

They also postulate that data-miners and statisticians should collaborate more closely in mining big data sets and in generating knowledge within a sound theoretical framework. They believe that statisticians should stop making unrealistic assumptions that remain unchecked, and that data-miners should work with statisticians in helping discover sound knowledge that will help manage the future. They also ask the question of whether data mining methods may help construct an appropriate empirical model. They argue that statisticians need to be more pragmatic and nicely refer to Breiman's paper on two statistical cultures [4], which discusses a kind of shift from model driven approaches to algorithm modeling-based approaches. Finally, they list examples of successful non-parametric methods, such as ensembles or Bayesian approaches, that have been developed by statisticians in the last few decades and represent this new methodological paradigm, and are, simultaneously, of particular interest to data miners.

An interesting integration of statistical and machine learning approaches is presented in chapter "Discovering Networks of Interdependent Features in High- Dimensional Problems" by M. Draminski, M. Dabrowski, K. Diamanti, J. Koronacki and J.Komorowski, where methods for feature selection taking into consideration feature interdependencies in genomic data are developed.

Finally, we direct the reader to Sect. 2 in chapter "A Machine Learning Perspective on Big Data Analysis" of our introductory chapter where we survey the literature on the above issues and present different arguments for and against the claim that a Big Data revolution is underway. In particular, regarding the role of statistics in Big Data Analysis, we summarize, in Sect. 2 of chapter "A Machine Learning Perspective on Big Data Analysis", the arguments of several researchers concerning sample and selection biases that cannot be eliminated, illusions of working with the complete population, unknowns in the data, needs to understand causality, and the fact that correlations are not always sufficient to take actions in the real world.

2.2 The Need to Develop New Methodological Frameworks for Complex Problems

The call for the development of new methodological frameworks in the context of distributed data sets and complex interaction systems is expressed by A. Skowron, A. Jankowski and S. Dutta in chapter "Toward Problem Solving Support Based on Big Data and Domain Knowledge: Interactive Granular Computing and Adaptive Judgement". These authors notice that big data sets are often distributed and their parts are linked together. Moreover, computations are performed on quite complex objects and often affected by uncertainty. They argue that such computations are distributed over networks of agents involved in complex interactions. Agents perform computations on complex objects of very different natures, e.g., (behavioral) patterns, classifiers, clusters, structural objects, sets of rules, aggregation operations, reasoning schemes, etc. Moreover, implementing the process of mining such complex data sets in distributed networks is related to modeling the complex analytical systems with some basic high level primitives for composing and building complex analytical pipelines over Big Data. Such primitives are very often expressed in natural language, and they should be approximated using other low-level primitives, accessible from raw data or from domain expert knowledge.

To model such complex systems at the higher level, A. Skowron, A. Jankowski and S. Dutta propose to exploit the paradigm of Granular Computing. Granulation of information should be considered when precision of information is too costly and not very meaningful in modeling and controlling complex systems. Moreover, data are incomplete, uncertain, and vague. The granular computing paradigm is based on soft computing, such as fuzzy or rough sets theory. Although basics of Granular Computing have already been proposed, also by A. Skowron in his earlier papers [44], here authors extend it by introducing new complex granules. They also show how to build such granules using data and approximating expert's ontologies, in particular for hierarchical and multi-domain approaches. Moreover, they present a new unified methodology for modeling and controlling computations with these complex granules in case of an interaction between agents. They argue that it could support users in solving problems of Big Data, as granules may represent computational building blocks for approximating (or inducing models of) the high-level primitives used by researchers to compose complex analytical pipelines over Big Data.

2.3   Dynamic and Evolving Data

Data streams are one of the most challenging forms of Big Data, in particular if data evolve over time. In supervised machine learning, such unexpected changes in the underlying data distribution over time are referred to as concept drift. Such changes deteriorate the predictive accuracy of classifiers learned from past examples and they require new learning algorithms that could detect and adapt to concept drifts.

I. Zliobaite, M. Pechenizkiy and J. Gama provide an application-oriented overview of research in this field. The original contribution of their chapter "An Overview of Concept Drift Applications" includes a detailed survey of concept drift handling methods and focuses the reader's attention to the new research tasks driven by the typical categories of applications found in the context of data streams.

First, the authors overview and categorize application tasks for which the problem of concept drift is particularly relevant. Then, they introduce a special reference framework for describing application-oriented tasks in a systematic way. Their original proposal for the framework includes three main components:

1. The main properties of the application tasks with concept drift (data and learning tasks, characterization of changes and operational setting for availability of the ground truth as class labels).
2. A categorization of application areas and tasks based on those properties (they distinguish mainly between monitoring and control, information management, and analytics and diagnostics applications).
3. Links between tasks and applications.

The authors noticed that their categories of applications differ in terms of the data types they use. Monitoring and control applications typically use streaming sensory data as inputs, and concept drifts typically happen fast and suddenly. Information management applications work with documents and concept drifts happen more slowly than in the previous case. Diagnostic applications typically use time-stamped observations and concept drifts are even slower—typically incremental, or evolving. Then, they survey application-oriented published works on adaptive learning and focus on a few application examples that represent different types of tasks. Using these examples, they illustrate how the prediction task is formulated, and how concept drifts are handled.

The other interesting lesson from this chapter is related to the implications of evolving data on current research in data mining. Although the problem of concept drift has been recognized as an interesting one, the current state of research is still in an early stage and the many proposals that have been formulated were examined in artificial and theoretical settings. Indeed, these approaches have been tested primarily on simulated data or real data with simulated drifts. Assumptions behind expected type of changes and reasons for the changes are not always precisely explained and studied. I. Zliobaite, M. Pechenizkiy and J. Gama postulate that more studies should highlight the peculiarities of particular applications and give intuition and/or empirical evidence as to why traditional general-purpose concept drift handling techniques are not expected to perform well. They also suggest more research on specialized techniques suitable for a particular application type in the real world context.

Their other lessons from several real-life projects are the following: seasonal effects with vague periodicity for a certain subgroup of object occur in some problems, external contextual information (which could be available) or extraction of hidden contexts from the predictive features may help handle recurrent concept drift better, mining temporal relationships can be used to identify related drifts, domain experts should play an important role in acceptance of Big Data solutions. These experts will slowly move from non interpretable black-box models towards control systems that support an understanding of how these changes are detected and what adaptation would happen.

They expect changes in research on concept drift and hope that these changes will be helpful in improving utility, usability and trust in the adaptive learning systems being developed for many of the Big Data applications.

R. Sarmento, M. Oliveira, M. Cordeiro, S. Tabassum and J. Gama also consider the analysis of real-time streaming data in chapter "Social Network Analysis in Streaming Call Graphs". They discuss the challenges encountered in the analysis and visualization of the network data collected by mobile network operators. As the conventional data analysis performed by telecom operators is slow and implies heavy costs in data warehouses, the authors have modeled these time changing graphs as a data stream. This modeling combined with a special sampling has helped the visualization of mobile graphs. To sum up, this chapter nicely illustrates the authors' research in network sampling, visualization of streaming social networks, stream analysis and online exploratory data analysis.

Postulates for developing new types of adaptive learning algorithms that should lead to incremental models from asynchronous streams coming from financial application are also discussed by E. Paquet, H. Viktor and H. Guo in chapter "Data Mining in Finance: Current Advances and Future Challenges". Finally, M. Shah, in chapter "Big Data and the Internet of Things" also argues that the speed and scale at which the smart devices of the Internet of Things produce data require new streaming algorithms.

2.4 Information Network Analysis

Nowadays many of Big Data applications come from the Web, social media, or more generally, from networks. As such, these applications are connected with the analysis of information networks. In chapter "Analysis of Text-Enriched Heterogeneous Information Networks", J. Kral, A. Valmarska, M. Grcar, M. Robnik-Sikonja and N. Lavrac present a brief history of this research starting from sociologists like Zahary and progress toward the current state-of-the-art in network analysis. Issues of graph mining and social network analysis were also surveyed in the introductory chapter of this book.

J. Kral, A. Valmarska, M. Grcar, M. Robnik-Sikonja and N. Lavrac focus the readers attention on heterogeneous information networks [46]. These types of networks describe heterogeneous types of entities and different types of relations. Moreover, in enriched heterogeneous information networks, nodes of certain types contain additional information. The authors introduce us to this newer research area. They also claim that the methods that take the heterogeneous nature of the networks into account are capable of solving tasks that cannot be defined on homogeneous information networks (like clustering two disjoint sets of entities). They show how to merge the network analysis with the analysis of other data formats, either in the form of text documents or results obtained from various past experiments. The novel contribution of the authors chapter is to present a method for mining text-enriched heterogeneous information networks, which combine the information stored in a heterogeneous network with textual data. Unlike the related approaches, the new method combines two separate sources (network structure and text) and joins them into a single representation.

Their chapter also includes two case studies illustrating this method. The results obtained on the VideoLectures.NET data show that using this method increases classification accuracy as compared to using only texts or only structural information about the instances. Moreover, the results obtained by their other study on psychology paper bibliographies show that the relational information hidden in the network structure is particularly useful.

Chapter "Social Network Analysis in Streaming Call Graphs" by R. Sarmento, M. Oliveira, M. Cordeiro, S. Tabassum and J. Gama describes a real life case study of telecommunication data transformed into graphs, where nodes represent subscribers and edges represent the phone calls. The authors discuss which aspects of the analysis of the social networks underlying call graphs can deliver valuable business insights to mobile telecom operators (e.g., topological aspects of the networks in terms of degree distribution, average path length, clustering, connected components and finding the key nodes in the network based on the position they occupy in the network structure, community detection, etc.). They also show other challenges pertaining to the network data collected by mobile network operators: data are more complex, they are continuously generated by the communication activity among subscribers, and in addition to the large volume, this data arrives at high rates. Therefore, the authors point out requirements for developing new methods that should be able to cope with data speed and volume and operate under the one-pass paradigm. This led them to model these call graphs as data streams, generate specialized sampling for them, as well as new techniques for visualization which were discussed in the previous subsection.

Combinations of various data types are also discussed in chapter "Industrial-Scale AdHoc Risk Analytics Using MapReduce" by E. Paquet, H. Viktor and H. Guo where the advantages of combing financial data analysis with non-traditional data (social, tweets, etc.) and its impact for trend predictions are considered.

2.5 Mining Sensing Data and Exploiting the Internet of Things

Advances in sensing and information technologies are making it possible to embed increasing computing power in small devices and open up new opportunities for both collecting and processing large–scale data. Combined with advances in communication, this results in a system of highly interconnected devices referred to as the Internet of Things. It is claimed that the Internet of Things will be a growing source contributing to Big Data Analysis in the nearest future [37].

In order to mine sensing data, the existing data mining techniques have to be adapted to dealing with constraints in resources and to performing an analysis in real–time. The underlying focus of ubiquitous systems is to perform computationally intensive analysis techniques on mobile device environments that are constrained by limited computational resources and varying network characteristics. Furthermore, it becomes necessary to perform synthesis and knowledge integration from multiple data streams in a resource constrained environment.

These problems are discussed in chapter "Big Data and the Internet of Things", where M. Shah presents several important aspects of the intersection of Big Data analytics and the Internet of Things. The brief review of the connectivity, communication and data acquisition issues is not the main focus of the chapter. Instead, the author focuses on the novel opportunities and challenges that the new world of interconnected devices offer, along with some advancements that are being made on various fronts to realize them.

In this chapter, M. Shah discusses how Big Data technologies and the Internet of Things are playing a transformative role in society. In his view, the ubiquitous nature of such technologies will profoundly change the world as we know it, just as the industrial revolution and the Internet did in the past. He expects that they will change the context in which predictive analytics is performed in many application problems (examples of real-time diagnostics of air-engines and electrical turbines are considered to illustrate this statement). The author predicts that some devices will take corrective actions, thus making themselves self-aware and self-maintaining.

Other lessons from M. Shah's chapter include recommendations for business organizations or companies developing applications at the intersection of Big Data and the Internet of things. Besides several computational and technological needs, he postulates that it will also become necessary to facilitate interfacing between engineering or domain experts and data scientists for efficient and productive knowledge transfer, agreed-upon validation, as well as adoption and integration mechanisms for analytics.

Finally, he argues that researchers and company have to pay more attention to societal aspects of the new technologies. He lists the following areas that need to receive more attention and efforts than they currently do: privacy, security, and interpretability of models and data quality issues. Other challenges pertain to issues resulting from the difficulties associated with the adoption of data mining analytics in various domains, e.g. the limitations of model validation and testing, the integration of the Internet of things devices with the human physical understanding of the world, the risks of systemic errors and failures.

2.6 The Need to Deal with Heterogeneous Representations, Vagueness and Unknown Data

Variety as it refers to heterogeneous data is one of the essential properties of Big Data. These different data forms are usually greatly interconnected, interrelated and may also be inconsistently represented which creates challenges for their integration and cleaning. Heterogeneity also forces analysts to deal with structured, semi-structured and unstructured data simultaneously, which is another difficult task to approach when using standard knowledge discovery tools.

These issues are discussed in a few chapters of this book. J. Kral, A. Valmarska, M. Grcar, M. Robnik-Sikonja and N. Lavrac in chapter "Analysis of Text-Enriched Heterogeneous Information Networks" present a new method, which combines structural information separately calculated from homogeneous networks with the text vector representation (obtained from textual information contained in network nodes). Their case study illustrates that combining these two different heterogeneous representations is feasible and is more powerful than standard methods that handle them independently.

Dealing with heterogeneous data is a major challenge in the integration phase of knowledge discovery. M. Shah in chapter "Big Data and the Internet of Things" describes the current efforts to standardize data protocols for data exchanges between various measurement devices and computer systems. However, he warns readers that the high resolution and temporal nature of such data makes it difficult to align multiple sources as well as devise strategies to learn from them in conjunction with static data sources. The protocols for obtaining the quantities from different measurement systems are still not uniform or standardized even within a given domain. The data integration becomes more difficult since it requires the transformation of such derived quantities and the solving of many conflict situations.

Moreover, in chapter "An Insight on Big Data Analytics", R. Sparks, A. Ickowicz and H. Lenz discuss the usefulness of statistical tools for integrating and reducing large data sets. The complexity of basic data elements, their vague description and several problems of using imprecise natural language are also mentioned in A. Skowron, A. Jankowski and S. Dutta in chapter "Toward Problem Solving Support Based on Big Data and Domain Knowledge: Interactive Granular Computing and Adaptive Judgement", where they postulate the development of new data mining methods for dealing with such data.

Similarly, E. Paquet, H. Viktor and H. Guo consider unknowns (data, parameters, etc.) associated with financial data. The authors show how analyzing and understanding which attributes and parameters are not known is crucial in order to create accurate and meaningful predictions.

2.7 Process of Knowledge Discovery from Data

Although Big Data projects may concern various data sets and involve quite different techniques, some of our authors suggest that more investigations into the systematic process approach to discovering knowledge and deploying final models are needed. Recall that in the practice of Knowledge Discovery from Data, such a way of thinking has resulted in useful standards, such as the CRISP-DM model [6]. This is also a leitmotif in chapter "Implementing Big Data Analytics Projects in Business" of F. Fogelman-Soulie and W. Lu, where the opportunities created by Big Data analytics for companies and the challenges associated with the practical implementation of such projects are discussed. In their view, the process of implementing Big Data Analysis projects in companies includes a number of stages that were inferred from earlier data mining projects, however, they believe that more efforts need to be put into integrating, cleaning and pre-processing the data.

They also claim that appropriate feature engineering is very meaningful for the business domain since such data sets are often high dimensional. Reports from various business or industrial projects show that working with at least 1,000 features is common, but some projects may generate even more features. However, the feature engineering stage is a very difficult step to perform given that it requires lots of data, large computation time and more complex models.

In Big Data Analysis problems, some additional features can be obtained from outside data sources, such as open data sources or private data obtained from partners or data providers. These new data may bring additional value. However, as they are of different formats and semantics—a problem reflected in the Variety of data issue—they need careful realizations of many transformation steps in pre-processing. Compared to earlier machine learning applications, these steps require new models and software tools. F. Fogelman-Soulié and W. Lu review some open-source tools in the section entitled "Architectures for Big Data" in their chapter. These authors nicely illustrate how feature engineering, especially with different semantics, can increase the performance of the final model by describing a real project of credit-card fraud detection on the Internet.

M. Draminski, M. Dabrowski, Kl. Diamanti, J. Koronacki and J. Komorowski also consider in chapter "Discovering Networks of Interdependent Features in High-Dimensic Problems" new methods for the identification of the most important and independent features in bio-informatics data. They argue that higher numbers of relevant features may be more challenging to obtain than increasing the number of observations.

An additional issue raised by F. Fogelman-Soulié and W. Lu in other parts of their chapter is that choosing appropriate learning algorithms from the many existing ones is not a trivial task. Like other researchers before them, they suggest that a practical lesson drawn from recent Big Data projects is that simple models with lots of data could perform better than complex models on less data. They propose an incremental strategy where the analyst should choose a relatively simple algorithm and work with increasing data volumes with feature engineering. Simpler algorithms are also easier to explain than more complex ones, so sometimes, domain experts or users will prefer simpler models to more accurate algorithms such as ensembles, due to their better interpretability. Finally, they warn readers of the importance and difficulty of choosing appropriate procedures for evaluating learning algorithms, in particular if bigger data are divided into smaller samples or when data sets are progressively increased (either by adding observation, or features).

Quite similar practical observations on the interpretability and evaluation of proposed models can be found in M. Shah's chapter—see the previous Sect. 2.5 in chapter "Big Data and the Internet of Things".

Finally, I. Zliobaite, M. Pechenizkiy and J. Gama present yet another process approach in chapter "An Overview of Concept Drift Applications". They start by discussing the classical model of the data mining process (the CRISP-DM standard), where the life cycle of a data mining project spans over six phases: business understanding, data understanding, data preparation, modeling, evaluation and deployment. As this model assumes that most of the data mining steps are performed offline, it is not appropriate for data streams. Therefore, they generalize it to the streaming settings, where concept drifts and changes of models are expected. The main differences between their proposed model and the standard process is that the data preparation, mining, and evaluation steps are completely automated, there is no manual data exploration, and there is an automated monitoring of performance, including change detection and alert services.

2.8 Architectural Support for the Efficient Mining of Big Data

Big Data requires new technologies to efficiently process huge amounts of data within a tolerable time. Standard storage disk systems may be too slow and limited for new tasks. Therefore, new storage infrastructures, suitable for parallel processing nodes have recently been developed [43]. Other technologies commonly applied to Big Data include massively parallel and distributed processing. Real, or near-real, time information processing and delivery of results is one of the requirements for Big Data Analytics in many applications. Massive, evolving and complex data characteristics lead to the development of new scalable algorithms allowing for data processing and analyzing. New architectures (software or hardware) for the efficient management of complex and dynamic data streams and their analysis (sometimes in an approximate way) are required as well.

Many of the chapters in this book consider these issues. They note that standard relational database management may be insufficient for the storage and management of big data sets. For instance, F. Fogelman-Soulié and W. Lu refer to efforts by many companies to integrate various data repositories into data warehouses. They discuss the difficulties and cost of constructing ETL models (i.e. Extraction, Transformation, Load of data into data warehouses) and their implementations in financial companies. However, considerations of heterogeneous representations, dynamic, constantly emerging data sources and other characteristics of Big Data have led them to conclusions that fixed static and structured data warehouse models are not adequate. To cope with these limitations they propose to use a new architecture, called "Data Lake" which is a special repository of all the data collected by an organization, where the data is stored in its original raw form. Because no a-priori structure or data model is imposed at collection time, all further usage should be possible without having to modify a pre-existing model.

Furthermore, M. Shah, in Sect. 2 of chapter "Big Data and the Internet of Things" surveys the problems of data integration and management in the context of mining data from mobile and sensing devices. He also explains why classical relational databases are no longer sufficient for dealing with such diversified data sources. NoSQL databases are the answer to these limitations. He advocates the use of columnar data stores such as BigTable, Cassandra, Hypertable, HBase (inspired by the BigTable); key-value and document databases such as MongoDB, Couchbase server, Dynamo and Cassandra; stream data stores such as Eventstore; graph based data- stores such as Neo4j and so on. A slightly more comprehensive description of these systems is also available in our introductory chapter.

The next issue concerns processing platforms. F. Fogelman-Soulié and W. Lu present a nice historical discussion of the tradeoff between traditional big servers (with a scaling-up mechanism) and clusters of less costly simpler machines.

Other authors of this book refer to the Hadoop distributed file system and MapReduce as solutions for running large-scale distributed Big Data processing applications. M. Szczerba, M. Wiewiórka, M. Okoniewski and H. Rybiński present an overview of cloud-based Big Data analytic tools that are currently used and developed for genomic data analysis and that are based on tools coming from the Hadoop system. M. Shah briefly discusses Hadoop relevance to dealing with data coming from the Internet of Things.

An interesting example of programming in the MapReduce framework is described in chapter "Industrial-Scale Ad Hoc Risk Analytics Using MapReduce" by A. Rau-

Chaplin, Z. Yao, and N. Zeh in the context of performing large- scale Monte Carlo simulations to approximate the portfolio-level in their risk analysis system.

Several other authors also notice the limitations of distributed systems such as Hadoop, with respect to time delays in performing analytics. Many machine learning algorithms require multiple passes on the data that are too costly in terms of communication with the underlying system. They direct our attention to newer frameworks such as Spark that were developed to address these issues and are more suitable for intensive machine learning and data mining scenarios (see, e.g., Sect. 3 of the chapter "Scalable Cloud-Based Data Analysis Software Systems for Big Data from Next Generation Sequencing" by M. Szczerba, M. Wiewiórka, M. Okoniewski and H. Rybiński). Then, F. Fogelman-Soulié and W. Lu describe the use of Spark tools inside the idea of a Big Data platform (see Sect. 5 of chapter "Implementing Big Data Analytics Projects in Business").

2.9   Domain-Specific Cases of Big Data Analysis

The chapters of this book also include the description of several Big Data Analysis applications to various problems. The three dominant application areas considered by the authors are life science (mainly biomedicine and genomics), business (mainly finance) and technology.

Life Science

M Szczerba, M. Wiewiórka, M. Okoniewski and H. Rybiński discuss in chapter "Scalable Cloud-Based Data Analysis Software Systems for Big Data from Next Generation Sequencing" problems of mining sequenced data coming from various molecular biology laboratory technologies (e.g., applications pertaining to DNA genotyping, RNA expression profiling, genome methylation searches, and many others). Due to the decreasing costs of the sequencing machines, the amount of collected biological data has significantly increased. The next generation of sequencing technology should consequently contribute much more to Big Data and will influence new diagnostics in medicine. The results of analyzing genomic data can be used in many stages of diagnosing and treatment procedures, especially for personalized medicine, as well as for constructing new functional knowledge bases. However, it causes challenges for efficient storage and data analysis. Discussing these challenges and dedicated software and architectural solutions are the main contributions of their paper. First, the authors present a very interesting overview of Big Data analytic cloud tools that are currently used, tested or are adapted for genomic data analysis. They describe examples of tools developed on the basis of Hadoop and Spark platforms. Moreover, their chapter gives a detailed case study of a special tool, called SparkSeq. It is the dedicated genomic big data processing system, which has already been applied in a number of biological sequencing analysis projects. Perspectives for similar system applications in biology and medicine are also discussed. The final sections of this chapter includes the authors view on the next generation sequencing big data architectures and open problems of developing new scalable software tools for bioinformatics.

Genomic applications are also considered in chapter "Discovering Networks of Interdependent Features in High-Dimensional Problems" by M. Draminski, M. Dabrowski, K. Diamanti, J. Koronacki and J. Komorowski. Their new methodology for selecting features and discovering their interactions is validated on a large, fairly complex real data set concerning gene expression levels in some human cells. The authors showed that their Monte-Carlo Feature Selection MCFS-ID algorithm returned a limited number of highly informative features, which could also support learning accurate classifiers. They also showed the usefulness of their other method for constructing Inter Dependent Graphs (for detecting strong interactions between features, and using a special approach to analysing rules discovered from data) on the same kind of the gene expression data set. These graphs and underlying rules provide experts with a refined view of biological results and support their interpretations. To sum up, this chapter shows that new methods for feature engineering are necessary in Life Science (where data sets are often highly dimensional) and the combination of such methods with the construction of graphs of interactions between features may help in understanding complex relations in bio-medical data.

Business and Financial Analysis

A few other authors considered the context of financial or more general economic problems.

For instance, A. Rau-Chaplin, Z. Yao, and N. Zeh discuss problems of risk analysis for reinsurance companies in chapter "Industrial-Scale Ad Hoc Risk Analytics Using MapReduce". They showed that typical systems for aggregate risk analysis are efficient at generating a small set of key portfolio metrics required by rating agencies and other regulatory organizations. However, these systems are not able to deal with ad hoc queries that provide a better view of the many dimensions of risks that can impact a reinsurance portfolio. To ensure better financial planning, the insurance companies need to carry out large-scale Monte Carlo simulations to estimate the probabilities of the losses incurred due to catastrophic or critical events. These more advanced risk-analysis queries and simulations require stronger computing power and are both data-intensive and time demanding. The main contributions of their chapter include: discussing new distributed and parallel solutions for such risk estimation with references to Big Data techniques, and presenting the authors' system which uses the MapReduce framework and carefully engineers data structure implementations.

Chapter Data Mining in Finance: Current Advances and Future Challenges by E. Paquet, H. Viktor, and H. Guo also addresses the issue of making predictions and building trading models for financial institutions. These authors provide a short overview of the current development of Big Data in this sector. Then, they focus on particular characteristics that occur in Big Data sets in the financial sector: unknown values and parameters, and randomness in the financial models. In their opinion, traditional data mining techniques are too limited to deal with such data characteristics. They describe stochastic predictive models for financial data, Although the major part of chapter "Big Data and the Internet of Things" by M. Shah concerns Big Data and the Internet of Things, the author also discusses many application domains impacted by Big Data analytics. He expects changes in the manufacturing sector, asset and fleet management, operations management, resource exploration, energy sector, healthcare, retail and logistics. Section 3 of chapter "Big Data and the Internet of Things" includes an illustrative case study, and a discussion of the opportunities that may arise from mining Big Data by showing its impact on organizations focusing on these domains. The next sections of this chapter are of great interest as well as they include a discussion of the necessary changes an organization is willing or capable to make in order to implement Big Data projects (see Sect. 4 in chapter "Big Data and the Internet of Things"), and the author's opinion on more general societal impact and areas of concerns (Sect. 5 of chapter "Big Data and the Internet of Things") which should be more appropriate for the high Volume and Variety of Big Data encountered in their area of application. The other part of their interesting discussion concerns the evolving aspect of financial data. These include highly fluctuating data, data arriving at a fast rate, late-arriving data, etc. (see Sect. 6 of chapter "Data Mining in Finance: Current Advances and Future Challenges").

Finally, F. Fogelman-Soulié and W. Lu illustrate their considerations with a real life project of credit-card fraud detection on the Internet, funded by the ANR (the French National Research Agency). This is an important area of applications for new data mining methods. It becomes more critical due to the increases in Internet transactions and in the activity of crime groups. The authors discuss the volume of collected transaction data, the specific limits of the recorded data items and their dynamic characteristics. The important part of their case study is to construct appropriate feature representation and to describe their experiences with building and evaluating good prediction models.

Technological Applications

Although the major part of chapter "Big Data and the Internet of Things" by M. Shah concerns Big Data and the Internet of Things, the author also discusses many application domains impacted by Big Data analytics. He expects changes in the manufacturing sector, asset and fleet management, operations management, resource exploration, energy sector, healthcare, retail and logistics. Section 3 of chapter "Big Data and the Internet of Things" includes an illustrative case study, and a discussion of the opportunities that may arise from mining Big Data by showing its impact on organizations focusing on these domains. The next sections of this chapter are of great interest as well as they include a discussion of the necessary changes an organization is willing or capable to make in order to implement Big Data projects (see Sect. 4); and the authors opinion on more general societal impact and areas of concerns (Sect. 5 of chapter "Big Data and the Internet of Things").

Finally, in chapter "Social Network Analysis in Streaming Call Graphs" R. Sarmento, M. Oliveira, M. Cordeiro, and J. Gama describe some of the problems that are encountered in the particular sector of telecommunications services. Their paper concerns the analysis of the very large and dynamic telecommunication networks graphs, looking for patterns of interactions between users. The authors also propose innovative visualization techniques and describe their implementation. Results of the analysis of such graphs provide useful insights into the social behaviors of users. These behavioral patterns provide significant gains to telecom service providers, e.g., maximizing profits by customer segmentation, profiling, churn and fraud detection etc. Apart from this, they also provide benefits to society in terms of users or subscribers.

3 Other Research Challenges of Big Data Analytics

In this section we very briefly discuss a few other issues, which have an impact on society and research.

3.1 Privacy and Ownership of Data

Privacy issues have become very important with the advent of Big Data and may have a great societal impact. Stan Matwin, as a matter of fact, is one of the first data mining researchers who have recognized this very dangerous side-effect of learning methods, warned researchers about it and looked for solutions to counter it. He came to that problem from moral and ethical concerns. In his words [33]:

> My interest in data privacy is a little different. I am concerned about the fact that modern computers may become a tool that can be used to breach and violate people's privacy easier and on a much larger scale than it was possible, say, 30 years ago. I believe that since the computer research community invented the tools that make it possible—databases, the internet, image and voice recognition, barcodes, etc.—it is then our moral obligation to at least think about tools that would make privacy easier and that would avoid many privacy-averse incidents

He has been working on developing methods that make it nearly impossible to identify a given individual in a data set [35, 53, 54].

We noticed our authors awareness of these problems as well. For instance, the reader can have a look at Sect. 8 of chapter "An Insight on Big Data Analytics" where the authors asked several important questions concerning the ownership of data sets, confidential agreements, new views on intellectual property of the data, unsolved limits of sharing data sets and integrating them from different sources. Moreover, these authors discuss various consequences of applying data mining results. M. Shah warns, in chapter "Big Data and the Internet of Things", that the current methods for privacy preserving data mining are still at a preliminary phase and that efforts to deal with that issue, to-date, have focused mainly on the data and basic analytics stage. He argues that the Internet of things applications have more specific requirements that should be properly addressed in future research.

Looking more widely in the literature, one can find more opinions saying that we still do not know how to share private data while ensuring that the data remains useful. The current techniques for maintaining privacy are too weak to allow the mining of Big Data with high quality results [9, 39]. It is believed that certain paradigms such as differential privacy reduce the information content too much to be useful in practical situations [50]. At the other extreme, as previously mentioned, data may be adequate for mining algorithms but, in such cases, privacy is not always properly considered.

Another related issue concerns the right of people to their own electronic records, and the understanding that their data is often used for analytical aims other than those they envisioned. The majority of users of on-line systems do not go beyond their basic level of data control, and they do not know what it means to share data or that their data (even web search phrases) will be linked to other data sources and mined to provide new results. Yet another ethical problem is using the results of mining personal data to predict the actions of other people.

All these and other issues open up many additional challenging problems. Some of them are more algorithm–oriented, while others are open law questions. Teen and Polonetsky call for new models balancing benefit for researchers and individual privacy rights [47]. As suggested in [38] the "foundations of data mining need to be reformulated in such a way that privacy protection and discrimination prevention are embedded in the foundations themselves, dealing with every moment in the data-knowledge life cycle, from data capture to data mining and analytics, up to the deployment of the extracted models".

3.2 Tracking the Accuracy, Trustworthiness and Provenance of the Data

As we have pointed out in the introductory chapter, the exploration of Big Data involves checking the quality of the data and its trustworthiness. Recall that some data sources produce low quality or uncertain data, see e.g. tweets, blogs, and social media. Earlier lessons of mining real data sets have clearly showed that the accuracy of the results strongly depends on the quality of the data and the appropriateness of the pre-processing. Moreover, if the final models interact with the environment and/or are applied to critical domains of human activities, then a good verification of the input data and their pre-processing as well as the deployment of data mining results all become much more crucial than in earlier information systems.

Some of the authors of this book mention these issues in the context of the process of knowledge discovery see, e.g., chapter "Big Data and the Internet of Things" by M. Shah (in Sect. 9 where he presents his concerns about the limitations of current solutions for the Internet of Things). Moreover, we have briefly described the provenance challenges for Big Data chapter "A Machine Learning Perspective on Big Data Analysis", Sect. 2.

It is important to note that more efforts should be done and new innovative approaches are needed. Some authors argue that new methods are necessary due to the complexity of Big Data. We can refer the reader to such papers as [10, 11, 19] for more information on new methods considered in the context of Big Data provenance. The authors of [22] describe approaches that attempt to track the provenance of workflows for MapReduce jobs. Recording provenance in distributed environments is also considered in [32].

Provenance also opens up additional topics for machine learning research. For instance, in the case of dynamic and changing data, the evolutionary history and the origins of data items become more complicated. The authors of [7] claim that trust measures are not static and that learning approaches could be applied to discover new measures of interesting data sources using others sources. In particular, new unsupervised methods have been proposed in [52]. Other research [51] has also shown the usefulness of semi-supervised learning methods that start with a portion of ground truth data. It was also advocated in [7] that developing new innovative methods, which can run on parallel platforms and deal with scalable data and numerous heterogeneous sources is one of the highly desired future research directions in the field.

3.3 Data Visualization and Visual Data Mining

Data analysts use visualization tools to understand the unknown structure of data and underlying patterns. Many tools have been developed for multidimensional data or more structured data. The reader is referred to [20] for their review. These authors also describe several visual data mining tools that may facilitate interactive mining based on the user's judgment of intermediate data mining results. Some of them use special methods to visualize mining results, e.g. clustering or classifiers. Interaction mechanisms for filtering, querying, and selecting data are also available. However, it is claimed that such visual exploration is too often available as a separate tool while it should be more tightly coupled with analytical methods into one knowledge discovery system.

R. Sarmento, M. Oliveira, M. Cordeiro, and J. Gama discuss the practical usefulness of visualizing large telecommunication networks in chapter "Social Network Analysis in Streaming Call Graphs". To efficiently handle very large and dynamic graphs, the authors have to model them as a kind of data stream and use special sampling techniques.

However, one could notice that many visualization methods and software tools have been developed in the context of standard, static and smaller data sets and that they are limited when it comes to exploring big data sets. The scale and complexity of Big Data may be too critical a challenge for current techniques and their implementations.

Reports like [23] list other requirements to make new visualization systems suitable for Big Data. These are:

- Enabling real-time data analysis (computationally cost-effective),
- Using in-memory compression to enable the handling of large-scale data,
- Supporting the interactive exploration of the data at different stages and the fast presentation of reports,
- Showing meaningful results (e.g., with appropriate context information and special presentation techniques to overcome the difficulties associated with too many results),
- Allowing users to share their presentations and reports with others and to collaborate in a sufficient secure way.

Then, DeGeer in [12] noticed that traditional visualization tools are too oriented toward the presentation of what a user may already know about the data. Instead, they should be exploring unknown aspects - which is more characteristic for data mining or even previously for Exploratory Data Analysis in statistics [48]. Furthermore, DeGeer presents a postulate of what a stronger visual interactivity means: the user has to be able to explore the data "on the fly", change its interests, filter out irrelevant information, deal with outliers and isolate unexpected patterns. He also notices that existing visualization tools are good for static information but that they generally fail to work with dynamic data.

Real-time visualization is particularly useful in data streams. Systems should handle a large number of very fast updates and offer innovative ideas on how to present changes in the data structure. The authors of the comprehensive survey on the topic present a similar opinion [31]. They also give an example of an open problem concerning the quick detection of breaking news events from huge amounts of streaming tweets. Following more recent papers by data stream researchers [24], the visualization of concept drifts and the graphical evaluation of model reactions to them are still open problems. Moreover, Gaber et al. claim that there are currently no on-line real-time visualization tools to complement the Ubiquitous Data Stream Mining algorithms [17]. A final postulate is to construct efficient visualization-based data discovery tools for mobile devices.

Other research reports [23] show other opportunities for applying visualization techniques to the protection of data quality (helping to find errors [1]) and supporting tracks of data provenance (graphical display of user activity records, characteristics of data sources).

3.4 User Feedback Integration and Result Interpretation

Since the beginnings of knowledge discovery from data, it has been stressed that users/decision makers should be able understand the analysis and the results of the machine learning algorithms. These postulates are also valid for many Big Data applications. For instance, [40] describes the real world successful application of data mining to predict manhole explosions and fires in the New York electrical network. Black-box (non-transparent) predictive models were treated as neither useful nor convincing. Every step of the process had to be verified by both scientists and company engineers. Therefore, the research team designed several software tools that allowed transparency of the main operations and provided reasons for the predictions made by the final system. This allowed the integration of domain expertise (by company specialists) into the modelling process, data verification, and system evaluation.

In [34], Stan Matwin pointed out that appropriate interpretation of the results may be more important than better accuracy of the models, in particular when results are used for making decisions concerning people, like medical diagnostics or administrative decisions. However, he also noted that a good interpretation is still a research challenge for the machine learning and data mining fields. A limited number of popular approaches mainly trees, rules, Bayesian networks—offer, so called, symbolic knowledge representations, which could be directly inspected and interpreted by humans. Measuring and evaluating the interpretation abilities offered by various learning algorithms is still less studied than other criteria. In his view, this question should be brought to the fore and treated in an inter-disciplinary manner. Visualization methods could partly support users in interpretation tasks.

Another issue is that, data sources may contain erroneous data, or applied algorithms may not meet all the assumptions and, as a result, may produce inaccurate results. Responsible users will not rely exclusively on computer calculations but, instead, will try to verify the results—which again should be supported by new developed techniques.

However, these expectations are real challenges for Big Data—due to data complexity, sophisticated workflow of data transformations, distributed processing, and the application of many algorithms. Similarly to studying data provenance, there is a need for capturing adequate metadata reports, and powerful visualization tools that could involve human experts into the analysis could help interpret analytical results.

This type of use for data mining systems calls for more adequate users' interaction facilities which would allow humans to provide feedback or guidance. Interactiveness has been relatively under-emphasized in the context of data mining [7]. However, it will become more important when dealing with Big Data properties, such as all "V" characteristics. For instance, user guidance can help narrow the massive data into reduced, promising sub-spaces and accelerate the processing. Users can also evaluate and interpret intermediate results, search for hypotheses directly, and repeat certain steps with different assumptions or parameters if necessary.

This means that beside designing good visualization tools, it is necessary to develop special infrastructures and carry out more advanced research on evaluation measures and validation procedures. In particular, this refers to situations where algorithms may produce too many results and where finding a limited number of interesting patterns is not an obvious task [2, 21].

4   Stan Matwin's Contributions to Big Data Analytics

Stan Matwin's contributions to Big Data Analytics are many and quite significant. They have impacted the field in many ways.

Although the issue was only briefly discussed in chapter "A Machine Learning Perspective on Big Data Analysis", the class imbalance problem has been and will remain a confounding problem for machine learning, data mining and Big Data Analysis for years to come. Matwin and his colleagues were some of the first researchers to address the issue in [25, 26]. The approach they proposed remains a popular way of solving the problem close to 20 years later. Their work also helped popularize the use of the geometric mean (G-Mean) in class imbalance problems [27]. This was important since, on the one hand, this measure is still used today and on the other hand, it was an early attempt to challenge the usefulness of accuracy as the sole criterion in all situations. This led to its gradual replacement by (or at least competition with) the AUC, Precision/Recall Curves, etc.

Another of Matwin's important contribution is in the area of Text Mining. As seen in Sect. 1 of this chapter, data will increasingly be coming from the Internet and, in particular, from Social Media. This means that text processing has been and will continue to be an extremely important area of research in Big Data Analysis. Matwin's most important contribution in this area has been in feature engineering— as discussed in Sect. 2.7 of this chapter[5]. Feature Engineering remains an important topic of research both in text mining and in biomedical applications—but he also contributed interesting results in the areas of co-training, name entity recognition, word sense recognition, etc. [30, 41, 42].

As discussed in Sect. 3.1 of this chapter, Matwin also became interested in the problem of Privacy in Data Mining long before it became a popular issue [54]. As early as 2002, he developed, together with students and colleagues, privacy-oriented Data Mining algorithms [14].

Matwin's interest in practical applications led him to work on a wide variety of problems, including predicting who in a hospital emergency room will need hospitalization, recognizing oil spills in the ocean, categorizing medical articles, detecting emerging trends in a political campaign or in public opinion. Overall, he has contributed to solving problems in such wide-ranging fields as neuro-ophthalmology, forestry, electronics, and many others.

In 2013, with this experience in hand, Matwin established the Institute for Big Data Analytics at Dalhousie University. The institute is thriving and currently includes 7 research professors (including 6, on the executive board), 3 postdoctoral fellows, 6 Ph.D. students and 8 M.Sc. students. Ongoing projects span the domains of global telecommunications services, home care, retirement living and nursing homes, Marine Ecology, Text, anesthetics and post-operative care, to name only a few. The Institute will also be hosting the prestigious Conference on Knowledge Discovery and Data Mining in 2017.

Acknowledgments The work of Jerzy Stefanowski was partially supported by the Polish National Science Center under Grant No. DEC-2013/11/B/ST6/00963. The work of Nathalie Japkowicz was supported by a Discovery Grant from the Natural Sciences and Engineering Research Council of Canada (NSERC).

References

1. ASA—Discovery with Data: Leveraging statistics and computer science to transform science and society. A report of a Working Group of the American Statistical Association (July 2, 2014)
2. Bayardo, R., Agrawal, R.: Mining the most interesting rules. In: Proceedings of the 5th ACM SIGKDD Conference on Knowledge Discovery and Data Mining, pp. 145–154 (1999)
3. Borne, K.: Scientific data mining in astronomy. In: Next Generation of Data Mining, pp. 91–114. Taylor & Francis, CRC Press (2009)
4. Breiman, L.: Statistical modeling: the two cultures. Statistical Sciences, pp. 199–231 (2001)
5. Caropreso, M., Matwin, S., Sebastiani, F.: A learner-independent evaluation of the usefulness of statistical phrases for automated text categorization. In: Text databases and document management: Theory and practice, pp. 78–102 (2001)
6. Chapman, P., Clinton, J., Kerber, R., Khabaza, T., Reinartz, T., Shearer, C., Wirth, R.: CRISP-DM 1.0 step-by-step data mining guide. Technical report, The CRISP-DM consortium (2000)
7. Che, D., Safran, M., Peng, Z.: From Big Data to Big Data mining: challenges, issues and opportunities. In: Hong B. et al. (eds) DASFAA Workshops, Springer, LNCS, vol. 7827, pp. 1–15, (2013)
8. Chen, M., Mao, S., Liu, Y.: Big data. A survey. Mob. New Appl. **19**, 171–209 (2014)
9. Crawford, K., Schultz, J.: Big data and due process: toward a framework to redress predictive privacy harms. Boston College Law Rev. **55**(1), 93–128 (2014), http://lawdigitalcommons.bc.edu/bclr/vol55/iss1/4
10. Dai, C., Lin, D., Bertino, E., Kantarcioglu, M.: An approach to evaluate data trustworthiness based on data provenance. In: Proceedings of the 5th VLDB Workshop on Secure Data Management, pp. 82–98 (2008)
11. Davidson, S., Freire, J.: Provenance and scientific workflows: challenges and opportunities. In: Proceedings of SIGMOD'08, (2008)
12. DeGeer, W.: What is Next in Big Data. Wired, 12 Feb (2014)
13. Dwork, C., Mulligan, D.: It is not privacy and it is not fair. Stanford Law Review, online 35, 3 Sept (2013)
14. Felty, A., Matwin, S.: Privacy-oriented data mining by proof checking. In: Proceedings of the 6th European Conference on Principles of Data Mining and Knowledge Discovery—PKDD 2002, Springer LNAI, pp. 138–149, (2002)
15. Gaber, M., Stahl, F., Gomes, J.: Pocket Data Mining. Big Data on Small Devices. Series: Studies in Big Data (2014)
16. Giannotti, F., Nanni, M., Pedreschi, D., Pinelli, F., Rinzivillo, S., Trasarti, R.: Unveiling the complexity of human mobility by querying and mining massive trajectory data. VLDB J. **20**(5), 695–719 (2011)
17. Gillick, B., Gaber, M., Krishnaswamy, S., Zaslavsky, A.: Visualisation of cluster dynamics and change detection in ubiquitous data stream mining. Proc. IWUC'2006, 29–38 (2006)
18. Ginsberg, J., Mohebbi, M.H., Patel, R.S., Brammer, L., Smolinski, M.S., Brilliant, L.: Detecting influenza epidemics using search engine query data. Nature **457**(7232), 1012–1014 (19 Feb 2009)
19. Glavic, B.: Big Data provenance: challenges and implications for benchmarking. In: Specifying Big Data Benchmarks, Springer, pp. 72–80, (2014)
20. Han, J., Gao, J.: Research challenges for data mining in science and engineering, In: Next Generation of Data Mining London: Chapman & Hall, pp. 1–18 (2009)
21. Hilderman, R.J., Hamilton, H.J.: Knowledge Discovery and Measures of Interest. Kluwer Academic, Boston (2002)

22. Ikeda, R., Park, H., Widom, J.: Provenance for generalized map and reduce workflows. In Proc. of CIDR, 273–283 (2011)
23. Intel White Paper: Big Data Visualization: Turning Big Data Into Big Insights—The Rise of Visualization-based Data Discovery Tools, (March 2013)
24. Krempl, G., Zliobaite, I., Brzezinski, D., Hullermeier, E., Last, M., Lemaire, V., Noack, T., Shaker, A., Sievi, S., Spiliopoulou, M., Stefanowski, J.: Open challenges for data stream mining research. ACM SIGKDD Explor. 16(1), 1–10 (2014). June
25. Kubat, M., Holte, R., Matwin, S.: Machine learning for the detection of oil spills in satellite radar images. Mach. Learn. 30(2–3), 195–215 (1998)
26. Kubat, M., Holte, R., Matwin, S.: Addressing the curse of imbalanced training sets: one-sided selection. Proc. ICML 97, 179–186 (1997)
27. Kubat, M., Holte, R., Matwin, S.: Learning when negative examples abound. In: Proc. ECML '97, pp. 146–153 (1997)
28. Lally, A., et al.: Question analysis: how Watson reads a clue. IBM J. Res. Dev. 56(3/4), (2012)
29. Lazer, D., Kennedy, R., King, G., Vespignani, A.: The parable of google flu: traps in big data analysis. Science, 343, 1203–1205 (14 March 2014)
30. Li, X., Szpakowicz, S., Matwin, S.: A WordNet-based algorithm for word sense disambiguation. In Proc. IJCAI-95, pp. 1368–1374, (1995)
31. Liu, S., Cui, W., Wu, Y., Liu, M.: A survey on information visualization: recent advances and challenges. Vis. Comput. 30(12), 1373–1393 (2014). December
32. Malik, T., Nistor, L., Gehani, A.: Tracking and sketching distributed data provenance. In: eScience, pp. 190–197 (2012)
33. Matwin's opinions on data privacy issues: http://www.dal.ca/faculty/computerscience/research-industry/researchchairs/stan_matwin.html (Retrieved 2015)
34. Matwin, S.: Machine learning: four lessons and what is next? Bull. Polish AI Soc. 2, 2–7 (2013)
35. Matwin, S.: Privacy-preserving data mining techniques: survey and challenges. In Custers, B., Calders, T., Schermer, B., Zarsky T. (eds.) Discrimination and Privacy in the Information Society. Springer Series on Studies in Applied Philosophy, Epistemology and Rational Ethics, vol. 3, pp. 209–221 (2013)
36. Mayer-Schonberger, V., Cukier, K.: Big data: a revolution that will transform how we live, work and think. Eamon, Dolan/Houghton Mifflin Harcourt (2013)
37. Musolesi, M.: Big mobile data mining: good or evil? IEEE Internet Computing, pp. 2–5 (2014)
38. Pederschi, D., Calders, T., Custer, B.: Big Data mining, fairness and privacy a vision statement towards an interdisciplinary roadmap of research. KDnuggest Rev. 11(26) (2011)
39. Richards, N., King, J.: Three paradoxes of big data. Stanford Law Rev. Online 66, 41–46 (2013)
40. Rudin, C., Passonneau, R., Radeva, A., Jerome, S., Issac, D.: 21st century data miners meet 19-th century electrical cables. IEEE Computer, 103–105, (June 2011)
41. Scott, S., Matwin, S.: Text classification using WordNet hypernyms. In: Procedings of the Conference—Use of WordNet in Natural Language Processing Systems, pp. 38–44 (1998)
42. Scott, S., Matwin, S.: Feature engineering for text classification. Proc. ICML'99, 379–388 (1999)
43. Singh, D., Reddy, C.: A survey on platforms for big data analytics. J. Big Data 1(8), 2–20 (2014)
44. Skowron, A., Stepaniuk, J., Swiniarski, R.: Modeling rough granular computing based on approximation spaces. Inf. Sci. 184, 20–43 (2012)
45. Smailovic, J., Grcar, M., Lavrac, N., Znidarsic, M.: Stream-based active learning for sentiment analysis in the financial domain. Inf. Sci. 285, 181–203 (2014)
46. Sun, Y., Han, J., Yan, X., Yu, P.: Mining knowledge from interconnected data: a heterogeneous information networks analysis approach. VLDB Endowment 5(12), 2022–2023 (2012)
47. Teen, O., Polonetsky, J.: Privacy in the age of big data. A time for big decisions. Stanford Law Rev. Online 64, 63–69 (2012)
48. Tukey, J.: Exploratory Data Analysis. Addison Wesley, Reading (1970)
49. Weisburd, D., Telep, C.: Hot spot policing: what we know and what we need to know. J. Contemp. Crim. Justice 30, 200–220 (2014)

50. Working Paper on Big Data and Privacy—Privacy principles under pressure in the age of Big Data analytics—55th Meeting of International Working Group on Data Protection in Telecommunications, vol. 5, 6 May 2014, Skopje (2014)
51. Yin, X., Tan, W.: Semi-supervised truth discovery. In: Proceedings of the 20th International Conference on WWW, pp. 217–226 (2011)
52. Yin, X., Han, J., Yu, P.: Truth discovery with multiple conflicting information providers on the Web. In: Proceedings of the 13th ACM SIGKDD Conference on KDD, pp. 1048–1052 (2007)
53. Zhan, J., Chang, L., Matwin, S.: Privacy-preserving multi-party decision tree induction. In: Research Directions in Data and Applications Security, vol. XVIII, pp. 341–355 (2004)
54. Zhan, J., Matwin, S., Chang, L.: Privacy-preserving collaborative association rule mining. J. Netw. Comput. Appl. **30**(3), 1216–1227 (2007)